# New Advances in Soil Pollution and Remediation

# New Advances in Soil Pollution and Remediation

Editors

**Fayuan Wang**
**Liping Li**
**Lanfang Han**
**Aiju Liu**

MDPI • Basel • Beijing • Wuhan • Barcelona • Belgrade • Manchester • Tokyo • Cluj • Tianjin

*Editors*

Fayuan Wang
Qingdao University of
Science and Technology
China

Liping Li
Henan University of
Technology
China

Lanfang Han
Guangdong University of
Technology
China

Aiju Liu
Shandong University of
Technology
China

*Editorial Office*
MDPI
St. Alban-Anlage 66
4052 Basel, Switzerland

This is a reprint of articles from the Special Issue published online in the open access journal *International Journal of Environmental Research and Public Health* (ISSN 1660-4601) (available at: https://www.mdpi.com/journal/ijerph/special_issues/soil_pollution_re).

For citation purposes, cite each article independently as indicated on the article page online and as indicated below:

LastName, A.A.; LastName, B.B.; LastName, C.C. Article Title. *Journal Name* **Year**, *Volume Number*, Page Range.

**ISBN 978-3-0365-5165-4 (Hbk)**
**ISBN 978-3-0365-5166-1 (PDF)**

© 2022 by the authors. Articles in this book are Open Access and distributed under the Creative Commons Attribution (CC BY) license, which allows users to download, copy and build upon published articles, as long as the author and publisher are properly credited, which ensures maximum dissemination and a wider impact of our publications.

The book as a whole is distributed by MDPI under the terms and conditions of the Creative Commons license CC BY-NC-ND.

# Contents

**About the Editors** . . . . . . . . . . . . . . . . . . . . . . . . . . . . . . . . . . . . . . . . . . . **vii**

**Preface to "New Advances in Soil Pollution and Remediation"** . . . . . . . . . . . . . . . . . . . **ix**

**Liping Li, Lanfang Han, Aiju Liu and Fayuan Wang**
Imperfect but Hopeful: New Advances in Soil Pollution and Remediation
Reprinted from: *Int. J. Environ. Res. Public Health* **2022**, *19*, 10164, doi:10.3390/ijerph191610164 . **1**

**Lei Tang, Yiyue Zhang, Shuai Ma, Changchun Yan, Huanhuan Geng, Guoqing Yu, Hongbing Ji and Fei Wang**
Potentially Toxic Element Contaminations and Lead Isotopic Fingerprinting in Soils and Sediments from a Historical Gold Mining Site
Reprinted from: *Int. J. Environ. Res. Public Health* **2021**, *18*, 10925, doi:10.3390/ijerph182010925 . **5**

**Kui Cai and Chang Li**
Ecological Risk, Input Flux, and Source of Heavy Metals in the Agricultural Plain of Hebei Province, China
Reprinted from: *Int. J. Environ. Res. Public Health* **2022**, *19*, 2288, doi:10.3390/ijerph19042288 . . . **23**

**Wei Chen, Faming Zeng, Wei Liu, Jianwei Bu, Guofeng Hu, Songshi Xie, Hongyan Yao, Hong Zhou, Shihua Qi and Huanfang Huang**
Organochlorine Pesticides in Karst Soil: Levels, Distribution, and Source Diagnosis
Reprinted from: *Int. J. Environ. Res. Public Health* **2021**, *18*, 11589, doi:10.3390/ijerph182111589 . **47**

**Yitong Yin, Ximing Luo, Xiangyu Guan, Jiawei Zhao, Yuan Tan, Xiaonan Shi, Mingtao Luo and Xiangcai Han**
Arsenic Release from Soil Induced by Microorganisms and Environmental Factors
Reprinted from: *Int. J. Environ. Res. Public Health* **2022**, *19*, 4512, doi:10.3390/ ijerph19084512 . . **63**

**Yuan Meng, Liang Zhang, Zhi-Long Yao, Yi-Bin Ren, Lin-Quan Wang and Xiao-Bin Ou**
Arsenic Accumulation and Physiological Response of Three Leafy Vegetable Varieties to As Stress
Reprinted from: *Int. J. Environ. Res. Public Health* **2022**, *19*, 2501, doi:10.3390/ijerph19052501 . . . **79**

**Ziwen Xu, Shiquan Lv, Shuxiang Hu, Liang Chao, Fangxu Rong, Xin Wang, Mengyang Dong, Kai Liu, Mingyue Li and Aiju Liu**
Effect of Soil Solution Properties and $Cu^{2+}$ Co-Existence on the Adsorption of Sulfadiazine onto Paddy Soil
Reprinted from: *Int. J. Environ. Res. Public Health* **2021**, *18*, 13383, doi:10.3390/ijerph182413383 . **95**

**Geng Li, Haibo Li, Yinghua Li, Xi Chen, Xinjing Li, Lixin Wang, Wenxin Zhang and Ying Zhou**
Stabilization/Solidification of Heavy Metals and PHe Contaminated Soil with β-Cyclodextrin Modified Biochar (β-CD-BC) and Portland Cement
Reprinted from: *Int. J. Environ. Res. Public Health* **2022**, *19*, 1060, doi:10.3390/ijerph19031060 . . . **107**

**Hongyang Lin, Yang Yang, Zhenxiao Shang, Qiuhong Li, Xiaoyin Niu, Yanfei Ma and Aiju Liu**
Study on the Enhanced Remediation of Petroleum-Contaminated Soil by Biochar/g-$C_3N_4$ Composites
Reprinted from: *Int. J. Environ. Res. Public Health* **2022**, *19*, 8290, doi:10.3390/ijerph19148290 . . . **125**

**Ningning Wang, Yucong Jiang, Tianxiang Xia, Feng Xu, Chengjun Zhang, Dan Zhang and Zhiyuan Wu**
Antimony Immobilization in Primary-Explosives-Contaminated Soils by Fe–Al-Based Amendments
Reprinted from: *Int. J. Environ. Res. Public Health* **2022**, *19*, 1979, doi:10.3390/ijerph19041979 . . . **139**

**Shuang Li, Liao He, Bo Zhang, Yan Yan, Wentao Jiao and Ning Ding**
A Comprehensive Evaluation Method for Soil Remediation Technology Selection: Case Study of Ex Situ Thermal Desorption
Reprinted from: *Int. J. Environ. Res. Public Health* **2022**, *19*, 3304, doi:10.3390/ijerph19063304 . . . **155**

# About the Editors

**Fayuan Wang (PhD)**

Fayuan Wang is a full professor of College of Environment and Safety Engineering, Qingdao University of Science and Technology, China. He received his M.A. degree in Soil Microbiology from Laiyang Agricultural College (2002) and his PhD degree in Plant Nutrition from Institute of Soil Science, Chinese Academy of Sciences (2005). He completed his postdoctoral studies at Tsinghua University (China) in 2012. In 2013–2014, he was a Visiting Scholar at Harvard University (USA). His research interests include ecotoxicology of heavy metals and emerging contaminants, environmental stress on plants and soil microbiota, remediation of polluted environments, and mycorrhizal diversity and environmental applications. He has published more than 100 journal papers. He currently serves as editor or editorial board member of several international journals, including International Journal of Environmental Research and Public Health (MDPI), Toxics (MDPI), Frontiers in Environmental Science (Frontiers), Ecotoxicology (Springer), Applied Soil Ecology (Elsevier), and Plant Physiology and Biochemistry (Elsevier) and reviewer for more than 50 international journals.

**Liping Li (PhD)**

Liping Li is a full professor of environmental soil science at Henan University of Technology in Zhengzhou City, Henan Province, China. Liping Li received his PhD in Soil Science at the Institute of Soil Science, Chinese Academy of Sciences, in 2004. The present research conducted in his group is focused on potentially toxic metal (PTM) pollution in the environment and the risk in Pb smelting contaminated areas in northern China. The distribution of PTM in soil and particulate matter, the migration of PTM from soil to crop, and particulate matter and their risk to humans are areas of focus in our present research. Specific examples include the processing and cooking of wheat flour as conducted in China and their effect on the PTM health risk, soil PTM bioavailability to free-range chickens in contaminated areas, Cd reduction in the wheat grain of contaminated areas, and reduction of Cd risk in wheat flour during processing. Dr. Li has published about 100 papers, including 30 in English. Dr. Li serves as a reviewer for more than 30 international journals. He is a member of Chinese Society of Soil Science.

**Lanfang Han (PhD)**

Lanfang Han received her PhD in Environmental Science from Beijing Normal University, China, in 2019, and has served as a professor in Guangdong University of Technology since 2021. Her research group focuses on the fate of pollutants in soils and recycling of agricultural solid wastes. She was selected into the "Young Elite Scientists Sponsorship Program" by China Association for Science and Technology and won second prize of the Environmental Protection Science and Technology Award from the Ministry of Environmental Protection of China (ranked 10th), Excellent Doctoral Dissertation Award of Beijing Normal University, and other honors. She was invited to participate in the editing of 2 books and has published 79 papers (70 SCI papers), of which she is the first/corresponding author of 33 papers (30 SCI). She is also the review editor of Frontiers in Soil Science, an editorial board member of Geochemistry, and guest editor of Frontiers in Environmental Science.

**Aiju Liu (PhD)**

Aiju Liu received a PhD in Environmental Science from the Nanjing University, China, in 2006. She has served as a full professor at the Shandong University of Technology since 2018. She is the current director of Shandong Tailing Resource Utilization Engineering Technology Development and Research Center. Her research group focuses on soil contamination and remediation, mainly including the effects of heavy metals on soil microbial ecology, improvement of acidified soil, and soil amelioration of tailings, such as bauxite residues. She has published more than 30 journal papers and numerous conference communications. She is a member of China Soil Society and of the Chinese Ceramic Society, a frequent reviewer for many journals, and a member of numerous grant evaluation panels for national institutions.

# Preface to "New Advances in Soil Pollution and Remediation"

Soil pollution has been identified as a global environmental issue, posing potential risks for soil ecosystems, food security, and human health. Recently, the Global Assessment of Soil Pollution report, jointly issued by FAO and UNEP, has drawn our attention to this topic again. As the report points out, "Soil pollution hampers the achievement of Sustainable Development Goals (SDGs), including achieving zero hunger, ending poverty, ensuring healthy lives and human well-being, halting and reversing land degradation and biodiversity loss, and making cities safe and resilient." Therefore, effective soil remediation technologies, including physical, chemical, and biological approaches, are required to make polluted soils safer for humans and other organisms. In addition to traditional pollutants, such as heavy metals and persistent organic pollutants, soils are also being contaminated by emerging contaminants, such as microplastics, nanoparticles, antibiotics, and antibiotic resistance genes. Due to their emerging state, little information is available on the occurrence and contamination status of these emerging contaminants in the soil environment, which need to be urgently investigated. Meanwhile, soil contamination by these contaminants calls for more novel and effective soil remediation techniques.

This book focuses on new advances in soil pollution due to both traditional and emerging contaminants as well as novel and green remediation technologies. A total of ten manuscripts are presented, six reporting on the occurrence, environmental behaviors, and risks of contaminants and the other four focusing on techniques for the remediation of polluted soils. A list of contaminants are targeted, including toxic metal(loid)s (e.g., Pb, As, Sb, and multi-metals), organic contaminants (e.g., organochlorine pesticides, phenanthrene, and petroleum), and antibiotics (e.g., sulfadiazine). Special focus is given to the remediation of polluted soils using new materials. These new findings can provide insights for researchers, technicians, teachers, and students in environmental science and technology in addition to helpful suggestions for policy-makers and environmental managers.

These studies were accomplished by a total of 71 researchers from 21 universities or research institutes in China. This fact reflects the significant progress in research on soil pollution and remediation in China as well as its future potential. The efforts and contributions of these Chinese researchers are commendable. Nevertheless, we call for more work from researchers in various countries, particularly those facing similar environmental issues. We have only one Earth, and we encourage all efforts aimed at its preservation from the global scientific community.

Finally, we would like to thank all the authors who shared with us their findings, and all the reviewers who provided constructive comments on the manuscripts included in this book. We also thank MDPI and the Editorial Staff for their professional support in the successful publication of this book.

**Fayuan Wang, Liping Li, Lanfang Han, and Aiju Liu**
*Editors*

*Editorial*

# Imperfect but Hopeful: New Advances in Soil Pollution and Remediation

Liping Li [1], Lanfang Han [2], Aiju Liu [3] and Fayuan Wang [4,*]

[1] School of the Environment, Henan University of Technology, Zhengzhou 450001, China
[2] School of Ecology, Environment and Resources, Guangdong University of Technology, Guangzhou 510006, China
[3] School of Resources and Environmental Engineering, Shandong University of Technology, Zibo 255049, China
[4] College of Environment and Safety Engineering, Qingdao University of Science and Technology, Qingdao 266042, China
\* Correspondence: wangfayuan@qust.edu.cn

Soil is the most important resource for plant growth and human survival, supporting agricultural production and human habitation. However, due to unreasonable anthropogenic activities, soil receives a large amount of hazardous and toxic materials, including common contaminants such as toxic metal(loid)s and organic contaminants, and emerging contaminants with uncertain hazards and toxicity, posing potential risks for the soil ecosystems, food security, and human health. Thus, it is of great importance to recognize the environmental behaviors and fate of contaminants, and to improve remediation techniques. However, the environmental behaviors of contaminants have high complexity and spatial heterogeneity in soil environments. It is really challenging to illustrate soil pollution processes and to successfully remediate polluted soil. In particular, highly efficient, low-cost, and environmentally friendly soil remediation techniques need to be developed urgently.

In this Special Issue (SI), new advances in soil pollution and remediation were reported. Ten papers were published, six ones reporting the occurrence, environmental behaviors, and risks of contaminants, and the other four focusing on the remediation techniques of polluted soils. A series of contaminants were targeted, including toxic metal(loid)s (e.g., Pb, As, Sb, and multi-metals), organic contaminants (e.g., organochlorine pesticides, phenanthrene (PHe), and petroleum), and antibiotics (e.g., sulfadiazine). Both single and combined pollution scenarios were involved. Below, we summarized the main findings of these studies.

Non-ferrous metal mining and smelting is one of the main sources of heavy metal pollution in the environment. Because of the co-existence of trace elements in the mining ore, non-ferrous metal mining and smelting often results in the accumulation of multiple heavy metals in the soil. It is necessary to identify the sources of different metals. For sources with different Pb isotopic fingerprints, the Pb isotopic ratios of the sample and end member materials (such as ore and unpolluted soil) can be used for source apportionment of the Pb in the soil samples. Tang et al. [1] investigated the distribution of Cu, Pb, Zn, Cr, Ni, Cd, As, and Hg in soils and sediments near former gold mines in Beijing and Pb source apportionment with sample Pb isotopic information. They found that Cd and Pb had the greatest pollution indexes (as contamination factor in this paper) among all the metals. The results highlight the need for reasonable management of such deserted sites with high heavy metal accumulation. Cai and Li [2] studied the potential ecological risk, source, and input flux of heavy metals in the Hebei plain of northern China, and detected the accumulation of As, Cu, Cd, and Zn. With the input flux calculation, atmospheric dry and wet deposition were interestingly found to contribute more to soil pollution than the usage of fertilizer or irrigation water. Their results provide valuable data for controlling and remedying heavy metals polluted with agricultural soils.

**Citation:** Li, L.; Han, L.; Liu, A.; Wang, F. Imperfect but Hopeful: New Advances in Soil Pollution and Remediation. *Int. J. Environ. Res. Public Health* **2022**, *19*, 10164. https://doi.org/10.3390/ijerph191610164

Received: 14 August 2022
Accepted: 16 August 2022
Published: 16 August 2022

**Publisher's Note:** MDPI stays neutral with regard to jurisdictional claims in published maps and institutional affiliations.

**Copyright:** © 2022 by the authors. Licensee MDPI, Basel, Switzerland. This article is an open access article distributed under the terms and conditions of the Creative Commons Attribution (CC BY) license (https://creativecommons.org/licenses/by/4.0/).

Differently, Chen et al. [3] investigated the occurrences of organochlorine pesticides (OCPs) in karst soil by analyzing 25 OCPs in the karst soils near the Three Gorges Dam, China. In this karst area, p,p′-DDT and mirex were the most abundant OCPs, with different concentrations in spatial distribution as the land use type and the water transport. These OCPs were mainly derived from current agrochemical use and current veterinary use in the study area. OCPs are among typical persistent organic pollutants (POPs) with high toxicity. More attention should be paid to the issues of their illegal uses and bioaccumulation via the food chain.

Arsenic is one of the most toxic and widespread metalloids in the natural environment, which was targeted by two studies in this SI. Yin et al. [4] analyzed the effects of microorganisms, low-molecular-weight organic acid salts, and phosphates on the migration of As in unrestored and nano zero-valent iron (nZVI)-restored soil. They showed that microorganisms suppressed As release in both unrestored and restored soil. Meng et al. [5] explored As stress-induced response of three leafy vegetables, i.e., garland chrysanthemum (*Chrysanthemum coronarium* L.), spinach (*Spinacia oleracea* L.), and lettuce (*Lactuca sativa* L.), and found that the high tolerance of garland chrysanthemum was mainly ascribed to the low transport of As from the roots to the shoots, the high activity of antioxidant enzymes (*superoxide dismutase, glutathione peroxidase,* and *catalase*), and the abundant phytochelatins in the roots.

Antibiotics are identified as an emerging contaminant. The co-existence of antibiotics with heavy metals in the soil environment is attracting increasing attention. Xu et al. [6] investigated the adsorption behavior of sulfadiazine (SDZ) in paddy soils, and found that the changes in soil pH and ion concentration decreased the adsorption of SDZ on soil components, while dissolved organic carbon (DOC) facilitated the adsorption of SDZ in paddy soils, but the effect of co-existent $Cu^{2+}$ was greatly dependent on the type of soil components. Their results confirmed that complexation may not be the only form of $Cu^{2+}$ and SDZ co-adsorption in paddy soils.

This SI also brings some exciting advances in soil remediation. Biochar is a promising amendment used for the remediation of soils polluted with heavy metals and organic contaminants. Li et al. [7] reported the simultaneous stabilization/solidification of heavy metals and PHe by β-cyclodextrin-modified biochar. This functional material has abundant oxygen-containing functional groups on the surface and a porous structure with a large specific surface area, thus greatly increasing the retention of Cd, Cr, Cu, Pb, and PHe in soil. Lin et al. [8] reported an environmentally friendly soil remediation method with biochar/graphite carbon nitride (BC/g-$C_3N_4$), and indicated its technical feasibility of remediating petroleum-contaminated soil. BC/g-$C_3N_4$ facilitated the degradation by reducing recombination and better electron–hole pair separation. After treatment with BC/g-$C_3N_4$, the removal rates of $nC_{13}$-$nC_{35}$ were above 90% in the contaminated soil. In sum, the BC/g-$C_3N_4$ composites can effectively remedy organic contaminated soil.

In recent years, Sb pollution in the soil environment is drawing more attention. For the immobilization of Sb in polluted soils, a dilemma is that Sb existed as anions, while co-existing Cd and Pb ions are positively charged; thus, when a positively charged amendment can immobilize Sb in soil, it can also increase the availability and mobility of cationic contaminants in the soil. Thus, methods which can reduce the availability of both cations and anions are needed for Sb-contaminated soils. For this purpose, a combination of different additives may be an ideal choice. Wang et al. [9] studied the immobilization of Sb, Cu, and Zn with $FeSO_4$ + $Al(OH)_3$, and found that 5% $FeSO_4$ + 4% $Al(OH)_3$ was effective for the stabilization of Sb and co-occurring metals. However, for practical purposes, these dosages are still too high. In addition to the high cost, a high dosage of amendments may alter soil properties (such as texture, soil permeability, and phyto-availability of microelements), causing unexpected ecological risks for soil biota and plants.

A comprehensive evaluation of soil remediation technologies is of critical importance to the optimization and selection of proper technology for contaminated sites. Thermal desorption is an effective method to remediate sites contaminated with organic contaminants.

Li et al. [10] developed a methodology to comprehensively evaluate three ex situ thermal desorption processes. The evaluation indicators included 20 qualitative and quantitative indicators covering technical, environmental, resource, economic, and social aspects. Their study provides a novel evaluation approach for the application of soil remediation technology. However, the applicability of this method needs to be verified in more case studies.

Due to their complexity of the soil environment, many contaminants are difficult to remove or degrade readily. Current techniques based on physical, chemical, and biological principles have some disadvantages, e.g., high price, environment unfriendliness, secondary pollution, and long processing time. While optimizing traditional techniques, new remediation materials and techniques should be developed. In addition, diverse emerging contaminants have been observed in the soil environment, including agricultural ecosystems [11,12]. However, their occurrence, environmental behaviors, fate, ecotoxicology, and potential risks are yet to be fully recognized, not to mention the remediation technology for their removal. Thus, we call for more efforts on emerging contaminants in the soil, particularly developing effective remediation techniques.

**Author Contributions:** Writing—original draft, L.L., L.H., A.L. and F.W.; Writing—review and editing, L.L., L.H., A.L. and F.W. All authors have read and agreed to the published version of the manuscript.

**Funding:** This research received no external funding.

**Acknowledgments:** We thank all the reviewers who provided valuable comments on the manuscripts submitted to this Special Issue.

**Conflicts of Interest:** The authors declare no conflict of interest.

**References**

1. Tang, L.; Zhang, Y.; Ma, S.; Yan, C.; Geng, H.; Yu, G.; Ji, H.; Wang, F. Potentially toxic element contaminations and lead isotopic fingerprinting in soils and sediments from a historical gold mining site. *Int. J. Environ. Res. Public. Health* **2021**, *18*, 10925. [CrossRef] [PubMed]
2. Cai, K.; Li, C. Ecological risk, input flux, and source of heavy metals in the agricultural plain of Hebei Province, China. *Int. J. Environ. Res. Public. Health* **2022**, *19*, 2288. [CrossRef] [PubMed]
3. Chen, W.; Zeng, F.; Liu, W.; Bu, J.; Hu, G.; Xie, S.; Yao, H.; Zhou, H.; Qi, S.; Huang, H. Organochlorine pesticides in karst soil: Levels, distribution, and source diagnosis. *Int. J. Environ. Res. Public. Health* **2021**, *18*, 11589. [CrossRef] [PubMed]
4. Yin, Y.; Luo, X.; Guan, X.; Zhao, J.; Tan, Y.; Shi, X.; Luo, M.; Han, X. Arsenic release from soil induced by microorganisms and environmental factors. *Int. J. Environ. Res. Public. Health* **2022**, *19*, 4512. [CrossRef] [PubMed]
5. Meng, Y.; Zhang, L.; Yao, Z.-L.; Ren, Y.-B.; Wang, L.-Q.; Ou, X.-B. Arsenic accumulation and physiological response of three leafy vegetable varieties to As stress. *Int. J. Environ. Res. Public. Health* **2022**, *19*, 2501. [CrossRef] [PubMed]
6. Xu, Z.; Lv, S.; Hu, S.; Chao, L.; Rong, F.; Wang, X.; Dong, M.; Liu, K.; Li, M.; Liu, A. Effect of soil solution properties and $Cu^{2+}$ co-existence on the adsorption of sulfadiazine onto paddy soil. *Int. J. Environ. Res. Public. Health* **2021**, *18*, 13383. [CrossRef] [PubMed]
7. Li, G.; Li, H.; Li, Y.; Chen, X.; Li, X.; Wang, L.; Zhang, W.; Zhou, Y. Stabilization/solidification of heavy metals and PHe contaminated soil with β-cyclodextrin modified biochar (β-CD-BC) and Portland cement. *Int. J. Environ. Res. Public. Health* **2022**, *19*, 1060. [CrossRef] [PubMed]
8. Lin, H.; Yang, Y.; Shang, Z.; Li, Q.; Niu, X.; Ma, Y.; Liu, A. Study on the enhanced remediation of petroleum-contaminated soil by biochar/g-$C_3N_4$ composites. *Int. J. Environ. Res. Public. Health* **2022**, *19*, 8290. [CrossRef] [PubMed]
9. Wang, N.; Jiang, Y.; Xia, T.; Xu, F.; Zhang, C.; Zhang, D.; Wu, Z. Antimony immobilization in primary-explosives-contaminated soils by Fe–Al-based amendments. *Int. J. Environ. Res. Public. Health* **2022**, *19*, 1979. [CrossRef] [PubMed]
10. Li, S.; He, L.; Zhang, B.; Yan, Y.; Jiao, W.; Ding, N. A comprehensive evaluation method for soil remediation technology selection: Case study of ex situ thermal desorption. *Int. J. Environ. Res. Public. Health* **2022**, *19*, 3304. [CrossRef] [PubMed]
11. Snow, D.D.; Cassada, D.A.; Larsen, M.L.; Mware, N.A.; Li, X.; D'Alessio, M.; Zhang, Y.; Zhang, Y.; Sallach, J.B. Detection, occurrence and fate of emerging contaminants in agricultural environments. *Water Environ. Res.* **2017**, *89*, 897–920. [CrossRef] [PubMed]
12. Wang, F.; Wang, Q.; Adams, C.A.; Sun, Y.; Zhang, S. Effects of microplastics on soil properties: Current knowledge and future perspectives. *J. Hazard. Mater.* **2022**, *424*, 127531. [CrossRef] [PubMed]

Article

# Potentially Toxic Element Contaminations and Lead Isotopic Fingerprinting in Soils and Sediments from a Historical Gold Mining Site

Lei Tang [1,2,3,†], Yiyue Zhang [1,2,†], Shuai Ma [1,2], Changchun Yan [1,2], Huanhuan Geng [1,2], Guoqing Yu [3], Hongbing Ji [1,2,*] and Fei Wang [1,2,*]

1. School of Energy & Environmental Engineering, University of Science and Technology Beijing, 30 Xueyuan Road, Beijing 100083, China; tangtom1220@163.com (L.T.); yiyue.zhang@hotmail.com (Y.Z.); shuai.ma@yale.edu (S.M.); y19801286607@163.com (C.Y.); genghuanhuan0325@outlook.com (H.G.)
2. Beijing Key Laboratory of Resource-Oriented Treatment of Industrial Pollutants, 30 Xueyuan Road, Beijing 100083, China
3. Beijing Geo-Exploration and Water Environment Engineering Institute Co., Ltd., 9 Linglong Road, Beijing 100142, China; 13366833331@163.com
* Correspondence: ji.hongbing@hotmail.com (H.J.); wangfei@ustb.edu.cn (F.W.); Tel./Fax: +86-10-62333305 (F.W.)
† L. Tang and Y. Zhang contributed equally to this work.

**Abstract:** Lead (Pb) isotopes have been widely used to identify and quantify Pb contamination in the environment. Here, the Pb isotopes, as well as the current contamination levels of Cu, Pb, Zn, Cr, Ni, Cd, As, and Hg, were investigated in soil and sediment from the historical gold mining area upstream of Miyun Reservoir, Beijing, China. The sediment had higher $^{206}Pb/^{207}Pb$ ratios (1.137 ± 0.0111) than unpolluted soil did (1.167 ± 0.0029), while the soil samples inside the mining area were much more variable (1.121 ± 0.0175). The mean concentrations (soil/sediment in mg·kg$^{-1}$) of Pb (2470/42.5), Zn (181/113), Cu (199/36.7), Cr (117/68.8), Ni (40.4/28.9), Cd (0.791/0.336), As (8.52/5.10), and Hg (0.168/0.000343) characterized the soil/sediment of the studied area with mean $I_{geo}$ values of the potentially toxic element (PTE) ranging from −4.71 to 9.59 for soil and from −3.39 to 2.43 for sediment. Meanwhile, principal component analysis (PCA) and hierarchical cluster analysis (HCA) coupled with Pearson's correlation coefficient among PTEs indicated that the major source of the Cu, Zn, Pb, and Cd contamination was likely the mining activities. Evidence from Pb isotopic fingerprinting and a binary mixing model further confirmed that Pb contamination in soil and sediment came from mixed sources that are dominated by mining activity. These results highlight the persistence of PTE contamination in the historical mining site and the usefulness of Pb isotopes combined with multivariate statistical analysis to quantify contamination from mining activities.

**Keywords:** miyun reservoir; pollution assessment; binary mixing model; source appointment

Citation: Tang, L.; Zhang, Y.; Ma, S.; Yan, C.; Geng, H.; Yu, G.; Ji, H.; Wang, F. Potentially Toxic Element Contaminations and Lead Isotopic Fingerprinting in Soils and Sediments from a Historical Gold Mining Site. *Int. J. Environ. Res. Public Health* **2021**, *18*, 10925. https://doi.org/10.3390/ijerph182010925

Academic Editors: Fayuan Wang, Liping Li, Lanfang Han and Aiju Liu

Received: 16 September 2021
Accepted: 12 October 2021
Published: 18 October 2021

**Publisher's Note:** MDPI stays neutral with regard to jurisdictional claims in published maps and institutional affiliations.

**Copyright:** © 2021 by the authors. Licensee MDPI, Basel, Switzerland. This article is an open access article distributed under the terms and conditions of the Creative Commons Attribution (CC BY) license (https://creativecommons.org/licenses/by/4.0/).

## 1. Introduction

In the past few decades, potentially toxic element (PTE) contaminations from mining activities have become a serious global environmental problem [1–4]. Mining activity constitutes prominent sources of toxic, corrosive, radioactive, or nonradioactive metal contaminants from ore, smelting, mineral dressing, and the erosion of mine tailings [5]. Both historical and ongoing mining activities have a nonnegligible impact on the surrounding environment, resulting in significant increases in PTE loads in both aquatic and terrestrial ecosystems [6]. The release of large amounts of mine wastes from mining and transportation, acid mine drainage (AMD) from ore mineral dressing, and fly ash, as well as particulate matter, from metal smelting and coal combustion all potentially lead to significant increases in PTE loads in the surrounding environment. Previous works have found strong evidence of PTE contamination in the soils, water, sediments, atmosphere, and biota in proximity to mining activities [2,7–10]. Identifying anthropogenic sources of PTEs and the apportionment of the contributions of anthropogenic and natural sources has caused significant concern because it is of crucial importance to preventing and controlling PTE contamination [11,12]. Although the contamination of the surrounding environment by PTEs from mining activities has been extensively studied and highlighted, source interpretation of mining-impacted areas remains challenging, especially in historical small-scale polymetallic mining sites [2].

Multivariate statistical analysis is a traditional and useful tool to identify potential factors that may indicate or hint to sources of PTE concentration and to explore similarities and hidden patterns among the sample [13,14]. Principal component analysis (PCA) and cluster analysis (CA) are common widely used techniques and are often combined to identify the potential sources of PTEs in soils and sediments [15]. The multivariable statistics analysis of PTE concentration provides vital information on the interrelationships of elements. However, its source identification and apportionment usually rely strongly on statistical approaches, which have required large databases and sophisticated statistics [16]. Pb isotope fingerprinting has shown great advantages in the identification and quantification of various sources in environmental studies [17,18]. The four natural Pb stable isotopes ($^{204}$Pb, $^{206}$Pb, $^{207}$Pb, and $^{208}$Pb) in natural or anthropogenic origin sources (e.g., ore deposits, coal, and leaded gasoline) typically have their unique signatures and result in the distinguishable Pb isotope ratio [13,16,19]. No significant Pb isotopic fractionation occurs during natural and anthropogenic processes, implying that the final Pb isotopic composition in the environment reflects only the original source of Pb or a mixture of multiple sources, thus allowing us to evaluate the contributions from the different Pb sources [16,19–21]. Studies are increasingly using Pb isotope fingerprinting to trace the anthropogenic Pb sources in sediments, soils, coal fly ash, aerosols, and other environment archives [11,12,22–25]. Recently, Pb isotopes have been employed to trace sources of gold deposits [26].

Miyun Reservoir (40°31′ N, 116°51′ E) is the largest reservoir and the primary surface drinking water source for Beijing with a population exceeding 20 million [27,28]. It has a storage capacity of 438 GL. Bai and Chao River are the leading natural replenishments of Miyun Reservoir that contribute mean flows of 111 GL yr$^{-1}$ and 203 GL yr$^{-1}$, respectively [29]. Historically, long-term small-scale metal (gold, iron, copper, etc.) mining and smelting activities have had a nonnegligible impact on the environment surrounding the Miyun Reservoir [27]. Since 2005, numerous small-scale metal mine sites have been closed in this area. However, mine waste from mining operations is still deposited in abandoned tailings ponds, continuing to cause PTE contamination to the neighboring environment. Several studies have reported that the PTE contamination of soils, river sediments, and river water upstream of Miyun Reservoir is mainly caused by mining activities [27,30–34]. Some other studies have concluded that coal combustion and vehicle exhaust were identified as the primary source of PTEs in surface soils. It is difficult to infer the contribution of various sources from the elevated concentration of PTEs, especially in the mining area [35]. In addition, the levels and sources of PTEs in sediments have rarely been reported in this area compared to soils (Zhu et al., 2013), even though sediment is the appropriate indicator of PTEs in aquatic systems [36].

To investigate the impact of mining activities on the accumulation of PTEs in the surrounding environment, Pb isotopes, as well as the current contamination status of eight typical PTEs (Cu, Pb, Zn, Cr, Ni, Cd, As, and Hg) in soils and sediments, were determined surrounding the mining-impacted area upstream of Miyun Reservoir, Beijing, China. We aimed (i) to assess the pollution of PTEs; (ii) to identify and appoint the potential pollution sources using Pb isotope fingerprinting and multivariate statistical analysis; (iii) to quantify the Pb contribution from the potential sources using stable isotope mixing models.

## 2. Materials and Methods

### 2.1. Sample Collection and Preparation

As shown in Figure 1, a total of 35 sampling sites (three points at each site) were taken from the small-scale gold ores scatter areas upstream of the Miyun Reservoir, including 16 surface soil sites, 15 surface sediments sites (SD1–SD15), and four tailings dam sites. Of the 16 soil sampling sites, 12 sites (SI1–SI12) were taken from an area heavily impacted by mining activities in proximity to local and regional potential contaminant sources (e.g., mines and tailings dams), and four unpolluted sites (SO1–SO4) were taken from woodlands, villages, and agricultural lands that were far from the mining activities and had not been strongly impacted. In 2005–2013, mining activities in the sampling area were completely abandoned due to local government policies.

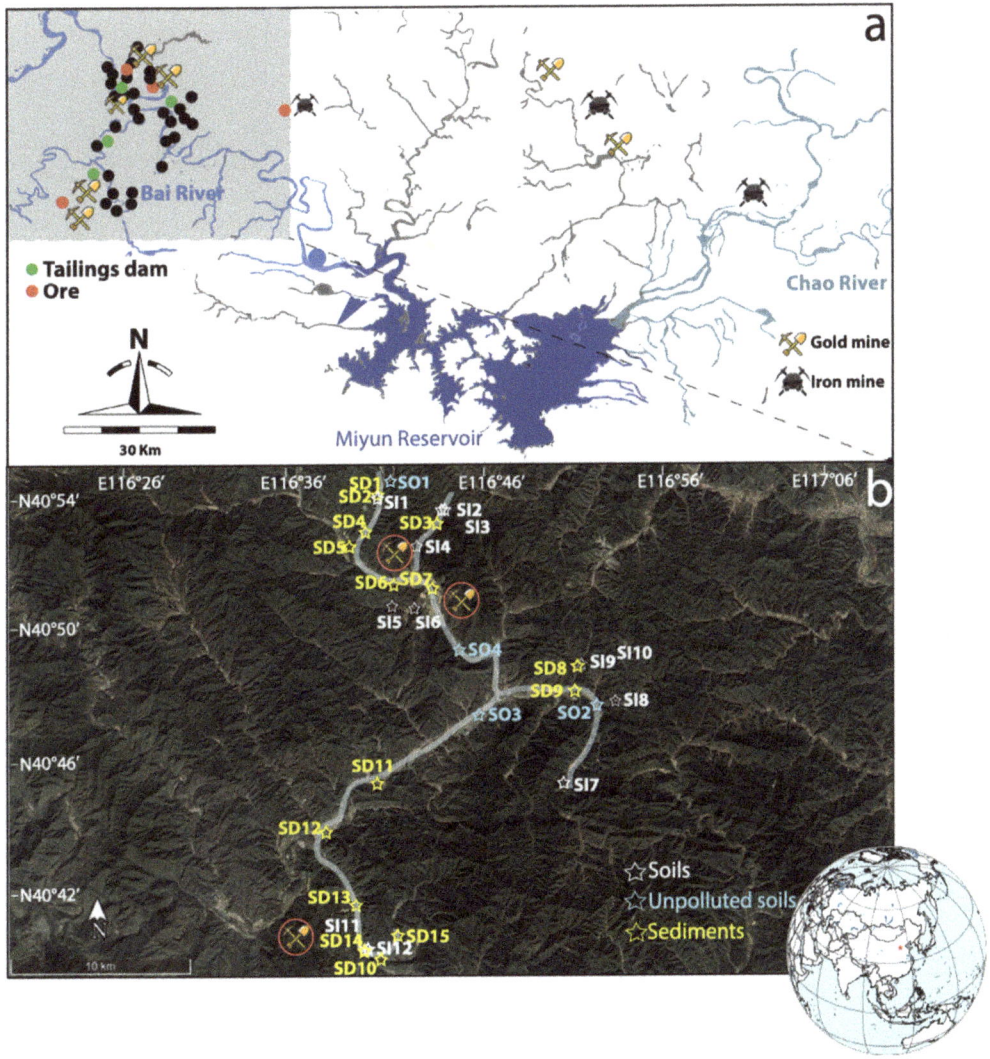

**Figure 1.** Map of the upstream of Miyun Reservoir (**a**); soils and sediments sampling sites (**b**).

## 2.2. Chemical Treatments and Analysis

Samples were first freeze-dried and sieved through the 200 mesh (<0.074 mm) stainless-steel sieve. Subsamples (0.1 g) were added with 4 mL of concentrated $HNO_3$ and 0.5 mL of $H_2O_2$ (30%) in a Teflon beaker before heating at 90 °C overnight until dryness. The samples were further digested with 2 mL of concentrated $HNO_3$, 2 mL of HF, and 1 mL of $HClO_4$ in a sealed beaker at 120 °C for 48 h. Upon evaporation until dryness, re-dissolved in 5% $HNO_3$, the digester was measured for total element concentration. Standard reference materials were processed with the same digestion procedure. Pb isotopic analyses ($^{204}Pb$, $^{206}Pb$, $^{207}Pb$, and $^{208}Pb$) were determined in a selection of soils and sediment samples using a Nu Instruments HR® double focusing MC-ICP-MS. Samples were calibrated against the National Institute of Standards and Technology (NIST) SRM 981 standard after each sample measurement. The measured isotopic ratios of the standard NIST SRM 981 were $^{204}Pb/^{206}Pb$ = 0.059 ± 0.001 (2SD), $^{206}Pb/^{207}Pb$ = 1.093 ± 0.002 (2SD), and $^{208}Pb/^{206}Pb$ = 2.166 ± 0.003 (2SD), which had good agreement with their respective certifications 0.059, 1.093, and 2.168 [17].

## 2.3. Quality Assurance and Quality Control (QA/QC)

The lab glassware and Teflon beakers were previously soaked in 50% $HNO_3$ (w/w) at 120 °C for at least 48 h, followed by rinsing with 18.2 MΩ/cm of Milli-Q water before usage. All analytical solutions were executed in triplicate, and the result was expressed as the mean value. The quality of the processing and analytical procedures was tested by measurements on the Chinese national geo-standard (GBW-07333 and GBW-07314) provided by the National Research Center for certified Reference Materials of China. The standard solutions (NIST) SRM 981 were measured after every ten samples in the analysis of PTE concentrations and after every signal sample in the analysis of Pb isotopic ratio. Instrument performance and analytical procedure reproducibility were determined by analyzing the United States Geological Survey (USGS) reference materials BCR-2 (Basalt, Columbia River) and AGV-2 (Andesite, Guano Valley). The BCR-2 standard resulted in $^{206}Pb/^{207}Pb$ = 1.209 ± 0.006 (2SD) and $^{208}Pb/^{206}Pb$ = 2.065 ± 0.003 (2SD), in agreement with the values reported [37]. The AGV-2 resulted in $^{206}Pb/^{207}Pb$ of 1.208 ± 0.001 (2SD) and $^{208}Pb/^{207}Pb$ of 2.468 ± 0.008 (2SD), also in agreement with previously published values [38]. The standard error of $^{207}Pb/^{206}Pb$ measurements was less than 0.3% RSD. The results of PTE concentrations were consistent with all reference values, and the differences were within ±7%. The relative error was lower than 10%, and the relative standard deviation (RSD) was lower than 5% for all tests.

## 2.4. Pollution Risk Assessment

The contamination factor (CF), degree of contamination and pollution load index (PLI), geo-accumulation index ($I_{geo}$), and potential ecological risk assessment (RI) were determined to assess the potential extent of PTE contamination in different sampling sites.

### 2.4.1. Contamination Factor (CF)

CF was used to assess environmental contamination caused by an excess of a single metal in a sample by calculating the ratio of the measured metal concentration to the natural background value of the metal [39], calculated according to Equation (1).

$$CF_i = C_f^i = C_m^i / C_b^i \qquad (1)$$

where $C_m$ represents the measured concentration of metal $i$, and $C_b$ represents the reference value for metal $i$. The reference value used here was the background value (BV) for PTEs in natural soils in Beijing [40]. Based on the calculated CF values, the degree of contamination was divided into four levels: low (CF < 1); moderate (1 ≤ CF < 3); considerable (3 ≤ CF < 6); and very high (CF ≥ 6).

### 2.4.2. Pollution Level Index (PLI)

The pollution load index (PLI) was used to assess the overall combined toxicity to the environment at each sampling site by standardizing the contribution of all the evaluated PTEs [41]. PLI was calculated as the nth root of the product of contamination factors ($CF_i$), calculated according to Equation (2).

$$PLI = \sqrt[n]{CF_1 \times CF_2 \times \cdots CF_i \times \cdots CF_n} \quad (2)$$

where n is the sum number of evaluated PTEs. PLI classifies six classes of metal contamination from low to high as follows [41]: unpolluted ($PLI \leq 1$); unpolluted to moderate ($1 < PLI \leq 2$); moderately polluted ($2 < PLI \leq 3$); moderately to highly polluted ($3 < PLI \leq 4$); highly polluted ($4 < PLI \leq 5$); and very highly polluted ($PLI \geq 5$).

### 2.4.3. Geo-Accumulation Index ($I_{geo}$)

The commonly used $I_{geo}$ is a geochemical criterion proposed by Muller [42] to quantify metal contamination from natural activities (geological and geographical processes) and anthropogenic activities in soils or sediments, calculated according to Equation (3).

$$I_{geo} = \log_2 \frac{C_m}{1.5 C_b} \quad (3)$$

The constant 1.5 is introduced as the background matrix correction factor for lithospheric effects. Igeo classifies seven classes of metal contamination from low to high as follows [43]: unpolluted ($I_{geo} < 0$); unpolluted to moderately polluted ($0 \leq I_{geo} < 1$); moderately polluted ($1 \leq I_{geo} < 2$); moderately to heavily polluted ($2 \leq I_{geo} < 3$); strongly polluted ($3 \leq I_{geo} < 4$); strongly to extremely polluted ($4 \leq I_{geo} < 5$); and extremely polluted ($I_{geo} \geq 5$).

### 2.4.4. Potential Ecological Risk Assessment

The potential ecological risk index of an individual element ($E_r^i$) and the comprehensive potential ecological risks (RI) and of PTEs in soils and sediments were evaluated following Equations (1), (4) and (5), as established by Hakanson [41].

$$E_r^i = C_f^i \times T_r^i \quad (4)$$

$$RI = \sum E_r^i = \sum (C_f^i \times T_r^i) \quad (5)$$

$T_r^i$ is the toxicity response factor of each metal, where Hg = 40, Cd = 30, As = 10, Cu = Pb = Ni = 5, Cr = 2, and Zn = 1 [44,45]. $C_f^i$ is calculated as Equation (1). The ecological risks of individual metal ($E_r^i$) were divided into five levels: low risk, ($E_r^i < 40$); moderate risk ($40 \leq E_r^i < 80$); considerable risk ($80 \leq E_r^i < 160$); high risk ($160 \leq E_r^i < 320$); and very high risk ($E_r^i > 320$). Based on $RI$ values, the comprehensive ecological risks of PTEs were divided into four levels: low risk ($RI < 150$); moderate risk ($150 \leq RI < 300$); considerable risk ($300 \leq RI < 600$), and high risk ($RI > 600$).

### 2.5. Data Analysis

XSLTAT software and R (version 3.6.3) were used for the statistical analysis, including Pearson's correlation analysis, HCA, and PCA. The statistical method was performed with a 95% confidence interval (significance $p < 0.05$). Due to a wide range of Pb and other metal concentrations in soils and sediments, the original data were standardized before carrying out HCA and PCA [15].

A binary mixing model was used to quantify the relative contributions of mining activity-associated Pb to the soils and sediments. The model was calculated from the values of $^{206}$Pb/$^{207}$Pb and $^{208}$Pb/$^{206}$Pb and the mean contribution was derived [46], calculated according to Equations (6)–(8).

$$X_1 = \frac{\left(\frac{^{206}Pb}{^{207}Pb}\right)_{Sample} - \left(\frac{^{206}Pb}{^{207}Pb}\right)_{Source\ A}}{\left(\frac{^{206}Pb}{^{207}Pb}\right)_{Source\ A} - \left(\frac{^{206}Pb}{^{207}Pb}\right)_{Source\ B}} \qquad (6)$$

$$X_2 = \frac{\left(\frac{^{208}Pb}{^{206}Pb}\right)_{Sample} - \left(\frac{^{208}Pb}{^{206}Pb}\right)_{Source\ A}}{\left(\frac{^{208}Pb}{^{206}Pb}\right)_{Source\ A} - \left(\frac{^{208}Pb}{^{206}Pb}\right)_{Source\ B}} \qquad (7)$$

$$\overline{X} = (X_1 + X_2)/2 \qquad (8)$$

where $X_1$ and $X_2$ are the percentages fraction of source A calculated from $^{206}$Pb/$^{207}$Pb and $^{208}$Pb/$^{206}$Pb, respectively. $\overline{X}$ is the average of $X_1$ and $X_2$.

## 3. Results and Discussion

### 3.1. Current PTEs Contamination in Soils and Sediments

The average concentrations of the studied PTEs in soils and sediments, in order of abundance, were as follows: Pb > Cu > Zn > Cr > Ni > As > Cd > Hg and Zn > Cr > Pb > Cu > Ni > As > Cd > Hg, respectively (Table 1). In general, the PTE concentration of unpolluted soils (outside mining area, SO1–SO4) was close to the background value of Beijing (Table 1), consistent with the previous studies for forest and grassland soils of Miyun Reservoir [30] and rural soils of Beijing [15]. The concentrations of all investigated PTEs were significantly higher in the mining-impacted soils (SI1–SI12) than in the unpolluted soils (SO1–SO4). For example, the mean concentration of Cd in the unpolluted soils (SO1–SO4) (0.13 mg·kg$^{-1}$) was similar to the Cd geochemical baseline concentration (0.12 mg·kg$^{-1}$) in Miyun Reservoir [47], whereas up to 0.79 mg·kg$^{-1}$ (mean value) of Cd was detected in the mining-impacted soils (SI1–SI12). The mean concentrations of Cu, Pb, Zn, Cr, Ni, Cd, As, and Hg in the sediments were 35, 39, 94, 67, 28, 0.3, 4.4, and 0.03 kg$^{-1}$, respectively, which are comparable to the levels in other river sediments in China [48,49]. However, the concentrations (in mg·kg$^{-1}$) of Zn (228), Pb (192), and Cu (92.1) were much higher than those of the upper continental crust [50] and Beijing background values [16]. Therefore, it can be predicted that metal-rich ores may be responsible for the significant increase in the concentration of some specific PTEs (Pb, Cu, and Cd) in the river sediments [48,51].

Spatially, the PTEs in the soils (32% < CV < 220%) were considerably more variable than in the sediments (29% < CV < 129%), although both were significantly higher than in unpolluted soils (10% < CV < 45%), as shown in Figure 2 by different colors. The variations may be caused by the complex geological and geographical features among the different sites and the surrounding anthropogenic activities [51]. Extremely high concentrations (in mg·kg$^{-1}$) of Pb (27,368), Cu (1582), Zn (792), and Cd (4.1) were found at Site SI6, which is geographically adjacent to the ore deposits (Figures 1 and 2). Cr (195) and Ni (62.3) exhibited the highest concentration at Site SI2, and the highest concentration of As (59.1) and Hg (1.14) was found at SI11. It can be seen from Table 1 that the maximum concentrations of each PTE were all detected in the soils (SI2, SI6, and SI11), and the mean concentrations of PTEs in the soils were also greater than in the sediments. Therefore, the overload of PTEs in soils and sediments is likely to have been disturbed by varying degrees of mining activities. Meanwhile, the soils in the study area were more significantly disturbed by mining activities. On the one hand, this phenomenon may originate from the fact that soils inside the mining area are closer to the source(s) point of contamination from various mining activities (Figure 1). The distance between sampling sites and point sources (e.g., mining, smelters, and tailings dams) may significantly affect metal accumulation through different diffusion intensities of anthropogenic activities [52,53]. On the other hand, the mobility and availability of PTEs in soils and sediments are influenced by several factors, including topography, oxic–anoxic interface, adsorption/desorption processes, pH, salinity, and organic matter [54]. The results of the current metal contamination in soils and sediments further confirm previous findings that there are varying degrees of metal contamination upstream of the Miyun Reservoir [27,33].

Table 1. Concentration of PTEs in soils, unpolluted soils, and sediments.

| Element | Cu | Pb | Zn | Cr | Ni | Cd | As | Hg |
|---|---|---|---|---|---|---|---|---|
| | | | | mg/kg | | | | |
| | | | | Soils (n = 36) | | | | |
| Max | 1580 | 27400 | 792 | 195 | 62.3 | 4.10 | 59.1 | 1.140 |
| Min | 13.0 | 4.77 | 18.7 | 44.3 | 18.2 | 0.0970 | 2.07 | 0.004 |
| Mean | 199 | 2470 | 181 | 117 | 40.4 | 0.791 | 8.52 | 0.168 |
| Meadium | 71.4 | 55.7 | 93.2 | 105 | 38.4 | 0.315 | 3.92 | 0.026 |
| SD | 438 | 785 | 211 | 52.7 | 13.8 | 1.12 | 16.0 | 0.329 |
| CV% | 220 | 31.8 | 116 | 45.2 | 34.1 | 141 | 188 | 197 |
| | | | | Unpolluted Soils (n = 12) | | | | |
| Max | 29.0 | 25.5 | 71.4 | 80.1 | 33.9 | 0.150 | 7.45 | 0.000019 |
| Min | 19.4 | 20.3 | 49.7 | 43.8 | 19.5 | 0.110 | 2.49 | 0.000010 |
| Mean | 23.6 | 23.1 | 64.2 | 55.7 | 24.9 | 0.128 | 4.71 | 0.000014 |
| Meadium | 23.0 | 23.3 | 67.9 | 49.5 | 23.2 | 0.125 | 4.46 | 0.000014 |
| SD | 4.79 | 2.21 | 9.83 | 17.0 | 6.87 | 0.0171 | 2.10 | 0.000004 |
| CV% | 20.3 | 9.57 | 15.3 | 30.5 | 27.6 | 13.4 | 44.5 | 28.0 |
| | | | | Sediments (n = 45) | | | | |
| Max | 92.1 | 192 | 325 | 129 | 45.5 | 0.910 | 15.0 | 0.000190 |
| Min | 16.7 | 16.1 | 46.9 | 44.7 | 18.4 | 0.100 | 2.46 | 0.000010 |
| Mean | 36.7 | 42.5 | 113 | 68.8 | 28.9 | 0.336 | 5.10 | 0.000034 |
| Meadium | 29.8 | 24.0 | 100 | 64.2 | 27.1 | 0.200 | 4.29 | 0.000017 |
| SD | 19.3 | 43.9 | 74.3 | 22.6 | 8.51 | 0.262 | 3.15 | 0.000044 |
| CV% | 52.7 | 103 | 66 | 32.8 | 29.4 | 77.8 | 61.7 | 129 |
| BBV [a] | 22.5 | 23.7 | 71.4 | 80.1 | 33.9 | 0.20 | 7.50 | 0.0700 |
| UCC [b] | 28.0 | 17.0 | 67.0 | 92.0 | 47.0 | 0.0900 | 4.80 | 0.0500 |

[a] Beijing Background Value; [b] Upper Continental Crust.

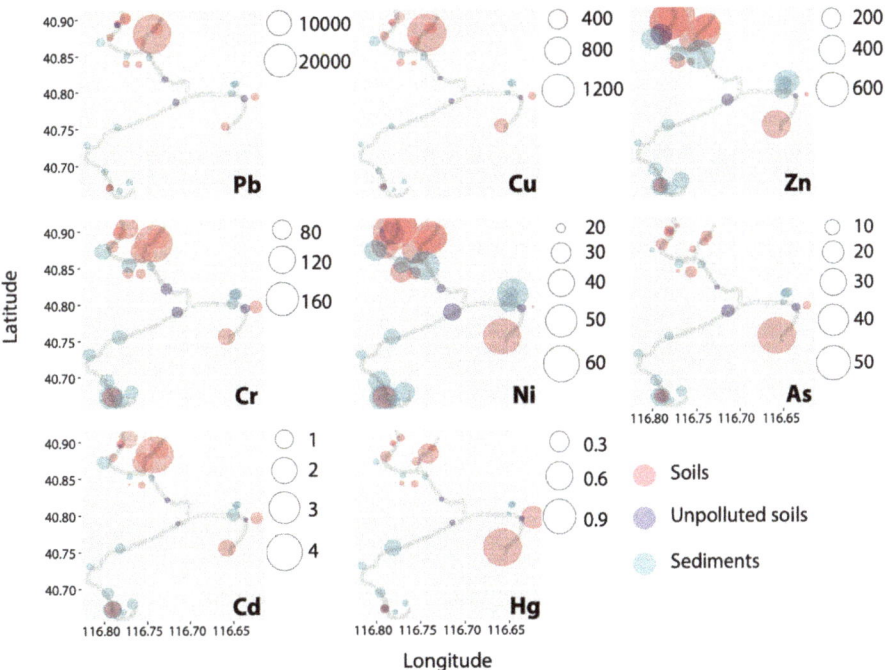

Figure 2. Spatial distribution of PTE contents (mg/kg) in soils and sediments.

## 3.2. Pollution Assessment of PTEs in Soils and Sediments

The pollution assessment methods, including CF, PLI, $I_{geo}$, and RI, generally reflected higher contamination levels in the soils than in the sediments, as shown in Figure 3 and Table 2. For example, average CF values of Cu, Pb, Zn, Cr, Ni, Cd, As, and Hg in sediments were 1.63, 1.79, 1.58, 0.86, 0.85, 1.68, 0.68, and 0.49, respectively, which belong to the unpolluted (CF < 1) and moderate-to-low ($1 \leq CF < 3$) polluted levels (Table 2). However, the average CF values of PTEs in soils were much higher with very high polluted levels ($CF \geq 6$) of Cu (8.84) and Pb (8.80); high polluted levels of Cd (3.96); and moderate-to-low polluted levels of Zn (2.53), Cr (1.46), Ni (1.19), As (1.14), and Hg (2.40). The PLI values of the sediment sites ranged from 0.57 to 1.92 with an average of 0.97, indicating almost no heavy pollution (PLI > 1), while seven of the 12 soil sampling sites had PLI values within 1.2–9.8 (Table 2). Similarly, the mean $I_{geo}$ value for each PTE in the sediments was below zero, while the mean $I_{geo}$ values in soils for PTEs ranged from −1.37 to 1.56 (Figure 2). The difference in $I_{geo}$ between soils impacted by mining activities (SI1-SI12) and unpolluted soils (SO1–SO4) can be clearly seen in Figure 2, with $I_{geo}$ values as high as 9 for the former and below 0 for the latter. Notably, for Pb, the Igeo values for SI1, SI5, and SI6 were classified as strongly contaminated ($4 \leq I_{geo} < 5$), and for SI6, it was classified as extremely contaminated ($I_{geo} \geq 5$). Notably, the highest PLI value was found at SI6 (9.8), followed by SI11 (4.4), SI1 (2.5), and SI5 (2.1), all located within 2 km of either of the tailings ponds, mines, or the smelters (Figure 1); thus, proximity to mining activities might account for the higher PLI at these soil sites. None of the PTEs in sediments showed a heavy or extreme pollution index with $I_{geo} < 3$. Sediment sampling sites did not have high or very high contamination indices, and most sampling sites were uncontaminated to moderately contaminated with PLI < 2, CD < 20, and RI < 300 (Table 2). One-third of the sediment sampling sites had uncontaminated levels ($I_{geo} < 0$) for all PTEs, indicating that these sites (SD5, SD6, SD9, SD12, and SD15) have not been impacted by mining activities.

**Table 2.** Pollution assessment of PTEs in soils and sediments.

| Type | Site | Contamination Factor (CF) | | | | | | | | Pollution Load Index (PLI) | | Potential Ecological Risk Index (RI) | |
|---|---|---|---|---|---|---|---|---|---|---|---|---|---|
| | | Cu | Pb | Zn | Cr | Ni | Cd | As | Hg | Values | Pollution Levels | Values | Pollution Levels |
| Soils | SI1 | 5.20 | 35.4 | 3.50 | 2.23 | 1.53 | 5.30 | 0.36 | 0.34 | 2.48 | Moderate | 381 | Considerable |
| | SI2 | 1.72 | 0.71 | 1.40 | 2.43 | 1.84 | 1.15 | 0.41 | 0.11 | 0.90 | Unpolluted | 66 | Low |
| | SI3 | 0.58 | 0.20 | 0.26 | 2.35 | 1.04 | 0.60 | 0.58 | 0.06 | 0.44 | Unpolluted | 38 | Low |
| | SI4 | 4.53 | 0.74 | 0.86 | 1.31 | 1.50 | 0.65 | 0.79 | 0.19 | 0.93 | Unpolluted | 65 | Low |
| | SI5 | 4.49 | 37.0 | 3.31 | 1.30 | 1.23 | 4.80 | 0.28 | 0.36 | 2.12 | Moderate | 366 | Considerable |
| | SI6 | 70.3 | 1150 | 11.1 | 1.92 | 1.06 | 20.5 | 0.55 | 3.97 | 9.75 | Very high | 6767 | High |
| | SI7 | 3.61 | 9.79 | 3.92 | 1.30 | 1.15 | 6.05 | 0.40 | 0.53 | 2.01 | Moderate | 265 | Moderate |
| | SI8 | 2.04 | 1.16 | 0.90 | 0.80 | 0.90 | 0.49 | 0.29 | 0.14 | 0.65 | Unpolluted | 40 | Low |
| | SI9 | 0.89 | 1.02 | 0.78 | 0.59 | 0.65 | 0.85 | 0.83 | 0.39 | 0.72 | Unpolluted | 49 | Low |
| | SI10 | 2.27 | 3.03 | 1.21 | 0.55 | 0.54 | 2.00 | 0.50 | 5.46 | 1.38 | Moderate to unpolluted | 98 | Low |
| | SI11 | 7.64 | 6.03 | 2.07 | 1.57 | 1.76 | 4.05 | 7.88 | 16.3 | 4.39 | High | 286 | Considerable |
| | SI12 | 2.73 | 1.67 | 1.08 | 1.13 | 1.11 | 1.05 | 0.79 | 0.93 | 1.21 | Moderate to unpolluted | 70 | Low |
| Unpolluted soils | SO1 | 1.29 | 0.86 | 0.94 | 1.00 | 1.00 | 0.75 | 0.67 | 0.27 | 0.78 | Unpolluted | 48 | Low |
| | SO2 | 0.88 | 1.08 | 1.00 | 0.55 | 0.58 | 0.65 | 0.33 | 0.14 | 0.56 | Unpolluted | 38 | Low |
| | SO3 | 1.16 | 1.01 | 0.96 | 0.68 | 0.79 | 0.60 | 0.99 | 0.21 | 0.73 | Unpolluted | 45 | Low |
| | SO4 | 0.86 | 0.95 | 0.70 | 0.56 | 0.58 | 0.55 | 0.52 | 0.17 | 0.56 | Unpolluted | 36 | Low |

Table 2. Cont.

| Type | Site | Contamination Factor (CF) | | | | | | | | Pollution Load Index (PLI) | | Potential Ecological Risk Index (RI) | |
|---|---|---|---|---|---|---|---|---|---|---|---|---|---|
| | | Cu | Pb | Zn | Cr | Ni | Cd | As | Hg | Values | Pollution Levels | Values | Pollution Levels |
| Sediments | SD1 | 4.09 | 8.10 | 1.99 | 1.05 | 0.89 | 2.50 | 0.33 | 0.24 | 1.37 | Moderate to unpolluted | 148 | Low |
| | SD2 | 1.77 | 0.88 | 1.61 | 0.82 | 0.75 | 0.75 | 0.539 | 0.23 | 0.78 | Unpolluted | 48 | Low |
| | SD3 | 2.30 | 0.68 | 0.66 | 1.61 | 1.27 | 0.75 | 0.49 | 0.21 | 0.80 | Unpolluted | 53 | Low |
| | SD4 | 2.59 | 1.00 | 0.89 | 1.31 | 1.34 | 1.00 | 0.59 | 0.19 | 0.90 | Unpolluted | 64 | Low |
| | SD5 | 1.22 | 0.93 | 0.83 | 0.62 | 0.66 | 0.90 | 0.57 | 0.33 | 0.71 | Unpolluted | 49 | Low |
| | SD6 | 0.74 | 0.85 | 0.83 | 0.75 | 0.54 | 0.50 | 0.55 | 0.20 | 0.57 | Unpolluted | 34 | Low |
| | SD7 | 1.23 | 1.89 | 1.07 | 0.84 | 0.90 | 2.30 | 0.34 | 0.14 | 0.81 | Unpolluted | 95 | Low |
| | SD8 | 1.32 | 1.85 | 1.62 | 0.69 | 0.73 | 1.55 | 0.53 | 2.64 | 1.20 | Moderate to unpolluted | 75 | Low |
| | SD9 | 1.03 | 1.01 | 1.24 | 0.58 | 0.61 | 0.85 | 0.60 | 0.51 | 0.77 | Unpolluted | 47 | Low |
| | SD10 | 1.70 | 1.35 | 1.57 | 0.75 | 0.90 | 1.95 | 0.78 | 0.89 | 1.16 | Moderate to unpolluted | 89 | Low |
| | SD11 | 1.90 | 3.00 | 4.55 | 0.94 | 1.20 | 4.55 | 2.00 | 0.70 | 1.92 | Moderate to unpolluted | 194 | Moderate |
| | SD12 | 1.04 | 0.96 | 0.74 | 0.56 | 0.63 | 0.50 | 0.65 | 0.14 | 0.57 | Unpolluted | 36 | Low |
| | SD13 | 1.54 | 2.07 | 3.19 | 0.80 | 1.00 | 4.45 | 1.19 | 0.46 | 1.45 | Moderate to unpolluted | 173 | Moderate |
| | SD14 | 1.18 | 1.44 | 1.40 | 0.86 | 0.87 | 1.75 | 0.67 | 0.29 | 0.94 | Unpolluted | 80 | Low |
| | SD15 | 0.81 | 0.92 | 1.44 | 0.73 | 0.62 | 0.90 | 0.38 | 0.19 | 0.65 | Unpolluted | 45 | Low |

### 3.3. Source Identification

Pearson's correlation coefficient (Figure 4) showed a strongly positive correlation between Pb with Cu (0.996), Zn (0.939), and Cd (0.995) in the soils ($p < 0.01$). Meanwhile, As and Hg (0.926) had a significantly positive correlation in the soil, as well as Cr and Ni (0.653). The results of PCA (Table 3) displayed that factor one (F1) captured Cu, Pb, Zn, Cd, and Hg (52.7%), and factor two (F2) captured Cr and Ni (23.4%) in the soils, accounting for 76.2% of the total variance. This evidence indicates that they have possible parallel geochemical behaviors, which means they are likely from the same source(s) [33,55,56]. HCA was used to group sample sites and PTE concentrations together (shown as 2D heatmap), which provides more information in terms of the point source distribution and potential sources in soils and sediments. As shown in Figure 5, the unpolluted soils (SO1–SO4) clustered together and exhibited relatively low levels of PTEs. With the exception of sample site SI9, all other soils inside the mining area exhibited contamination with different PTEs. For example, SI5, SI6, and SI7 exhibited relatively high levels of Pb, Zn, Cu, and Cd. In addition, SI11 showed high levels of Hg, As, Ni, and Cr. These pieces of evidence combining the higher loading of PTEs than the background value of Beijing pointed out that Cu, Pb, Zn, and Cd in the soils inside the mining area likely originated from mining activities. The results confirm previous research that mining activities are the major source of Cu, Zn, and Pb from Pb-Zn ores, atmosphere deposition, acidic mine drainage wastewater from smelters, erosion, and leaching of tailings [57,58]. In addition, Ni and Cr in soils were closely associated with natural sources likely originating from the soils' parent material (lithogenic origin). As and Hg are likely derived from the traditional extraction process of gold ore, amalgamation for gold extraction [27,59].

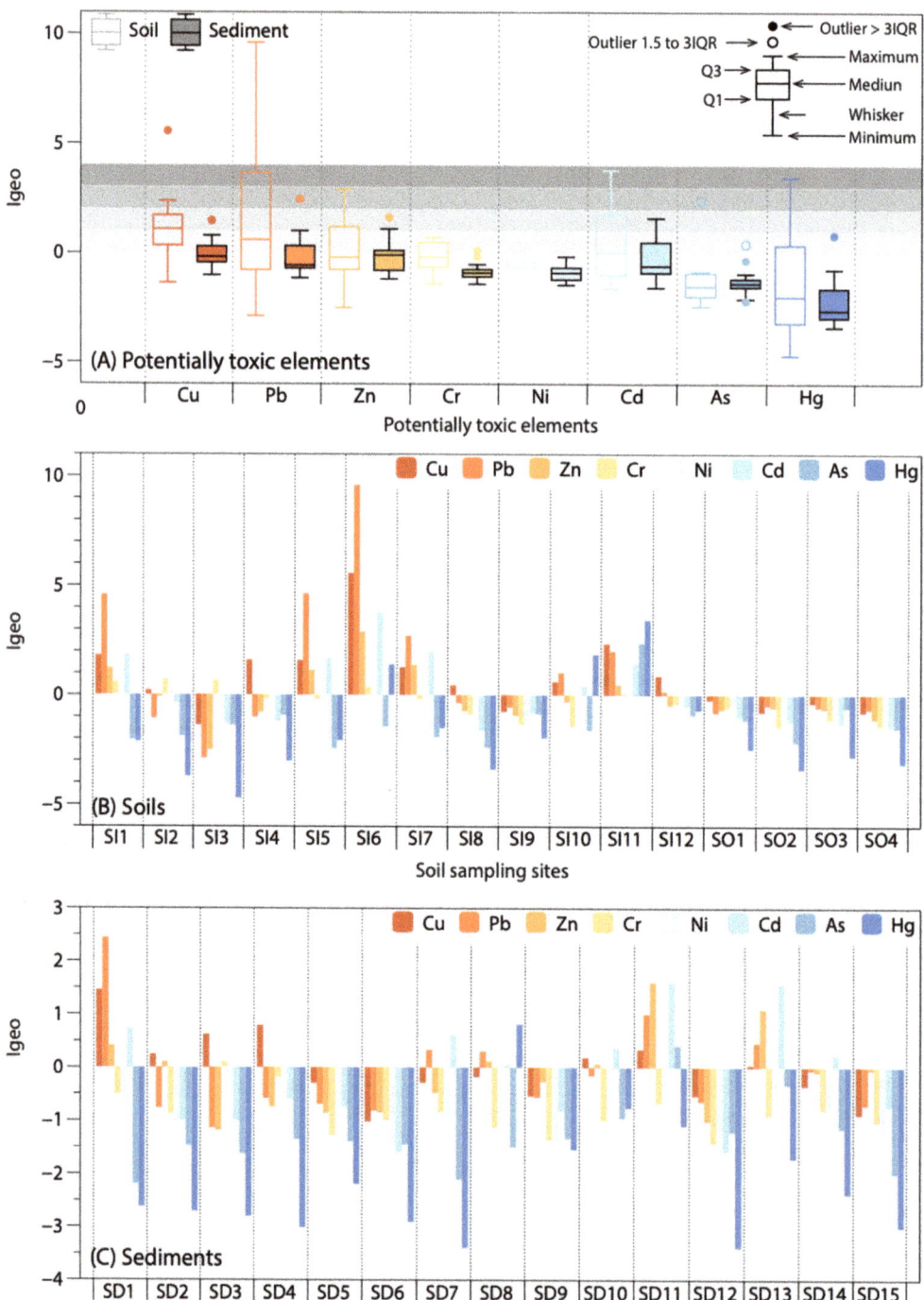

**Figure 3.** $I_{geo}$ of PTEs in soils and sediments.

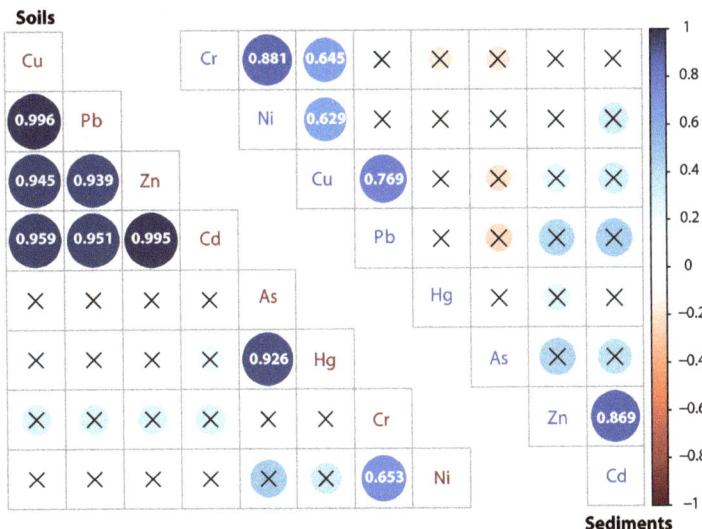

**Figure 4.** Pearson's correlation analysis of PTE concentration in soils (**Left**) and sediments (**Right**) inside the mining area.

The results obtained from PCA (Table 3) for PTEs in sediments showed that F1 explained 42.7% of the total variance with high positive loadings, Cd (0.850), Cu (0.820), Ni (0.753), Pb (0.722), Zn (0.685), and Cr (0.593), meaning a common source is possibly mining activities; F2 with a high value for Cr (0.733) described 26.6% of the total variance. Both F1 and F2 contained Cr, suggesting that Cr could originate from multiple sources. This speculation is further supported by the HCA and Pearson's correlation coefficient. The HCA results divided the PTEs into two groups in the sediments: As, Hg, Pb, Zn, and Cd in group 1; and Cu, Cr, and Ni in group 2, indicating that the same group may originate from the common source(s) (Figure 5). In addition, Cu-Cr-Ni and Zn-Cd were significantly correlated with each other ($r < 0.6$, $p < 0.01$) based on Pearson's correlation analysis. Thus, it can be observed that the correlation of PTEs in sediments is weaker compared to soils. At the same time, the overall content of PTEs is low, and fewer sampling sites are affected by mining activity disturbance.

**Table 3.** Results of principal component analysis.

| Element | Soils | | Sediments | |
|---|---|---|---|---|
| | F1 | F2 | F1 | F2 |
| Cu | **0.938** | 0.046 | **0.820** | 0.417 |
| Pb | **0.943** | −0.169 | **0.722** | −0.228 |
| Zn | **0.943** | −0.062 | **0.685** | −0.600 |
| Cr | 0.234 | **0.904** | **0.592** | **0.733** |
| Ni | 0.225 | **0.924** | **0.753** | 0.487 |
| Cd | **0.952** | −0.054 | **0.850** | −0.365 |
| As | 0.222 | 0.223 | 0.102 | −0.468 |
| Hg | **0.706** | −0.343 | 0.325 | −0.643 |
| Eigenvalue | 4.22 | 1.88 | 3.42 | 2.13 |
| Variability (%) | 52.7 | 23.4 | 42.7 | 26.6 |
| Cumulative (%) | 52.7 | 76.2 | 42.7 | 69.3 |

Greater than 0.5 are shown in bold.

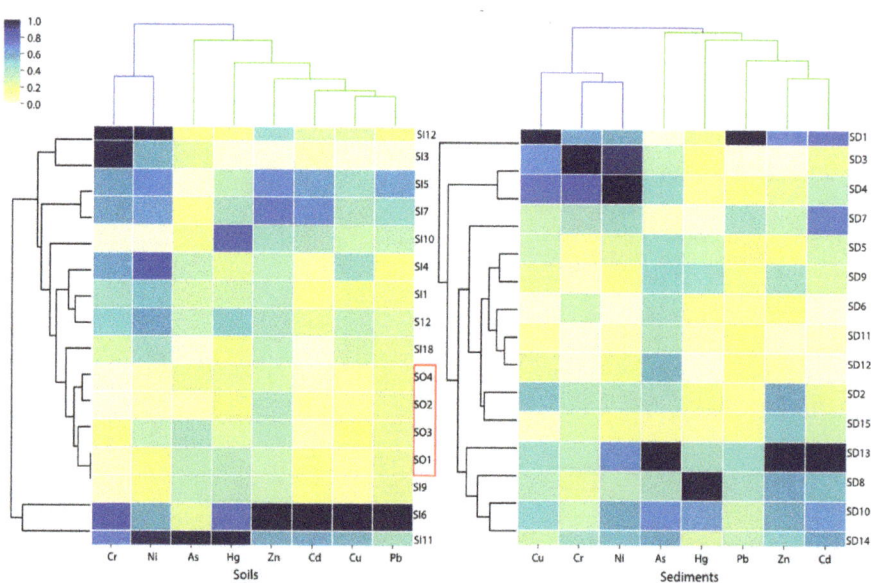

**Figure 5.** The HCA results are shown as 2D (Euclidean distance; agglomeration method: Ward's method).

*3.4. Pb Isotope Ratios and Source Apportionment*

As shown in Figure 6, the uncorrelated relationship between Pb concentration (1/Pb, kg/mg) and Pb isotopic ratio ($^{206}Pb/^{207}Pb$ and $^{208}Pb/^{206}Pb$) in both soils ($R^2 = 0.07$) and sediments ($R^2 = 0.03$) indicated a mixing of different Pb sources in soils and sediments (Xu et al., 2019b). The plot of $^{206}Pb/^{207}Pb$ vs. $^{208}Pb/^{206}Pb$ ratios of the soils, sediments, and tailings in this study is displayed in Figure 6c. The Pb isotopic composition of soils outside the mining area (SO1–SO4) received a significant input of adventitious Pb with high $^{206}Pb/^{207}Pb$ (1.167 ± 0.0029) and low $^{208}Pb/^{206}Pb$ (2.105 ± 0.0048), which is in line with China soils from the northeast geochemical region $^{206}Pb/^{207}Pb$ of 1.153–1.175, and $^{208}Pb/^{206}Pb$ ratios of 2.11 ± 0.005 [60]. Pb found in unpolluted sediments was usually derived from various natural sources, including weathering of catchment soils and bedrock or transported more directly within mineral matter eroded from catchment [61]. As listed in Table 4, there was a wider range of $^{206}Pb/^{207}Pb$ and $^{208}Pb/^{206}Pb$ ratios in soils (1.095–1.148 and 2.127–2.196) compared to sediments (1.120–1.154 and 2.122–2.167), which is consistent with data obtained in the corresponding Pb concentration. This reflected that soils within the mining area may be more severely disturbed by mining activities than sediments are. The dominating sources of Pb pollution in the Chinese mining area may originate from mining and industrial emissions such as metal processing and manufacturing, as well as coal combustion (transportation of aerosol deposition) and vehicle exhausts [30,47,53]. The contribution of leaded gasoline was not considered in this study, because, at the end of the last century, leaded gasoline was completely banned in China, and leaded gasoline has a quite low Pb concentration [16].

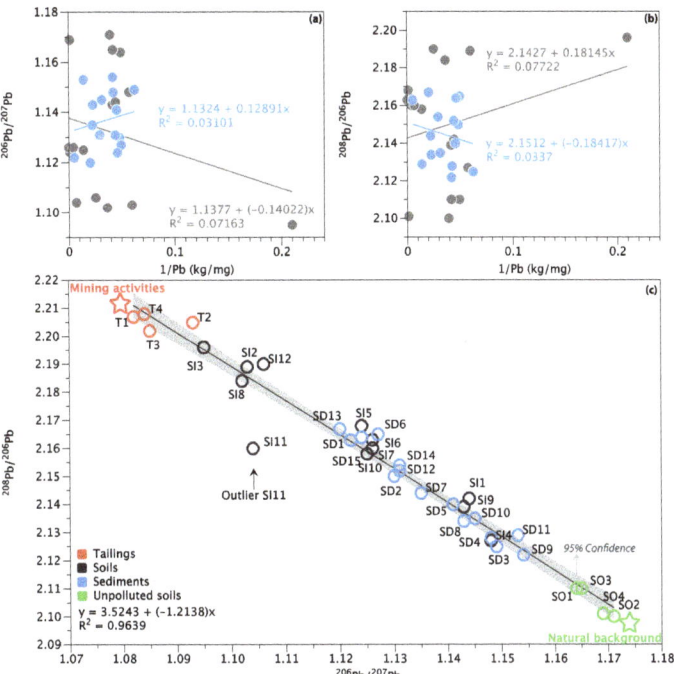

**Figure 6.** The plot of $^{206}Pb/^{207}Pb$ (**a**) and $^{208}Pb/^{206}Pb$ (**b**) versus the inverse of Pb concentration of the soils and sediments, as well as the plot of $^{206}Pb/^{207}Pb$ versus $^{208}Pb/^{206}Pb$ ratios of the sediments, soils, and tailings in this study (**c**).

Given a strong linear trend between the Pb isotope ratios ($^{208}Pb/^{206}Pb$ vs. $^{206}Pb/^{207}Pb$) of tailings, unpolluted soils, soils, and sediments ($R^2$ = 0.96), only the source contributions of Pb from mining activities (as source A) and natural background (as source B) were considered in this study, and the average relative Pb contributions were calculated for each site according to Equations (6)–(8). As shown in Figure 7, the results showed that mining activity contributes most of the mining activity-related Pb to soils, with an average relative contribution of 58.9%, ranging from 27.2% (SI4) to 86.7% (SI3). For the sediments, the natural background appeared to be the main source of Pb (58.8%), while contributions from mining activity ranged from 21.7% (SD9) to 60.2% (SD13). It confirmed that mining activity was the major source of Pb pollution in soils. This phenomenon is mainly due to the considerable contribution of long-term frequent mining activities, ore mining and smelting, and abundant small-scale mines distributed in the upper area of the Miyun Reservoir; therefore, most soil Pb likely represents locally emitted Pb [27,30,33,62]. In contrast, natural background sources are an important source of Pb in sediments. Nevertheless, some of the sample sites (e.g., SD13, SD1, SD15, and SD6) are still strongly disturbed by mining activities, with the contribution of mining activities greater than 50%. Several pieces of research have also suggested that mining activities were the dominant anthropogenic Pb source in reservoir sediments [63,64]. It is noteworthy that mining activity-related Pb sources account for 66% of the significant outliers (SI11) in Figure 6. It is speculated that the reason for the deviation may be the influence of other external sources at this sample site, which significantly changed its Pb isotopic composition. This is also supported by the HCA results that, although both sample sites SI6 and SI11 are heavily contaminated (Table 2), the dominant PTEs are significantly different (Figure 5).

Table 4. Pb isotopic component in soils and sediments.

| Type | Sample Site | $^{208}Pb/^{204}Pb$ | $^{207}Pb/^{204}Pb$ | $^{206}Pb/^{204}Pb$ | $^{206}Pb/^{207}Pb$ | $^{208}Pb/^{206}Pb$ |
|---|---|---|---|---|---|---|
| Soils | SI1 | 37.92 | 15.48 | 17.70 | 1.144 | 2.142 |
| | SI2 | 36.72 | 15.21 | 16.77 | 1.103 | 2.189 |
| | SI3 | 36.77 | 15.30 | 16.75 | 1.095 | 2.196 |
| | SI4 | 37.78 | 15.47 | 17.76 | 1.148 | 2.127 |
| | SI5 | 38.04 | 15.61 | 17.55 | 1.124 | 2.168 |
| | SI6 | 37.91 | 15.56 | 17.52 | 1.126 | 2.163 |
| | SI7 | 37.61 | 15.47 | 17.41 | 1.126 | 2.160 |
| | SI8 | 36.91 | 15.34 | 16.90 | 1.102 | 2.184 |
| | SI9 | 37.81 | 15.47 | 17.68 | 1.143 | 2.139 |
| | SI10 | 37.51 | 15.45 | 17.38 | 1.125 | 2.158 |
| | SI11 | 36.27 | 15.21 | 16.79 | 1.104 | 2.160 |
| | SI12 | 37.15 | 15.33 | 16.96 | 1.106 | 2.190 |
| | mean | 37.37 ± 0.558 | 15.41 ± 0.124 | 17.26 ± 0.382 | 1.121 ± 0.0175 | 2.165 ± 0.0210 |
| | median | 37.56 | 15.46 | 17.40 | 1.125 | 2.162 |
| Unpolluted soils | SO1 | 38.21 | 15.55 | 18.11 | 1.164 | 2.110 |
| | SO2 | 38.21 | 15.54 | 18.19 | 1.171 | 2.100 |
| | SO3 | 38.15 | 15.52 | 18.08 | 1.165 | 2.110 |
| | SO4 | 38.15 | 15.54 | 18.16 | 1.169 | 2.101 |
| | mean | 38.18 ± 0.030 | 15.54 ± 0.011 | 18.14 ± 0.043 | 1.167 ± 0.0029 | 2.105 ± 0.0048 |
| | median | 38.18 | 15.54 | 18.14 | 1.167 | 2.106 |
| Sediments | SD1 | 37.48 | 15.45 | 17.33 | 1.122 | 2.163 |
| | SD2 | 37.47 | 15.42 | 17.42 | 1.130 | 2.150 |
| | SD3 | 37.82 | 15.48 | 17.79 | 1.149 | 2.125 |
| | SD4 | 37.79 | 15.47 | 17.76 | 1.148 | 2.128 |
| | SD5 | 37.80 | 15.48 | 17.66 | 1.141 | 2.140 |
| | SD6 | 37.66 | 15.43 | 17.39 | 1.127 | 2.165 |
| | SD7 | 37.67 | 15.48 | 17.57 | 1.135 | 2.144 |
| | SD8 | 37.76 | 15.48 | 17.69 | 1.143 | 2.134 |
| | SD9 | 37.90 | 15.47 | 17.85 | 1.154 | 2.122 |
| | SD10 | 37.84 | 15.48 | 17.73 | 1.145 | 2.135 |
| | SD11 | 37.99 | 15.49 | 17.85 | 1.153 | 2.129 |
| | SD12 | 37.58 | 15.43 | 17.46 | 1.131 | 2.152 |
| | SD13 | 37.42 | 15.41 | 17.27 | 1.120 | 2.167 |
| | SD14 | 37.57 | 15.42 | 17.44 | 1.131 | 2.154 |
| | SD15 | 37.52 | 15.430 | 17.34 | 1.124 | 2.164 |
| | mean | 37.68 ± 0.168 | 15.45 ± 0.027 | 17.57 ± 0.196 | 1.137 ± 0.0111 | 2.145 ± 0.0152 |
| | median | 37.67 | 15.47 | 17.57 | 1.135 | 2.144 |
| Tailings | T1 | 36.63 | 15.30 | 16.60 | 1.085 | 2.207 |
| | T2 | 36.69 | 15.23 | 16.64 | 1.093 | 2.205 |
| | T3 | 36.53 | 15.30 | 16.60 | 1.085 | 2.202 |
| | T4 | 36.63 | 15.24 | 16.60 | 1.085 | 2.208 |
| | mean | 36.62 ± 0.057 | 15.27 ± 0.033 | 16.61 ± 0.017 | 1.087 ± 0.0035 | 2.205 ± 0.0023 |
| | median | 36.63 | 15.27 | 16.60 | 1.085 | 2.206 |

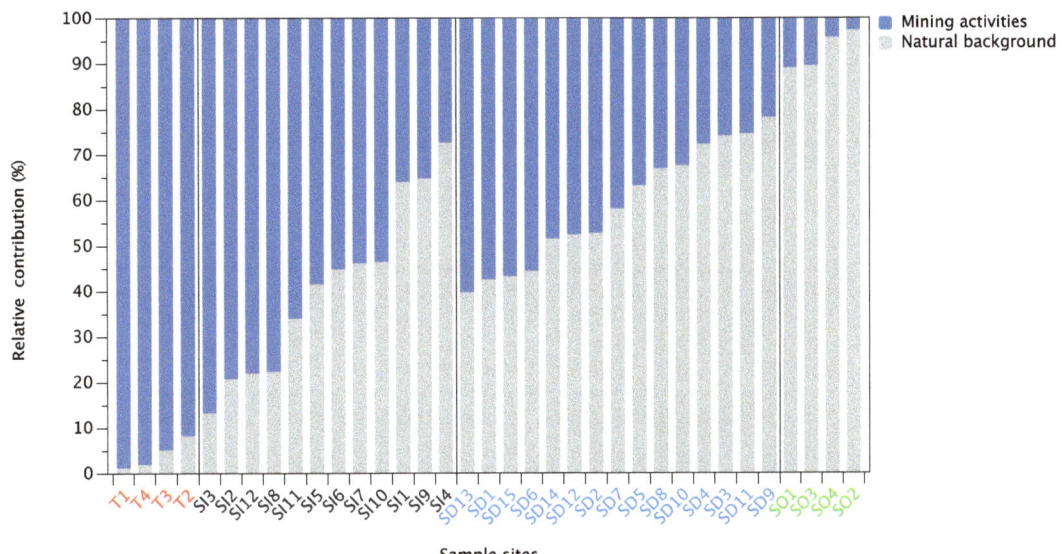

**Figure 7.** Average relative contribution (%) of Pb from different sources in soils and sediments.

## 4. Conclusions

Soils and sediments around the gold mine site have been affected to varying degrees by mining activities, and the soils have been more strongly disturbed. The average concentrations of PTEs in the soil markedly exceeded the local background values, and the Pb content in some sample sites was even several hundred times higher than the background values. The results of the multivariate statistical analysis suggested that the accumulation of Cu, Zn, Pb, and Cd may be mainly from mining activities, while Cr and Ni are from natural background sources in soils. Soils have much wider Pb isotopic ratio ($^{208}Pb/^{206}Pb$ and $^{206}Pb/^{207}Pb$) ranges than sediments do in the study area. The results of Pb isotopic fingerprinting with a binary mixing model indicated that the average relative contribution of mining activities to Pb accumulation accounts for 58.9% in soils and 41.2% in sediments. The mining activities were suggested to be the dominant contribution of Pb pollution in soils. The findings provide quantitative guidance for the environmental management of PTEs and control of the mining activities around the Miyun Reservoir.

**Author Contributions:** Conceptualization, L.T. and Y.Z.; methodology, S.M.; software, C.Y.; validation, H.G., G.Y. and L.T.; formal analysis, Y.Z.; investigation, H.G.; resources, C.Y.; data curation, S.M.; writing—original draft preparation, L.T. and Y.Z.; writing—review and editing, F.W. and H.J.; visualization, L.T.; supervision, H.J. and F.W.; project administration, F.W.; funding acquisition, F.W. All authors have read and agreed to the published version of the manuscript.

**Funding:** This research was funded by the National Natural Science Foundation of China, grant numbers 41822706, 41473096; Beijing Natural Science Foundation, grant number 8182034; and Fundamental Research Funds for the Central Universities, grant number FRF-TP-19-001C1.

**Institutional Review Board Statement:** Not applicable.

**Informed Consent Statement:** Not applicable.

**Acknowledgments:** We thank Zheng Gong for helping with the map.

**Conflicts of Interest:** The authors declare no conflict of interest. The funders had no role in the design of the study; in the collection, analyses, or interpretation of data; in the writing of the manuscript, or in the decision to publish the results.

## References

1. Murguía, D.I.; Bringezu, S.; Schaldach, R. Global direct pressures on biodiversity by large-scale metal mining: Spatial distribution and implications for conservation. *J. Environ. Manag.* **2016**, *180*, 409–420. [CrossRef] [PubMed]
2. Xiao, R.; Wang, S.; Li, R.; Wang, J.J.; Zhang, Z. Soil heavy metal contamination and health risks associated with artisanal gold mining in Tongguan, Shaanxi, China. *Ecotoxicol. Environ. Saf.* **2017**, *141*, 17–24. [CrossRef]
3. Purves, D. *Trace-Element Contamination of the Environment*; Elsevier: Amsterdam, The Netherlands, 2012.
4. Agudelo-Echavarría, D.M.; Olid, C.; Molina, F.; Vallejo-Toro, P.P.; Garcia-Orellana, J. Historical reconstruction of small-scale gold mining activities in tropical wetland sediments in Bajo Cauca-Antioquia, Colombia. *Chemosphere* **2020**, *254*, 126733. [CrossRef]
5. Zhang, Y.; Wang, F.; Hudson-Edwards, K.A.; Blake, R.; Zhao, F.; Yuan, Z.; Gao, W. Characterization of mining-related aromatic contaminants in active and abandoned metal(loid) tailings ponds. *Environ. Sci. Technol.* **2020**, *54*, 15097–15107. [CrossRef] [PubMed]
6. Masindi, V.; Muedi, K.L. Environmental contamination by heavy metals. *Heavy Met.* **2018**, *10*, 115–132. [CrossRef]
7. Hutchinson, T.C.; Whitby, L.M. Heavy-metal pollution in the sudbury mining and smelting region of canada, I. Soil and vegetation contamination by nickel, copper, and other metals. *Environ. Conserv.* **1974**, *1*, 123–132. [CrossRef]
8. Querol, X.; Alastuey, A.; Lopez-Soler, A.; Plana, F. Levels and chemistry of atmospheric particulates induced by a spill of heavy metal mining wastes in the Doñana area, Southwest Spain. *Atmos. Environ.* **2000**, *34*, 239–253. [CrossRef]
9. Resongles, E.; Casiot, C.; Freydier, R.; Dezileau, L.; Viers, J.; Elbaz-Poulichet, F. Persisting impact of historical mining activity to metal (Pb, Zn, Cd, Tl, Hg) and metalloid (As, Sb) enrichment in sediments of the Gardon River, Southern France. *Sci. Total Environ.* **2014**, *481*, 509–521. [CrossRef]
10. Sun, Z.; Xie, X.; Wang, P.; Hu, Y.; Cheng, H. Heavy metal pollution caused by small-scale metal ore mining activities: A case study from a polymetallic mine in South China. *Sci. Total Environ.* **2018**, *639*, 217–227. [CrossRef]
11. Huang, Y.; Zhang, S.; Chen, Y.; Wang, L.; Long, Z.; Hughes, S.S.; Ni, S.; Cheng, X.; Wang, J.; Li, T.; et al. Tracing Pb and possible correlated Cd contamination in soils by using lead isotopic compositions. *J. Hazard. Mater.* **2020**, *385*, 121528. [CrossRef]
12. Wang, P.; Li, Z.; Liu, J.; Bi, X.; Ning, Y.; Yang, S.; Yang, X. Apportionment of sources of heavy metals to agricultural soils using isotope fingerprints and multi-variate statistical analyses. *Environ. Pollut.* **2019**, *249*, 208–216. [CrossRef]
13. Hou, D.; O'Connor, D.; Nathanail, P.; Tian, L.; Ma, Y. Integrated GIS and multivariate statistical analysis for regional scale assessment of heavy metal soil contamination: A critical review. *Environ. Pollut.* **2017**, *231*, 1188–1200. [CrossRef] [PubMed]
14. Mostert, M.; Ayoko, G.A.; Kokot, S. Application of chemometrics to analysis of soil pollutants. *TrAC Trends Anal. Chem.* **2010**, *29*, 430–445. [CrossRef]
15. Wu, S.; Xia, X.; Lin, C.; Chen, X.; Zhou, C. Levels of arsenic and heavy metals in the rural soils of Beijing and their changes over the last two decades (1985–2008). *J. Hazard. Mater.* **2010**, *179*, 860–868. [CrossRef] [PubMed]
16. Cheng, H.; Hu, Y. Lead (Pb) isotopic fingerprinting and its applications in lead pollution studies in China: A review. *Environ. Pollut.* **2010**, *158*, 1134–1146. [CrossRef] [PubMed]
17. Bi, X.-Y.; Li, Z.-G.; Wang, S.-X.; Zhang, L.; Xu, R.; Liu, J.-L.; Yang, H.-M.; Guo, M.-Z. Lead isotopic compositions of selected coals, Pb/Zn ores and fuels in China and the application for source tracing. *Environ. Sci. Technol.* **2017**, *51*, 13502–13508. [CrossRef]
18. Deng, W.; Liu, W.; Wen, Y.; Li, X. A new inverse distance model to calculate the percentage contribution of various Pb sources. *Environ. Res.* **2020**, *185*, 109475. [CrossRef]
19. Komárek, M.; Ettler, V.; Chrastný, V.; Mihaljevič, M. Lead isotopes in environmental sciences: A review. *Environ. Int.* **2008**, *34*, 562–577. [CrossRef]
20. Hamilton, E.I. Book Review: Principles of isotope geology. *Earth-Sci. Rev.* **1978**, *14*, 190–191. [CrossRef]
21. ABollhöfer, A.; Rosman, K.J.R. Isotopic source signatures for atmospheric lead: The Southern Hemisphere. *Geochim. Cosmochim. Acta* **2000**, *64*, 3251–3262. [CrossRef]
22. Xu, D.; Wang, R.; Wang, W.; Ge, Q.; Zhang, W.; Chen, L.; Chu, F. Tracing the source of Pb using stable Pb isotope ratios in sediments of eastern Beibu Gulf, South China Sea. *Mar. Pollut. Bull.* **2019**, *141*, 127–136. [CrossRef]
23. Chakraborty, S.; Chakraborty, P.; Hathorne, E.; Sarkar, A.; Linsy, P.; Frank, M.; Nath, B.N. Evidence for increasing anthropogenic Pb concentrations in Indian shelf sediments during the last century. *Sci. Total Environ.* **2021**, *760*, 143833. [CrossRef]
24. Liu, J.; Luo, X.; Wang, J.; Xiao, T.; Yin, M.; Belshaw, N.S.; Lippold, H.; Kong, L.; Xiao, E.; Bao, Z.; et al. Provenance of uranium in a sediment core from a natural reservoir, South China: Application of Pb stable isotope analysis. *Chemosphere* **2018**, *193*, 1172–1180. [CrossRef]
25. Wang, Z.; Dwyer, G.S.; Coleman, D.S.; Vengosh, A. Lead isotopes as a new tracer for detecting coal fly ash in the environment. *Environ. Sci. Technol. Lett.* **2019**, *6*, 714–719. [CrossRef]
26. Xiong, L.; Zhao, X.; Wei, J.; Jin, X.; Fu, L.; Lin, Z. Linking mesozoic lode gold deposits to metal-fertilized lower continental crust in the North China Craton: Evidence from Pb isotope systematics. *Chem. Geol.* **2020**, *533*, 119440. [CrossRef]
27. Chen, X.; Ji, H.; Yang, W.; Zhu, B.; Ding, H. Speciation and distribution of mercury in soils around gold mines located upstream of Miyun Reservoir, Beijing, China. *J. Geochem. Explor.* **2016**, *163*, 1–9. [CrossRef]
28. Zhao, H.; Huang, Y.; You, S.; Wu, Y.; Zheng, F. A framework for assessing the effects of afforestation and South-to-North Water Transfer on nitrogen and phosphorus uptake by plants in a critical riparian zone. *Sci. Total Environ.* **2019**, *651*, 942–952. [CrossRef] [PubMed]
29. Su, M.; Andersen, T.; Burch, M.; Jia, Z.; An, W.; Yu, J.; Yang, M. Succession and interaction of surface and subsurface cyanobacterial blooms in oligotrophic/mesotrophic reservoirs: A case study in Miyun Reservoir. *Sci. Total Environ.* **2019**, *649*, 1553–1562. [CrossRef] [PubMed]
30. Han, L.; Gao, B.; Lu, J.; Zhou, Y.; Xu, D.; Gao, L.; Sun, K. Pollution characteristics and source identification of trace metals in riparian soils of Miyun Reservoir, China. *Ecotoxicol. Environ. Saf.* **2017**, *144*, 321–329. [CrossRef] [PubMed]

31. Zhu, X.; Ji, H.; Chen, Y.; Qiao, M.; Tang, L. Assessment and sources of heavy metals in surface sediments of Miyun Reservoir, Beijing. *Environ. Monit. Assess.* **2013**, *185*, 6049–6062. [CrossRef]
32. Li, Q.; Ji, H.; Qin, F.; Tang, L.; Guo, X.; Feng, J. Sources and the distribution of heavy metals in the particle size of soil polluted by gold mining upstream of Miyun Reservoir, Beijing: Implications for assessing the potential risks. *Environ. Monit. Assess.* **2014**, *186*, 6605–6626. [CrossRef] [PubMed]
33. Huang, X.; Zhu, Y.; Ji, H. Distribution, speciation, and risk assessment of selected metals in the gold and iron mine soils of the catchment area of Miyun Reservoir, Beijing, China. *Environ. Monit. Assess.* **2013**, *185*, 8525–8545. [CrossRef] [PubMed]
34. Gao, L.; Gao, B.; Zhou, Y.; Xu, D.; Sun, K. Predicting remobilization characteristics of cobalt in riparian soils in the Miyun Reservoir prior to water retention. *Ecol. Indic.* **2017**, *80*, 196–203. [CrossRef]
35. Zhou, X.; Xia, B. Defining and modeling the soil geochemical background of heavy metals from the Hengshi River watershed (southern China): Integrating EDA, stochastic simulation and magnetic parameters. *J. Hazard. Mater.* **2010**, *180*, 542–551. [CrossRef]
36. Kinimo, K.C.; Yao, K.M.; Marcotte, S.; Kouassi, N.L.B.; Trokourey, A. Distribution trends and ecological risks of arsenic and trace metals in wetland sediments around gold mining activities in central-southern and southeastern Côte d'Ivoire. *J. Geochem. Explor.* **2018**, *190*, 265–280. [CrossRef]
37. Sherman, L.S.; Blum, J.D.; Dvonch, J.T.; Gratz, L.E.; Landis, M.S. The use of Pb, Sr, and Hg isotopes in Great Lakes precipitation as a tool for pollution source attribution. *Sci. Total Environ.* **2015**, *502*, 362–374. [CrossRef] [PubMed]
38. Bohdalkova, L.; Novák, M.; Stepanova, M.; Fottová, D.; Chrastný, V.; Mikova, J.; Kuběna, A.A. The fate of atmospherically derived Pb in Central European catchments: Insights from spatial and temporal pollution gradients and Pb isotope ratios. *Environ. Sci. Technol.* **2014**, *48*, 4336–4343. [CrossRef]
39. Turekian, K.K.; Wedepohl, K.H. Distribution of the elements in some major units of the earth's crust. *Geol. Soc. Am. Bull.* **1961**, *72*, 175–192. [CrossRef]
40. Chen, X.; Xia, X.; Zhao, Y.; Zhang, P. Heavy metal concentrations in roadside soils and correlation with urban traffic in Beijing, China. *J. Hazard. Mater.* **2010**, *181*, 640–646. [CrossRef]
41. Tomlinson, D.L.; Wilson, J.G.; Harris, C.R.; Jeffrey, D.W. Problems in the assessment of heavy-metal levels in estuaries and the formation of a pollution index. *Helgoländer Meeresunters.* **1980**, *33*, 566–575. [CrossRef]
42. Muller, G. Index of geoaccumulation in sediments of the Rhine River. *Geojournal* **1969**, *2*, 108–118.
43. Muller, G. The heavy metal pollution of the sediments of Neckars and its tributary: A stocktaking. *Chem. Zeit* **1981**, *105*, 157–164.
44. Hakanson, L. An ecological risk index for aquatic pollution control. A sedimentological approach. *Water Res.* **1980**, *14*, 975–1001. [CrossRef]
45. Smith, K.S.; Huyck, H.L.; Plumlee, G.; Logsdon, M.; Filipek, L. An overview of the abundance, relative mobility, bioavailability, and human toxicity of metals. *Environ. Geochem. Miner. Depos.* **1997**, *6*, 29–70. [CrossRef]
46. Bird, G.; Brewer, P.A.; Macklin, M.G.; Nikolova, M.; Kotsev, T.; Mollov, M.; Swain, C. Quantifying sediment-associated metal dispersal using Pb isotopes: Application of binary and multi-variate mixing models at the catchment-scale. *Environ. Pollut.* **2010**, *158*, 2158–2169. [CrossRef] [PubMed]
47. Xu, D.; Gao, B.; Peng, W.; Gao, L.; Wan, X.; Li, Y. Application of DGT/DIFS and geochemical baseline to assess Cd release risk in reservoir riparian soils, China. *Sci. Total Environ.* **2019**, *646*, 1546–1553. [CrossRef] [PubMed]
48. Gao, L.; Han, L.; Peng, W.; Gao, B.; Xu, D.; Wan, X. Identification of anthropogenic inputs of trace metals in lake sediments using geochemical baseline and Pb isotopic composition. *Ecotoxicol. Environ. Saf.* **2018**, *164*, 226–233. [CrossRef] [PubMed]
49. Dai, L.; Wang, L.; Li, L.; Liang, T.; Zhang, Y.; Ma, C.; Xing, B. Multivariate geostatistical analysis and source identification of heavy metals in the sediment of Poyang Lake in China. *Sci. Total Environ.* **2018**, *621*, 1433–1444. [CrossRef] [PubMed]
50. Taylor, S.R.; Mclennan, S.M. The continental crust: Its composition and evolution. *J. Geol.* **1985**, *94*, 57–72.
51. Li, T.; Shi, Y.; Li, X.; Zhang, H.; Pi, K.; Gerson, A.R.; Liu, D. Leaching behaviors and speciation of cadmium from river sediment dewatered using contrasting conditioning. *Environ. Pollut.* **2020**, *263*, 114427. [CrossRef]
52. Tchounwou, P.B.; Yedjou, C.G.; Patlolla, A.K.; Sutton, D.J. Heavy metal toxicity and the environment. *NIH* **2012**, *101*, 133–164. [CrossRef]
53. Yu, Y.; Li, Y.; Li, B.; Shen, Z.; Stenstrom, M. Metal enrichment and lead isotope analysis for source apportionment in the urban dust and rural surface soil. *Environ. Pollut.* **2016**, *216*, 764–772. [CrossRef] [PubMed]
54. Du Laing, G.; Rinklebe, J.; Vandecasteele, B.; Meers, E.; Tack, F.M.G. Trace metal behaviour in estuarine and riverine floodplain soils and sediments: A review. *Sci. Total Environ.* **2009**, *407*, 3972–3985. [CrossRef] [PubMed]
55. Shajib MT, I.; Hansen HC, B.; Liang, T. Metals in surface specific urban runoff in Beijing. *Environ. Pollut.* **2019**, *248*, 584–598. [CrossRef]
56. Shi, G.; Chen, Z.; Xu, S.; Zhang, J.; Wang, L.; Bi, C.; Teng, J. Potentially toxic metal contamination of urban soils and roadside dust in Shanghai, China. *Environ. Pollut.* **2008**, *156*, 251–260. [CrossRef]
57. Cheng, X.; Danek, T.; Drozdova, J.; Huang, Q.; Qi, W.; Zou, L.; Yang, S.; Zhao, X.; Xiang, Y. Soil heavy metal pollution and risk assessment associated with the Zn-Pb mining region in Yunnan, Southwest China. *Environ. Monit. Assess.* **2018**, *190*, 194. [CrossRef]
58. Taylor, M.P.; Mackay, A.K.; Hudson-Edwards, K.A.; Holz, E. Soil Cd, Cu, Pb and Zn contaminants around Mount Isa city, Queensland, Australia: Potential sources and risks to human health. *Appl. Geochem.* **2010**, *25*, 841–855. [CrossRef]
59. Veiga, M.M.; Maxson, P.A.; Hylander, L.D. Origin and consumption of mercury in small-scale gold mining. *J. Clean. Prod.* **2006**, *14*, 436–447. [CrossRef]
60. Bing-Quan, Z.; Yu-Wei, C.; Xiang-Yang, C. Application of Pb isotopic mapping to environment evaluation in China. *Chem. Speciat. Bioavailab.* **2002**, *14*, 49–56. [CrossRef]

61. Renberg, I.; Brännvall, M.-L.; Bindler, R.; Emteryd, O. Stable lead isotopes and lake sediments—A useful combination for the study of atmospheric lead pollution history. *Sci. Total Environ.* **2002**, *292*, 45–54. [CrossRef]
62. Pan, L.; Fang, G.; Wang, Y.; Wang, L.; Su, B.; Li, D.; Xiang, B. Potentially toxic element pollution levels and risk assessment of soils and sediments in the upstream river, miyun reservoir, China. *Int. J. Environ. Res. Public Health* **2018**, *15*, 2364. [CrossRef] [PubMed]
63. Han, L.; Gao, B.; Wei, X.; Gao, L.; Xu, D.; Sun, K. The characteristic of Pb isotopic compositions in different chemical fractions in sediments from Three Gorges Reservoir, China. *Environ. Pollut.* **2015**, *206*, 627–635. [CrossRef] [PubMed]
64. Nassiri, O.; Rhoujjati, A.; EL Hachimi, M.L. Contamination, sources and environmental risk assessment of heavy metals in water, sediment and soil around an abandoned Pb mine site in North East Morocco. *Environ. Earth Sci.* **2021**, *80*, 96. [CrossRef]

Article

# Ecological Risk, Input Flux, and Source of Heavy Metals in the Agricultural Plain of Hebei Province, China

Kui Cai [1,2] and Chang Li [3,*]

1. Institute of Geological Survey, Hebei GEO University, Shijiazhuang 050031, China; kcai@hgu.edu.cn
2. Hebei Key Laboratory of Strategic Critical Mineral Resources, Hebei GEO University, Shijiazhuang 050031, China
3. School of Economics and Management, Hebei University of Science & Technology, Shijiazhuang 050018, China
* Correspondence: changli@hebust.edu.cn

**Abstract:** A large amount of heavy metal (HM) inputs exists in the farming areas of the Hebei plain of northern China. However, the potential ecological risk, source, and input flux of HMs in these areas have not been well-investigated. In this study, atmospheric deposition, fertilizer, irrigation water, and agricultural soil samples were collected from farming areas (~74,111 km$^2$) in Hebei Province, China. The HM index of geoaccumulation ($I_{geo}$) and potential ecological risk index (RI) of soil was calculated for eight HMs. The source and input flux of each element were predicted using the input flux and principal component score–multiple linear regression (PCS–MLR) methods. The results showed that Cd and Hg increased $I_{geo}$ values, and the maximum levels of As (29.5 mg/kg), Cu (228.9 mg/kg), Cd (4.52 mg/kg), and Zn (879.0 mg/kg) were greater than the health risk screening values in the soil quality standard of China. The potential ecological risk factor (Er) of Cd demonstrated a moderately potential ecological risk, accounting for 67.72%. The distribution map showed that Cd was mainly concentrated in eastern area of Baoding (BD) in the study area. The result of the atmospheric dry and wet deposition contributed more to soil pollution than the usage of fertilizer or irrigation water by calculating the input flux. The order was Zn (94%) > Cu (92%) > Pb (89%) > Cr (86%) > Cd (72%) > Hg = Ni (71%) > As (59%). Principal component analysis (PCA) results showed that there were four sources of HMs in soil. Geological sources contribute to the accumulation of As, Cr, and Ni in soil. Cu and Pb in the soil were attributable to the input from vehicular emissions and irrigation water. Cd and Zn in the soil were attributable to the farming activity, whereas Hg originates from the combustion of coal. The results of PCS–MLR demonstrated that the contribution rate of As, Ni, and Cr in the study area was 30.06%, 71.86%, 57.71% for the first group (natural source); Cu, Pb and Zn were 71.78%, 63.59%, and 30.72% for the second group (vehicle emissions); Zn was 60.93% for the third group (fertilizer application and irrigation water); and Hg was 85.16%, for the fourth group (coal combustion). These factors provide a valuable reference for remediating HM pollution.

**Keywords:** heavy metals; pollution assessment; input flux; source; management

Citation: Cai, K.; Li, C. Ecological Risk, Input Flux, and Source of Heavy Metals in the Agricultural Plain of Hebei Province, China. *Int. J. Environ. Res. Public Health* **2022**, *19*, 2288. https://doi.org/10.3390/ijerph19042288

Academic Editors: Fayuan Wang, Liping Li, Lanfang Han and Aiju Liu

Received: 2 January 2022
Accepted: 15 February 2022
Published: 17 February 2022

**Publisher's Note:** MDPI stays neutral with regard to jurisdictional claims in published maps and institutional affiliations.

**Copyright:** © 2022 by the authors. Licensee MDPI, Basel, Switzerland. This article is an open access article distributed under the terms and conditions of the Creative Commons Attribution (CC BY) license (https://creativecommons.org/licenses/by/4.0/).

## 1. Introduction

In agroecological environments, heavy metal pollution (HMP) is attributed to improper application of chemical fertilizers and pesticides, as well as irrigation water and industrial factors [1–3]. The dispersion of HMP over large areas has become a hotspot in environmental management. China is primarily a country having a large agricultural production; therefore, there are concerns about environmental problems related to agriculture [4,5]. Agricultural pollution influences food safety, which affects human health. HMP in agricultural soils, in addition to the quality and safety of agricultural products, has recently become an essential topic that should be considered to protect farmlands. In a soil pollution survey, the standards were breached at 19.4% of sites; i.e., 13.7%, 2.8%, 1.8%, and 1.1% of the considered area was slightly, mildly, moderately, and considerably polluted, respectively. Trace elements, such as Cd, Ni, Cu, As, Hg, and Pb, were the major pollutants [6]. Therefore, the overall situation was not good. The quality of arable land and soil is worrisome; moreover, pollution levels are high in soils near abandoned industrial sites and mines. China issued novel soil environmental quality standards and guidelines to mitigate risks on agricultural land (GB15618-2018) [7]. Based on the standards, measures must be taken to appropriately manage the soil environment on agricultural lands [8].

Industry and agriculture form the economic foundation of Hebei Province. However, industrial activity developments cause serious HMP, thereby limiting the agriculture economy. For example, Cd (0.92 mg/kg) and Hg (0.0725 mg/kg) were major toxic elements in the smelting of a middle area in the Hebei plain, China [9]. The heavy metal (HM) concentrations of Cd, Hg, Pb, Zn, Cu, Cr, and Ni were 1.86, 0.29, 154.78, 496.17, 91.06, 131.7, and 40.99 mg/kg, which exceeded the background values in street dust around an industrial zone in Shijiazhuang (SJZ) [10].

Based on a multiobjective geochemical survey conducted to identify pollutants from industrial, mining, and agricultural activities, the fertilizer usage rate, livestock, poultry manure, pesticides, irrigation water, atmospheric deposition of industrial waste and automobile exhaust, and other human activities were the primary causes of soil pollution [11,12].

The usage of chemical fertilizers has attracted considerable attention because fertilizers are major contributors to inorganic agricultural pollution. A 2018 statistical yearbook demonstrated that ~3.3 million tons of chemical fertilizers were used in Hebei Province from 2011 to 2017. Fertilizers are used inefficiently because their usage rate is only 30–40% [12]. Unused fertilizers are dispersed in soil and groundwater. The atmospheric deposition of pollutants is another concern. Atmospheric dry and wet deposition accounts for 43–85% of the total As, Cr, Hg, Ni, and Pb inputs. Note that >50% of Cd, Cu, and Zn inputs are attributable to livestock and poultry manures [13]. Moreover, a major water shortage can be observed in the Hebei plain, China (the study area); therefore, there is a major contradiction between supply and demand in agricultural production. The surface water is polluted; hence, groundwater is the primary source of irrigation water for agricultural production. Thus, agricultural pollution is a serious problem in the Hebei plain. Consequently, it is essential to establish annual inputs of HMs in the agricultural soil in Hebei Province for controlling and reducing HMP.

The primary pollutants of agricultural nonpoint sources (chemical fertilizer usage, livestock and poultry farming, and rural solid waste) can be understood using geographic information system [14], SWAT [15], AGNPS [16], export coefficient [17], input flux [18], and other HMP research models. To examine the soil pollution sources, the principal component score–multiple linear regression (PCS–MLR) method [19,20] and positive matrix factorization (PMF) method [21–23] are used for analyzing the contribution rate of HMs to soil pollutants. In this study, soil samples and three types of medium samples, including samples subjected to fertilizer, irrigation water, and atmospheric deposition, were obtained from the Hebei plain. An input flux method was used to determine the primary input fluxes associated with HMs for atmospheric deposition, fertilizer, and irrigation in agricultural soils. Moreover, the source apportionment (PCS–MLR) was used to analyze the contribution level of each pollution source with respect to the HMs. Hence, the aim of this study is to (1) analyze HMs concentration, the HMP level, and potential ecological risks of Hebei Plain, China, caused by industry and agriculture, (2) evaluate HMs spatial distribution characteristics, (3) calculate the HMs input fluxes of fertilizers, irrigation water, and atmospheric depositions to agricultural soils, (4) identify the pollution source and contribution rate of HMs observed in the study area via principal component score–multiple linear regression (PCS–MLR), and (5) provide valuable management to control the pollution source. We will provide valuable data for controlling and remediation HMP with respect to Hebei agricultural soils.

## 2. Materials and Methods

### 2.1. Study Area

The Hebei plain covers an area of ~$7.4 \times 10^4$ km$^2$ and encompasses both Jidong (JD) and Jizhongnan (JZN) plains (Figure 1). The Qinhuangdao (QHD) and Tangshan (TS) cities are located on the JD plain. The Langfang (LF), BD, Cangzhou (CZ), SJZ, Hengshui (HS), Handan (HD), and Xingtai (XT) cities are located on the JZN plain. The Hebei plain is bound by the Yellow River to the south, Yanshan (YS) Mountains to the north, Taihang (TH) Mountains to the west, and the Bohai Sea to the east. The Haihe and Luanhe River systems are located on the Hebei plain. The terrain varies in elevation from ~100 m in the west to ~3 m along the Bohai Coast. The region has a temperate monsoon climate, and climate changes are evident with a warm summer and a cold and dry winter. The regions between the foothills and coast contain moist and brown soils having clear variations. Moist soil can be observed in the middle of the plain as well as along the YS Mountains to the north and TH Mountains to the west. The farmland on the east side of the plain contains both Eutric Cambisols and Eutric Luvisols, whereas saline–alkali soil can be observed in the coastal region. The farmland covers an area of $6.5 \times 10^4$ km$^2$, of which $4.46 \times 10^4$ km$^2$ is irrigated. The grain output in 2019 was $3.41 \times 10^7$ tons, which primarily comprised wheat, rice, and maize [12]. The JD plain's main industry is steel and cement. The JZN plain's main industry is steel, coal, cement, and metallurgy. Meanwhile, the sewage irrigation is more serious.

**Figure 1.** The sampling site of soil and atmospheric deposition in Hebei plain, China.

### 2.2. Sample Collection and Treatment

#### 2.2.1. Soil Samples

A total of 287 soil samples were collected from June to September in 2012 from the top 20 cm of the arable layer in the farming areas. The surface soils, weeds, roots, gravels, bricks, fertilizer clumps, and other debris were removed during sampling. To ensure representativeness of samples, one point was considered to remain fixed during sampling. Furthermore, three to five subsamples were obtained at multiple points within a range of 20 m to obtain a sample with a weight of >1 kg. After natural air drying, gravels, biological debris, and plant roots were extracted. Subsequently, the samples were passed via a 20-mesh nylon screen, mixed, and ground to −200 mesh. Note that ~100 g of the samples was packed and sent for analysis [24].

2.2.2. Atmospheric Deposition

In this study, 60 sampling barrels were arranged in the farmland areas of the entire study area, including 12 samples in the JD plain and 3 counties in the north of LF and 48 samples in the south of the JZN plain. They were carried to the sampling sites after being treated using distilled water. When arranging the sampling sites, they were placed on a rural roof, ~5–8 m above the ground. Anhydrous ethanol was placed in a barrel at a 1-cm depth to prevent the deposition from producing secondary dust and avoid industrial areas and highways. The tank was placed in a room to allow natural evaporation, and then, the dry sediment was extracted. After weighing and recording, the samples were sealed, numbered, and sent to the laboratory for testing.

2.2.3. Chemical Fertilizer

Based on field surveys, well-known local chemical fertilizers were determined and used as samples. A total of 19 types of compound fertilizers, 17 types of urea, and 5 phosphate fertilizers were collected. Then, the application amounts of various types of chemical fertilizers were recorded, and the samples were sealed, numbered, and directly sent to the laboratory for analysis. A total of 41 samples were collected.

2.2.4. Irrigation Water

Two hundred and thirty-two samples were stored in a polythene plastic pot (the inner plug must be plastic) having a milky white rectangular stopper and a volume of 1 L. The polyethylene pot containing water samples was soaked in 10% $HNO_3$ for three days before being filled with groundwater samples. Subsequently, the pot was cleaned with tap water and distilled water. The sample bottles were washed three to five times using water samples before sampling. Then, the samples were sent to the laboratory for analysis.

The analysis and measurements were performed in strict accordance with standard GB/T 5750.6 [25]. For more details of the analysis and measurement methods, please refer to previous studies [24,26].

2.3. Chemical Analysis and Quality

We weighed a 0.25-g soil sample and placed it in a Teflon beaker, added 5 mL of HCL, heated it at low temperature on an electric heating plate, added 2 of mL $HNO_3$, HF, and $HClO_4$, continued heating and cooling, transferred it to a 25 mL colorimetric tube, diluted it with water to a scale, and stirred well. For more details of the process, please refer to DZ/T0279-2016 [26]. Cr, Cu, Ni, Pb, and Zn in the supernatant were analyzed via inductively coupled plasma atomic emission spectrometry (X Series II, Thermo Electron Corporation, Waltham, MA, USA). Cd was analyzed using a graphite furnace atomic absorption spectrometer (PerkinElmer PinAAcle 900T, Perkin Elmer Instruments (Shanghai) Co., LTD, Shanghai, China), and As and Hg in the supernatant were measured using a hydride generation atomic fluorescence spectrometer (AFS-2202E, Beijing HaiGuang Instrument Co., Ltd., Beijing, China). Moreover, for determining the pH value using an ion-selective electrode method, refer to DZ/T 0279.34-2016; for determining cation exchange capacity using a Hexamminecobalt trichloride solution/spectrophotometric method, refer to HJ 889-2017); and for determining organic carbon (orgC) using a potassium dichromate oxidation–external heating method, please refer to LY/T 1237-1999.

The precision and accuracy of experimental tests were evaluated using standard samples, recovery tests, indoor and outdoor repeat samples, and coded samples. The accuracy of deposition and solution samples can be controlled as per the national standard of substances GBW [26]. The average relative error ((Relative error/Absolute error) × 100%) of all samples was <4%. The precision of sample analysis was less than that specified in the standard; the precision rate was 100%. The precision and accuracy of all samples agreed with the requirement for Regional Ecogeochemical Evaluation [25].

### 2.4. Statistical Analysis

The HMs were subjected to descriptive statistical analysis using SPSS. In the analysis, the minimum, maximum, mean, correlation analysis, and input flux were determined in the study area. All data must be transformed via logarithmic transformations if a normal distribution is not obtained before Pearson's correlation analysis. The spatial distribution of the HMs was obtained using ARCGIS version 10.5. Principal component analysis (PCA) results were obtained via factor analysis (FA) using SPSS. Subsequently, the PCS–MLR results were obtained. The extraction of As, Cu, Pb, Cd, Ni, Cr, Hg, and Zn was 0.726, 0.846, 0.797, 0.978, 0.943, 0.898, 0.97, and 0.983, respectively.

### 2.5. Pollution Assessment Methodology

#### 2.5.1. Index of Geoaccumulation

The index of geoaccumulation ($I_{geo}$) was calculated for each metal to assess soil pollution [27]. $I_{geo}$ allows the impact of human activity on the soil environment to be distinguished from natural factors [28] and is essential for identifying the pollution source. $I_{geo}$ can be calculated as follows:

$$I_{geo} = log_2 \left[ \frac{C_n}{1.5 \times B_n} \right] \quad (1)$$

where $C_n$ is the concentration of metal $n$ in the farm soil (mg·kg$^{-1}$) and $B_n$ is the background value of metal $n$ (mg·kg$^{-1}$). A correction factor of 1.5 was used to consider natural fluctuations in the background value because of lithographic variations. The $B_n$ values for Hg, Cd, As, Cu, Ni, Pb, Cr, and Zn were 0.04, 0.11, 12.8, 21.8, 30.8, 21.5, 68.3, and 71.9 mg·kg$^{-1}$ [29], respectively. The contamination level based on $I_{geo}$ belonged to one of the following classes: $I_{geo} \leq 0$ (clean), $0 < I_{geo} \leq 1$ (slight), $1 < I_{geo} \leq 2$ (mild), $2 < I_{geo} \leq 3$ (moderate), $3 < I_{geo} \leq 4$ (moderate-heavy), $4 < I_{geo} \leq 5$ (heavy), and $I_{geo} > 5$ (severe).

#### 2.5.2. Potential Ecological Risk Index (RI)

Lars Håkanson—a Swedish scientist—proposed the potential ecological risk index RI [30], which quantitatively shows the response observed for basic elemental abundance and the synergistic effect of pollutants. Currently, it is primarily applied to assess the HMP in soil and RI. This factor considers the potential ecological risk of a single HM and the integrated ecological effect of various HMs [31]. Thus, the corresponding risk level could be obtained. Equation (2) was then used to calculate the RI for each metal.

$$RI = \sum_{i}^{m} E_r^i, \text{ where } C_f^i = C_i / C_n^i \text{ and } E_r^i = T_r^i \times C_f^i \quad (2)$$

where $C_f^i$ is the pollution coefficient of HM $i$, $C_i$ is the concentration of HM $i$ (mg·kg$^{-1}$), $C_n^i$ is the preindustrial reference value of the substance for HM $i$ [30], and $T_r^i$ is the toxicity coefficient of HM $i$. $T_r^i$ values were 40, 30, 10, 5, 5, 5, 2, and 1 for Hg, Cd, As, Cu, Ni, Pb, Cr, and Zn, respectively. $E_r^i$ is the potential ecological risk factor Er for HM $i$, which belongs to one of the five following categories: <40 (slight); 40–80 (moderate), 80–160 (strong), 160–320 (very strong), and >320 (extremely strong). RI shows the comprehensive index for all HMs. RI was classified into five: <150 (slight), 150–300 (moderate), 300–600 (strong), 600–1200 (very strong) and >1200 (extremely strong).

## 2.6. Input Flux Analysis

### 2.6.1. Atmospheric Deposition

Sixty sampling bottles were arranged in the study area. The average representative area of each bottle was $1.35 \times 10^9$ m$^2$. The amounts of HMs in the agricultural land due to atmospheric deposition can be calculated as follows:

$$Q_{a,i} = \frac{C_i \times W \times K}{S} \tag{3}$$

where $Q_{a,i}$ is the amount of HMs ($i$ = As, Cd, Cr, Cu, Ni, Pb, Zn, and Hg) present in the study area because of atmospheric deposition (mg·m$^{-2}$·y$^{-1}$), $Ci$ is the content of HM $i$ because of atmospheric deposition (mg·kg$^{-1}$), $W$ is the annual amount of deposition obtained based on the sampling bottle (kg), $K$ is the conversion coefficient (10,000 m$^2$), and $S$ is the area in which the individual soil sample can be observed (706.5 cm$^2$).

### 2.6.2. Irrigation Water

The amounts of HMs in agricultural land that can be attributed to the irrigation water were estimated based on the volume of irrigation water used annually and the heavy metal concentrations in water as per Equation (4).

$$Q_{w,i} = C_i \times V \times 10^{-4}, \tag{4}$$

where $Q_{w,i}$ is the annual input of the HM $i$ from irrigation water (mg·m$^{-2}$·y$^{-1}$), $C_i$ is the concentration of the HM $i$ in the irrigation water (g·L$^{-1}$), and $V$ is the volume of irrigation water used (m$^3$·ha$^{-1}$·y$^{-1}$).

### 2.6.3. Fertilizer

The content could not be easily calculated by homogenizing various fertilizers because of differences between various fertilizers and their complex formulations. Thus, we calculated the input flux of each fertilizer based on the amount of fertilizer applied per year and the sum using Equation (5).

$$Q_{f,i} = \sum_{\substack{i=1 \\ j=1}}^{n} M_{i,j} \times C_{i,j} \times 10^{-4}, \tag{5}$$

where $Q_{f,i}$ is the total HM input obtained from fertilizer $j$ (mg·m$^{-2}$·y$^{-1}$), $M_{i,j}$ is the total amount of fertilizer $j$ applied (kg·ha$^{-1}$·y$^{-1}$), and $C_{i,j}$ is the concentration of HM $i$ in fertilizer $j$ (mg·kg$^{-1}$).

The total input fluxes of HM $i$ ($Q_{t,i}$) from atmospheric deposition, irrigation water, and fertilizer were calculated as follows [32]:

$$Q_{t,i} = Q_{a,i} + Q_{w,i} + Q_{f,i} \tag{6}$$

## 2.7. Source Apportionment Methodology

PMF [21,22] and PCS–MLR [19,20] are examples of source apportionment methods [33]. Recently, the latter has been applied to soil. Because of the complexity of soil systems, the contributions of different sources to the overall concentration of a given element cannot be easily estimated quantitatively. PCS–MLR has enormous application for soil source; it may be applicable in local areas with similar geological conditions (such as diagenetic processes, parent materials, soil types, and landforms). Recently, researchers successfully tracked the origin of HMs in soil via PCS–MLR [19,20,34]. The fundamental assumption associated with PCS–MLR is that the total element concentration was linearly correlated with the contribution of different sources. The basic principle has been described by Thurston and Spengler (1985) in detail [33].

Therefore, in this study, PCS–MLR was used to assess the contribution of HMs from different sources to explain the variation in the HM concentration of the agricultural soil. The current PCS–MLR method has been modified based on Thurston and Spengler (1985) [33] (generally, considerable differences can be observed with respect to the concentrations of different elements). Therefore, data are converted in a dimensionless standard form as follows:

$$Z_{i,j} = \frac{C_{i,j} - \mu_i}{\sigma_i} \quad (7)$$

where $i = 1, 2, \ldots, n$ and $j = 1, 2, \ldots, m$ are the total numbers of HMs and samples, respectively; $Z_{i,j}$ is the standardized value of element $i$ for sample $j$; $C_{i,j}$ is the concentration of element $i$ for sample $j$; $\mu_i$ is the mean concentration of element $i$; and $\sigma_i$ is the standard deviation of the concentration distribution of element $i$ [34].

First, the normal distribution of variables was tested. The results showed that all variables were normally distributed before PCA. The PCA with a varimax rotation transform is applied to normalized data, and the PCA is calculated for the rotation. These PCA are related to multiple sources influencing the local soil position. The initial-rotated PCA of each component was used to estimate the absolute PCS (APCS) as per the method proposed by Thurston and Spengler (1985) [33] to obtain an improved proportional relation with the corresponding source contribution.

The regression of the normalized elemental data of the APCS provides a coefficient that can be used to convert APCS into the source contribution to the sample. The equation is as follows:

$$Y_j = X_0 + \sum_{i=1}^{a} X_i APCS_{i,j} \quad (8)$$

where $Y_j$ is the standardized value of the concentration of the HM $j$; $APCS_{i,j}$ is the rotated absolute component score for component $i$ of element $j$; $X_i$ $APCS_{i,j}$ is the contribution of element $j$ by the source identified with component $i$; and $X_0$ is the contribution by sources not considered in PCA. An $X_0$ approximately equal to 0 shows successful PCA–APCS [34]. Based on regression results, coefficients were employed to calculate the contribution of a pollution source with respect to the HMs in the study area.

## 3. Results and Discussion

### 3.1. Soil HM Concentration and Pollution Indices

Table 1 shows descriptive statistical data for pH, orgC, CEC, and HM concentrations in soil in the study area. The mean values of pH, orgC, CEC were 8.15, 1.03%, and 11.23 cmol/kg. The mean values of Cu, Pb, Cd, Hg, and Zn were higher than the local background values. In particular, Cd was twice the background value. The ranges of As, Cu, Pb, Cd, Ni, Cr, Hg, and Zn were 2.47–29.50, 5.60–228.90, 13.7–125.70, 0.05–4.52, 5.40–43.20, 25.00–112.10, 0.01–0.36, and 15.80–879.00 mg/kg, respectively. The maximum values of As, Cu, Cd, and Zn were greater than soil risk screening values for the Chinese Standard [35], indicating that certain samples were polluted in the study area. Especially for Cd, the maximum value was 90.4 times the minimum value. Moreover, the coefficient of variation of Cd was 183.02%. The main reason was that two samples with highest content (4.16 mg/kg and 4.52 mg/kg) originated from the eastern part of BD and led to a high coefficient of variation. The result of Cd concentration is similar with Zhou (2021) who reported that the maximum value of Cd in arable soil of eastern area of BD was 3.83 mg/kg, 96.67% of the samples that have exceeded the soil risk screening values [36].

Table 2 lists the $I_{geo}$ and RI mean values. When $I_{geo} > 0$, the HMs in the soil primarily originate from human activities rather than the natural source. When the $I_{geo}$ results showed that the mean values of the HMs were <0, no pollution owing to human activities can be observed in all samples. However, the $I_{geo}$ mean values of Cd and Hg were very close to 0. The sample numbers of Cd and Hg were considerably greater than zero and accounted for 43.86% and 32.63% of the total, respectively, indicating that the majority of Cd and Hg originate from human activities. However, the $I_{geo}$ values were >0 for Pb, Zn, and Cu at 9, 6, and 13 samples, respectively, indicating that the human effects must be seriously considered, although they were smaller than those observed for Cd and Hg. The mean $I_{geo}$ values of As, Cr, and Ni were 99.30%, 98.95%, and 100% lower than zero, indicating the natural origin of these elements. The result is similar to the result previously reported in Cai (2020). Moreover, Cai (2020) also showed that As was controlled by Fe and Mn oxide [23]. The mean $I_{geo}$ values of HMs in the soil of the study area were as follows: Cd (−0.008) > Hg (−0.05) > Pb (−0.14) > Cu (−0.15) > Zn (−0.18) > Cr (−0.19) > Ni (−0.24) > As (−0.34).

Table 1. Descriptive statistical analysis of the HMs in the study area (in mg/kg).

| Statistical | As | Cu | Pb | Cd | Ni | Cr | Hg | Zn | pH | orgC % | CEC cmol/kg |
|---|---|---|---|---|---|---|---|---|---|---|---|
| Mean | 9.38 | 24.77 | 24.48 | 0.19 | 27.62 | 66.65 | 0.06 | 75.62 | 8.15 | 1.03 | 11.23 |
| Median | 9.22 | 23.70 | 23.50 | 0.16 | 27.60 | 67.20 | 0.05 | 72.10 | 8.27 | 0.99 | 10.50 |
| Std. Deviation | 3.21 | 14.11 | 9.08 | 0.35 | 6.60 | 11.23 | 0.04 | 53.76 | 0.47 | 0.40 | 4.12 |
| coefficient of variation % | 34.18 | 56.98 | 37.07 | 183.02 | 23.88 | 16.85 | 64.65 | 71.08 | 5.77 | 38.83 | 36.69 |
| Skewness | 0.96 | 11.07 | 8.50 | 11.66 | −0.70 | −0.27 | 4.40 | 12.55 | −1.52 | 1.81 | 0.82 |
| Kurtosis | 4.87 | 156.19 | 88.87 | 137.08 | 1.29 | 3.65 | 27.34 | 180.81 | 3.06 | 7.98 | 0.66 |
| Minimum | 2.47 | 5.60 | 13.70 | 0.05 | 5.40 | 25.00 | 0.01 | 15.80 | 6.20 | 0.16 | 2.70 |
| Maximum | 29.50 | 228.90 | 125.70 | 4.52 | 43.20 | 112.10 | 0.36 | 879.00 | 9.05 | 3.66 | 27.40 |
| Local background [29] | 12.80 | 21.80 | 21.50 | 0.09 | 30.80 | 68.30 | 0.04 | 71.90 | - | - | - |
| Soil risk screening values [35] | 25.00 | 100.00 | 170.00 | 0.60 | 190.00 | 250.00 | 3.40 | 300.00 | - | - | - |

Ref. [29] Chinese soil element background value 1990. Ref. [35] Soil environmental quality GB15618-2018.

Table 2. The Igeo and RI mean values of the HMs in the study area.

| Heavy Metals | As | Cu | Pb | Cd | Ni | Cr | Hg | Zn |
|---|---|---|---|---|---|---|---|---|
| Igeo | −0.34 ± 0.16 | −0.15 ± 0.14 | −0.14 ± 0.09 | −0.008 ± 0.16 | −0.24 ± 0.13 | −0.19 ± 0.08 | −0.05 ± 0.19 | −0.18 ± 0.13 |
| Er | 9.52 ± 3.25 | 4.88 ± 2.78 | 5.62 ± 2.08 | 52.46 ± 96.02 | 4.35 ± 1.04 | 1.92 ± 0.32 | 133.01 ± 85.99 | 1.13 ± 0.80 |
| RI | 212.90 ± 142.55 | | | | | | | |

The Er and RI results indicated that Hg has the highest Er value, thus resulting in a (more) moderately potential ecological risk compared with that associated with the remaining HMs in the study area. This Er value accounts for 99.65% of that associated with the total sample. For Cd, the Er value showed a moderately potential ecological risk, accounting for 67.72% of that associated with the total sample. Meanwhile, the corresponding value of Er of the two samples (4.16 mg/kg and 4.52 mg/kg) was 1134 and 1233. It has reached extremely strong risk levels. The remaining HMs, such as As, Cr, Ni, Pb, and Zn, did not show potential ecological risk values except for one sample in which Cu achieved a moderate ecological risk. However, the comprehensive RI risk analysis demonstrated that 82.46% of the samples reached moderate risk levels. The risk levels were observed to be extremely strong, very strong, and strong in two, three, and seventeen samples, respectively.

*3.2. Soil HM Spatial Distribution*

3.2.1. As

Figure 2 shows the distribution map of As in the study area. The distribution trend demonstrated that the mountain front has lower As than the middle-plain and seaside areas. Higher soil As concentrations were reported in the northeast part of the study area.

The high-value areas were primarily distributed in the west and north part of CZ with (pH > 8.5) soil and the eastern parts of BD. A large number of scholars indicated that pH was one of the key factors affecting As [23,37]. For example, Shen (2020) reported that the content of bioavailable forms of As increased gradually with the increase of pH [37].

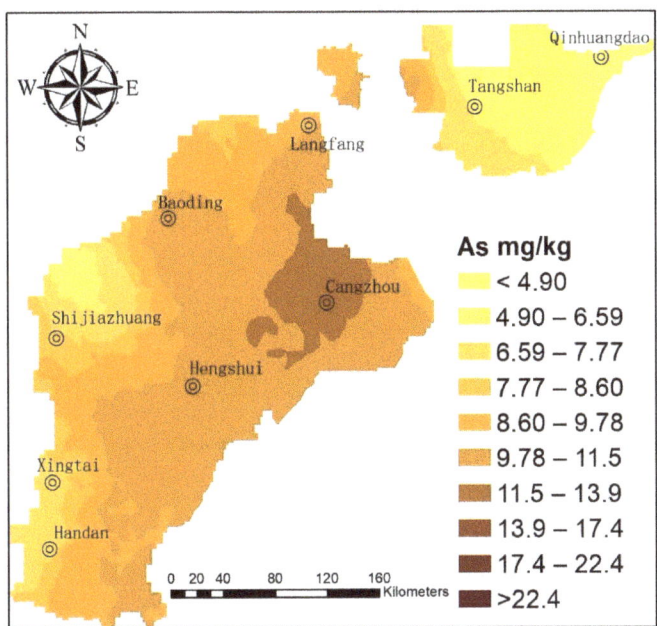

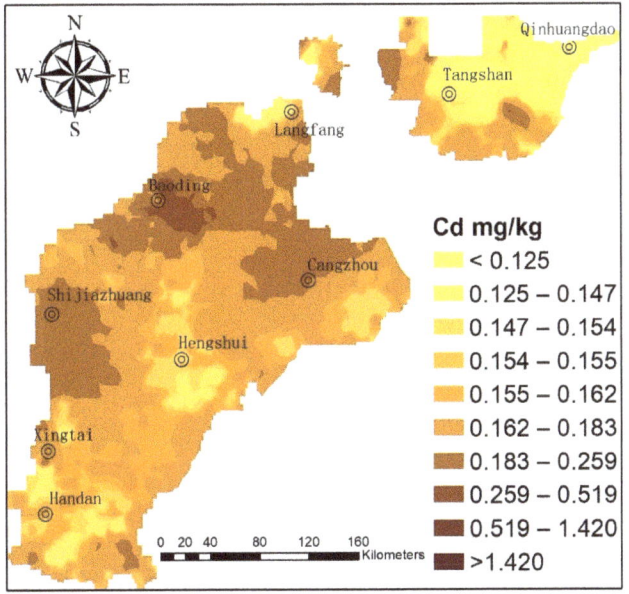

**Figure 2.** *Cont.*

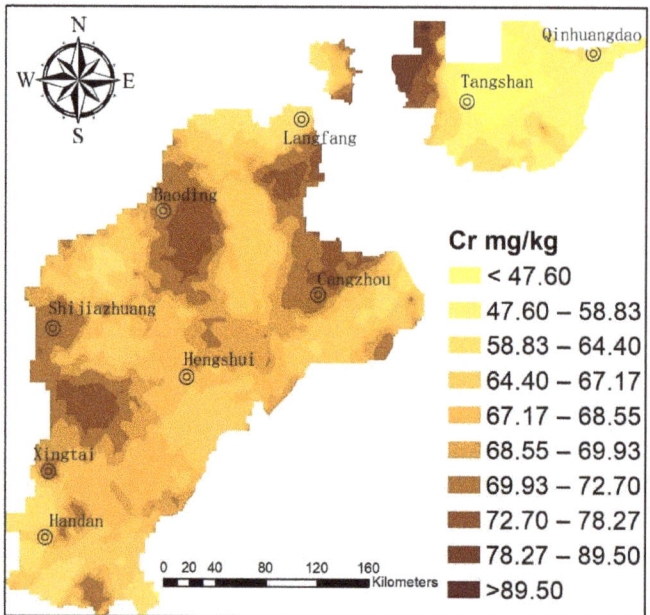

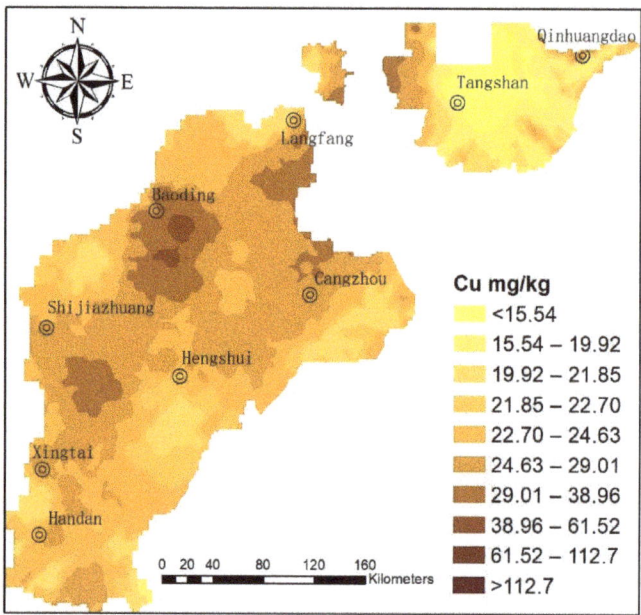

**Figure 2.** *Cont.*

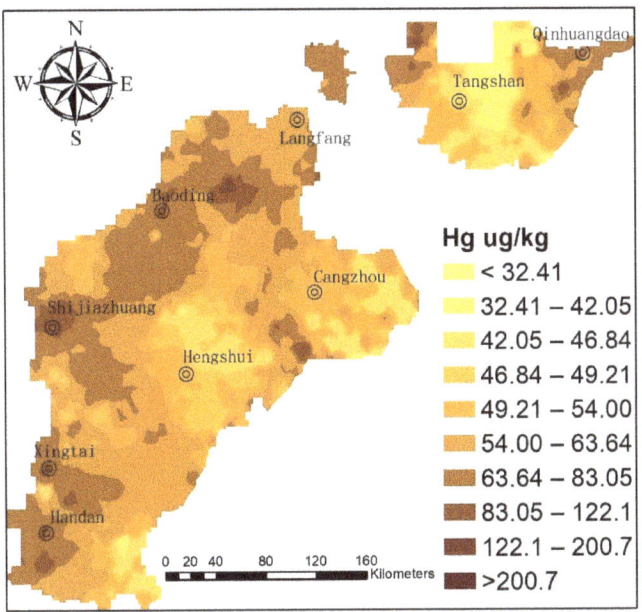

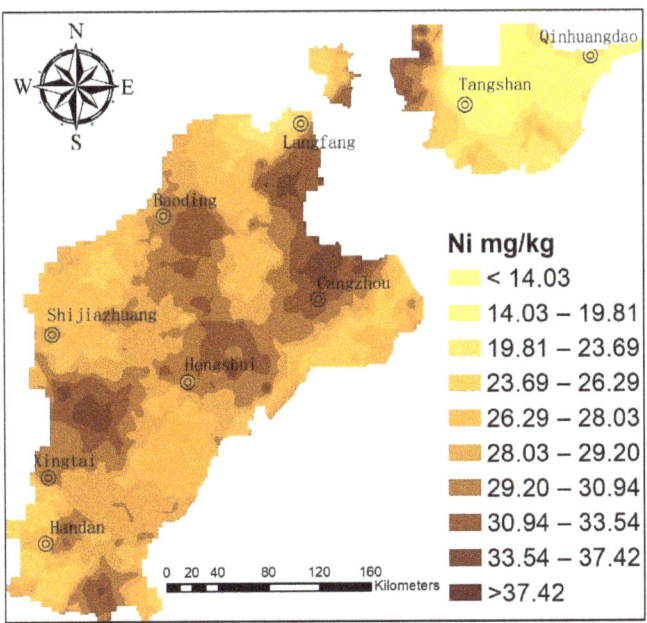

**Figure 2.** *Cont.*

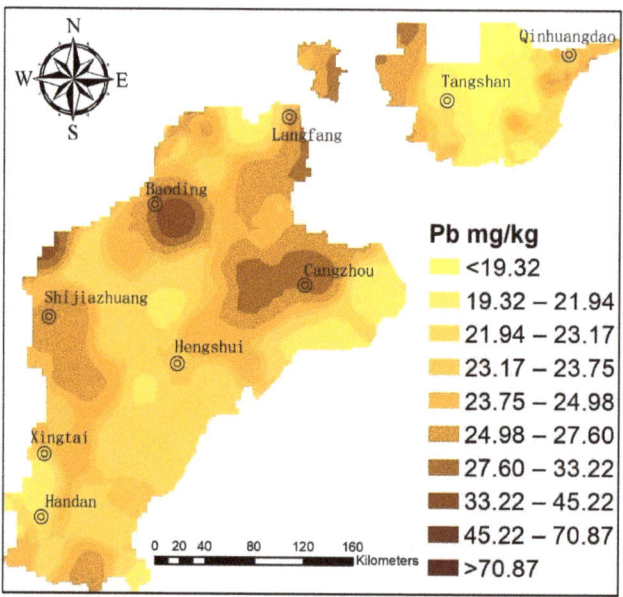

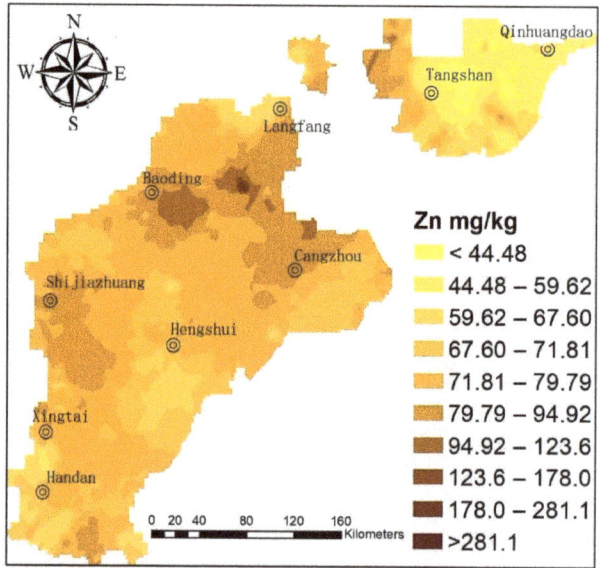

**Figure 2.** Prediction map of soil HMs in the Hebei plain prepared via an ordinary Kriging method.

In the eastern part of BD, the accumulation of As due to metallurgy and sewage irrigation has affected food security [9,38,39]. Wang (2021) reported that the abnormal high values of As were mainly caused by the sewage irrigation, the mean value of As has risen to 23.69 mg/kg [9]. Zhou (2021) showed that As was also affected by industrial emissions in addition to the weathering process of geological parent rock in the BD aera [36].

### 3.2.2. Ni

From Figure 2, the surface soil of the study area mostly contained low Ni content (Table 1), indicating that the spatial distribution of Ni in the piedmont of the TH Mountains was better than that of Ni in the YS Mountains. The distribution pattern of Ni in the northeast of the central plain was lower than that in the other areas. The low-value environment was primarily distributed in TS and certain areas toward the east of the study area. Ni was mainly affected by natural geological background factors in the Hebei plain [9,38].

### 3.2.3. Cr

According to Figure 2, the Cr content in the soil from the study area was low (Table 1). The Cr distribution in the front of the YS Mountains was lower than that in front of the TH Mountains. High-value zones were observed in BD, south of CZ, south of SJZ, and LF. The average value of Cr was close to the background value, primarily indicating a natural origin. Moreover, Cr and Ni had very similar distribution trends. The result shows that Cr and Ni have the same source. Wang (2021) and Guo (2021) examined HMP in BD and Xiong'an New District of Hebei plain and demonstrated that Ni and Cr varied in a small range, thus representing natural soil formation background [9,38].

### 3.2.4. Pb

As shown in Figure 2, the Pb content in most surface soil of the study area was relatively low. The high-value area was primarily distributed in BD, whereas the low-value area was located in TS and QHD. Many industrial and mining enterprises involved with the smelting, manufacturing, and usage of Pb products are located around BD and surrounding areas. Hence, the high Pb content in this area is attributable to the Pb-containing wastewater, waste gas, and waste residue discharged by such enterprises [39]. During gasoline combustion, Pb enters the atmosphere via the exhaust gas discharged by an automobile and then enters the surface soil through atmospheric settlement. Therefore, a high-value area of Pb can be observed in the jurisdiction of BD [40,41].

### 3.2.5. Cu

As shown in Figure 2, the Cu concentration in the surface soil of the study area was slightly higher than that in most areas. The Cu concentration in the surface soil of the YS Mountains was lower than that of the TH Mountains. The overall distribution pattern showed a low trend in the NNE belt to the north of HD. A high-value area could be observed in the BD area of Hebei Province. Metal processing, machinery manufacturing, steel production, and other such enterprises were distributed around the city in the high-value zone [9,40]. The waste residue, waste gas, and wastewater discharged by these enterprises were the primary reasons for the presence of Cu in the surface soil.

### 3.2.6. Hg

As shown in Figure 2, areas having high Hg content in the surface soil were primarily distributed in the surrounding regions of SJZ, BD, XT, and HD in Hebei Province. The mountain front showed higher Hg values than the coast. A higher value could be observed in front of the TH Mountains compared with that of the YS Mountains. Low-value areas were mostly distributed in the east area of TS. In the northern cities, coal-fired gas was used for heating during the winter. The deposition of Zn-containing compounds in dust and soot during combustion was the primary cause of Hg accumulation in the soils of the abovementioned areas [42,43].

### 3.2.7. Cd

The Cd content in southern Hebei was higher than that in northern Hebei (TS–QHD). Cd-rich zones were located in the east area of BD, and the surrounding regions. These areas may have been affected because of smelting and practice of wastewater irrigation that has been followed for decades [41]. The Cd content in the area was higher than the background value of the soil. Cui (2014) reported the wastewater irrigation as the main factor influencing higher Cd concentration in BD soils [39]. Zhou (2021) and Guo (2021) also have indicated that smelting was the main reason contributing to Cd and other HM concentration in BD soils [36,38].

3.2.8. Zn

Figure 2 shows that the Zn content in the surface soil of the study area was high in the south, low in the north, high in front of the TH Mountains, and low in the front of the YS Mountains, SJZ, Anyang, BD, and CZ. Industrial enterprises involved in various activities, including smelting processing, machinery manufacturing, galvanizing, instrumentation, organic synthesis, and paper-making, were located in the abovementioned high-value zone. The Zn accumulation in local soils is attributable to the metal smelting and waste emissions of these enterprises. Zhou (2021) considered that smelting contributed to Zn concentration in BD soils [36]. Guo (2021) reported similar results in the same area [38].

In summary, the Hebei province has been producing coal, iron and steel, metallurgy, and sewage irrigation for 40 years as a large industrial province [40]. Irrational development of industry and agriculture has resulted in the distribution characteristics of HMs in agricultural soil. The contents of As, Cr and Ni in agricultural soil were basically consistent with the background values of soil. However, due to the difference in pH value, leading to the high-value region of As, it was more inclined to the region with higher pH. The high-value distribution of Hg was concentrated in areas with concentrated coal-burning activities. The distribution of high values of Cd, Cu, Pb and Zn were mainly concentrated in the eastern and southern parts of BD due to wastewater irrigation and smelting.

3.3. Input Flux

Figure 3 shows the results of the input fluxes of HMs in the Hebei plain. The input fluxes contributions of HMs of atmospheric deposition in the Hebei plain were very significant. The order was Zn (94%) > Cu (92%)> Pb (89%)> Cr (86%) > Cd (72%)> Hg (71%) = Ni (71%) > As (59%). The contribution rate of 24% and 29% of irrigation water and fertilizer were higher for As and Hg. Hou (2014) reported that irrigation water contributing 60–71% of the total inputs was the main source of metals (As, Cd, Cu and Hg), atmospheric deposition account for 72% and 84% of the total inputs was an important source of Zn and Pb, in the Yangtze River delta, China [18]. Hence, the contribution rate of As and Hg is smaller than that in the Yangtze River delta, China. However, the contribution rate of Zn and Pb is higher than that in the Yangtze River delta, China. In addition, the input fluxes of As, Cd, Cr, Cu, Pb, Zn in the study area are much higher compared with the mean value of As 28.0 g/hm$^2$·a, Cd 4.0 g/hm$^2$·a, Cr 61.0 g/hm$^2$·a, Cu 108.0 g/hm$^2$·a, Hg 1.4 g/hm$^2$·a, Pb 202.0 g/hm$^2$·a, Zn 647.0 g/hm$^2$·a of China, except Hg 1.33 g/hm$^2$·a [44].

Moreover, Figure 2 shows that HMs have regional and industrial characteristics. In particular, the chemical industry base close to CZ results in large input fluxes of As, Cr, Cu, Hg, Ni, and Pb. The input fluxes of Cd were the largest in the north LF, and the largest input flux of Zn could be observed in the north of TS where mining and metallurgy industries could be observed.

A large high-input area can be observed with respect to Ni, Pb, and other elements. In addition to CZ, as can be observed in the urban area of LF, Hg can be observed in the east of SJZ, and Pb can be observed in the urban areas of HS and Wuyi county, thus forming a high-input area because of the high contribution rate of atmospheric deposition.

The atmospheric deposition resulted in a high Cd input zone toward the north of LF and a high Zn input zone in the south and north of TS. This was because the Cd input obtained via local atmospheric deposition reached 130 g/hm$^2$·a, which was ten times the average input flux (12.84 g/hm$^2$·a) in the remaining study areas. Meanwhile, the value was much higher than the mean value of Cd 4.0 g/hm$^2$·a of China [44]. The proportion of input flux of atmospheric deposition was 72%. Therefore, the atmospheric deposition had a major influence on the Cd distribution in the study area.

In all, the fertilizer input fluxes percentage for HMs ranged from 3% to 29%. The input flux percentages of irrigation water for HMs ranged from 2% to 24%, except for Hg. The input flux of Hg majorly comes from atmospheric deposition and fertilizer. Input flux percentages for different HMs also varied greatly (Figure 3). The input flux of most elements provided via irrigation water and fertilizer was smaller than the atmospheric deposition flux obtained via irrigation water and fertilizer, indicating that the groundwater quality in the Hebei plain was good. Furthermore, the high-value sites of the HM with respect to the input fluxes of irrigation water were scattered in the central and southern parts of the study area. The regions with high input fluxes of HMs in chemical fertilizers were observed in areas having developed agricultural production. Thus, the input fluxes of chemical fertilizers in the piedmont plain of the TH Mountains were considerably higher than that in CZ, TS, and other coastal areas. In the area, the use of fertilizers has led to HMs entering agricultural soils, and had a significant influence on the quality and safety of arable land. Compared to the results of Hou (2014) and Jiang (2014), the contribution rates and input fluxes for all HMs in the arable soil were significantly different in the study area. The main reasons might be: (1) other input pathways should be considered, such as agrochemicals, livestock manures, and sewage sludge. (2) The types and dosages of fertilizers were different in the three study areas. (3) The majority of farmland is wheat field in the study area, less irrigation water is applied in the agricultural production than in the other two study areas.

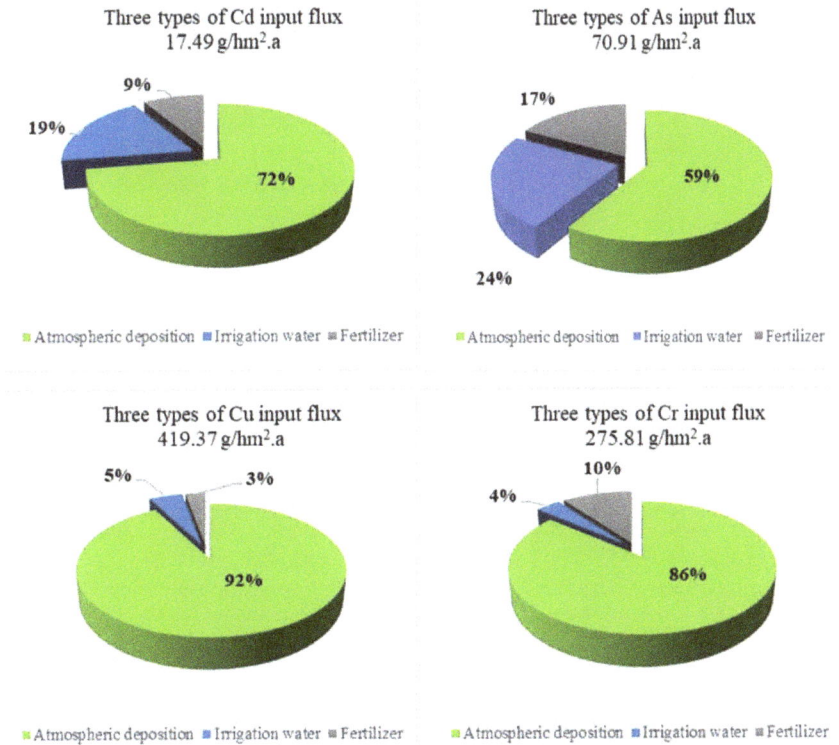

Figure 3. Cont.

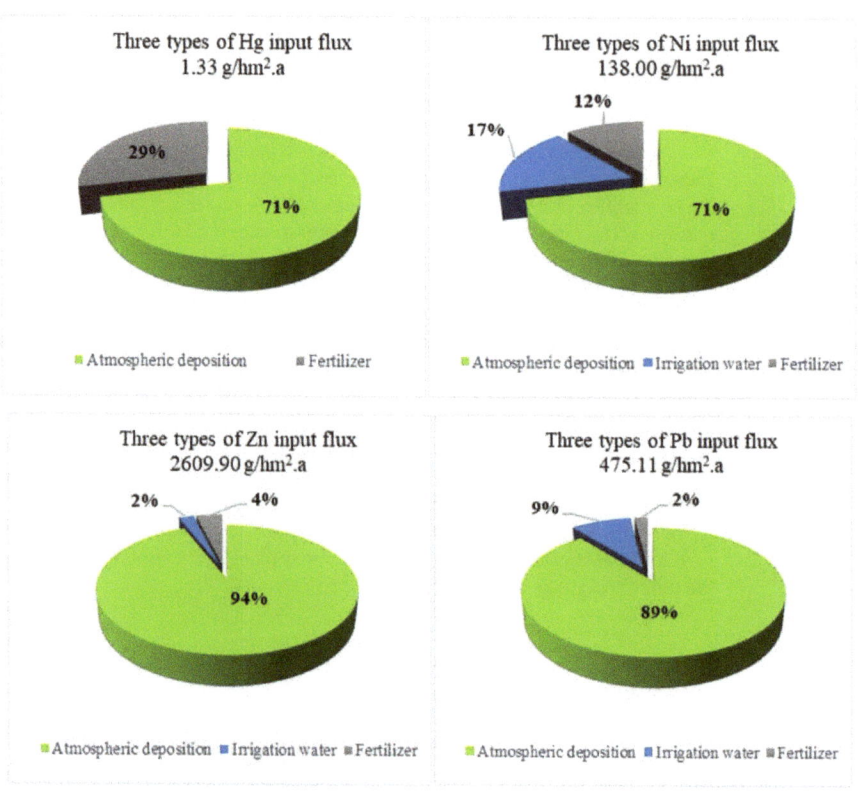

**Figure 3.** Contributions of three input types to agricultural land in the Hebei plain, China.

*3.4. Correlation Coefficient*

We evaluated the spatial distribution map of HMs as per the distribution trend of each element. Table S1 shows Pearson's correlation coefficients for HMs for the soil samples obtained from the agricultural soil of Hebei plain. All samples were observed, i.e., Zn–Cd > Cr–Ni > Cu–Pb > Ni–As > Cd–Cu with a value of >0.60 (Table S1). The result of Pearson's correlation coefficients indicated that they may have a similar source of Zn and Cd, Cr and Ni, Cu and Pb, Ni and As, Cd and Cu.

*3.5. Contribution Rate*

3.5.1. FA

FA was employed to assess the As, Cu, Pb, Cd, Ni, Cr, Hg, and Zn source identifications with respect to the 287 soil samples obtained from the agricultural soil of Hebei plain.

PCA was used as the extraction method. Varimax rotation was performed, and Kaiser normalization was employed. The Kaiser–Meyer–Olkin test value was ~0.60, indicating that FA was reasonable. Moreover, Bartlett's sphericity test result, which was <0.01, confirmed the suitability of data for FA. Table S2 shows the rotated component matrix. Four primary factors with eigenvalues of >1 could be observed, accounting for 89.26% of the total variance (Figure 4). The first factor primarily involves As, Ni, and Cr, indicating that they originate from natural sources because the mean value approached the background value. The second factor involves Cu and Pb, the third factor is dominated by Cd and Zn, and the fourth factor is Hg. All target HMs for PCA were obtained from natural and anthropogenic sources. Cu and Pb associated with the second factor were primarily released to the environment from the exhaust and nonexhaust traffic-related emissions, including fuel combustion, fuel additives, erosion of the asphalt material, tire attrition, smelting, and brake wear. Zhou (2021) and Guo (2021) demonstrated that the Cu and Pb pollution is caused by metal smelting, metal waste and debris processing, and battery manufacturing [9,43]. Cd and Zn, which dominate the third factor, were derived from other human activities [45–47]. The considerable variations of Cd and Zn concentrations in soil samples confirm the usage of anthropogenic resources, including phosphatic fertilizers. The fourth component, i.e., Hg, primarily originated from the burning of coal, as previously indicated by Zhou et al. (2011) [48] and Huang et al. (2018) [41]. Moreover, Hg is the primary contributor to coal combustion in China.

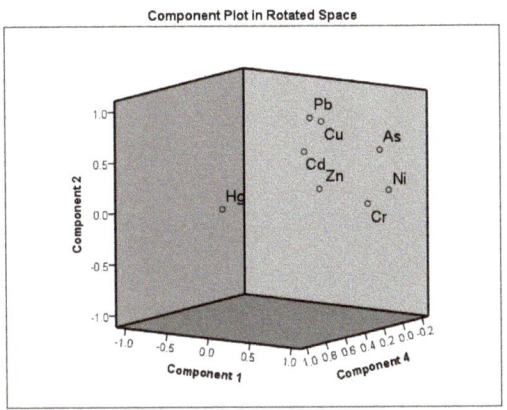

**Figure 4.** Each component plot in the rotated space map of HMs in the study area.

3.5.2. Source Distribution

As per the sources and contributions of the abovementioned pollutants, agricultural soil pollution shows obvious multisource characteristics (Figure 5). Table S3 summarizes the concentration of HMs in fertilizers, irrigation water, and atmospheric depositions. Phosphate fertilizers contribute the most to the Zn and Cd amount, followed by compound fertilizers. Urea was the lowest contributor; hence, the usage of phosphate fertilizers was considered to be another component. The amounts of HMs, such as Pb, Zn, Cu, Pb, Cd, and Hg, obtained via atmospheric deposition were considerably higher than other elements in the soil, except for As, Ni, and Cr. The agricultural atmospheric deposition of HMs is primarily attributable to industrial waste and coal burning. Therefore, S3 and S4 are defined as farming and coal-burning groups, respectively.

The PCS–MLR results demonstrated that the contribution rate of As, Cu, Pb, Cd, Ni, Cr, Hg, and Zn in the study area was 30.06%, 10.09%, 17.94%, 19.48%, 71.86%, 57.71%, 3.73%, and 8.29%, respectively, for the first group; 32.01%, 71.78%, 63.59%, 9.33%, 19.76%, 26.15%, 5.99%, and 30.72%, respectively, for the second group; 15.38%, 16.62%, 17.63%, 66.32%, 2.30%, 3.66%, 5.12%, and 60.93%, respectively, for the third group; and 22.46%, 0.69%, 0.84%, 4.87%, 6.09%, 12.48%, 85.16%, and 0.06%, respectively, for the fourth group.

The contribution of As to the first and second group exceeded 30%, which was attributed to the natural source based on the correlation between the mean concentration level (close to the background) and Cr and Ni. The contribution of As to the third group was >15%, indicating multiple pollutions caused by the contribution of As in the soil. The contributions of Cu and Pb to the second group were >60%, indicating the major role of transportation and sewage water. The contributions of Cd and Zn to the third group were >60%, indicating the high contribution of fertilizers and industrial waste emissions. The contributions of As and Pb were >15%, whereas the contribution of Hg to the fourth group was >85%, indicating the high contribution of the coal combustion. The contribution of As was >20%, indicating that coal combustion releases a considerable amount of As. The results are similar to those of Zeng (2001) and Tian (2009), who reported that As originates from coal combustion [49,50]. The results indicated that all investigated HMs in the agricultural soil of Hebei plain were contributed by S2, S3, and S4 pollution sources. Therefore, a study should be conducted to examine how their release can be reduced by controlling the sources.

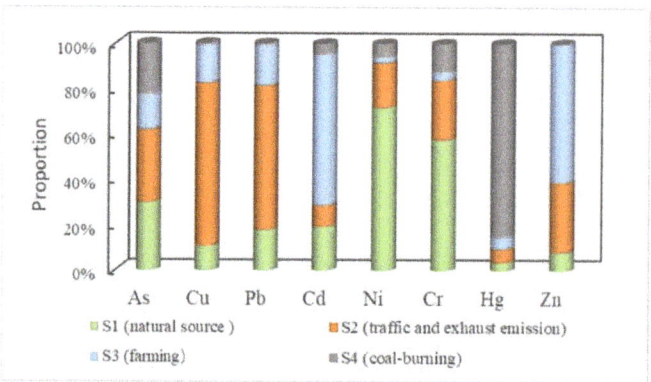

**Figure 5.** Contribution of HMs from four pollution sources estimated using PCS–MLR: S1 (natural source), S2 (vehicle emissions), S3 (fertilizer application and irrigation water), and S4 (coal combustion).

*3.6. Management*

As per the abovementioned research results, the agricultural HMP in the study area has become a serious problem, affecting the sustainable development of agriculture. Agricultural HMP usually refers to the large-area pollution caused by the unreasonable application of chemical fertilizers and pesticides, irrigation water, industrial and agricultural wastes, and household garbage during the agricultural production process. Therefore, the local environment department must take corresponding measures to solve this issue.

3.6.1. Fertilizer

According to statistics, 3.124 million tons of fertilizers were used in 2008, which increased to 3.356 million tons in 2014 and then decreased to 3.22 million tons in 2017 [12]. The application of chemical fertilizers has decreased, although the output of wheat and corn has increased, indicating that the usage of chemical fertilizers has increased. However, the current efficiency of domestic fertilizers was significantly different from that of foreign fertilizers. Therefore, the local area should scientifically and reasonably control the usage of chemical fertilizers and pesticides, particularly phosphate fertilizers [51,52]. We suggest focusing on new high-efficiency fertilizers and developing high-efficiency and low-toxicity products, such as pesticides, from agricultural waste. Furthermore, we must achieve green prevention as well as develop a control technology and an inexpensive treatment technology for livestock and poultry manure and agricultural straw [53–55]. The content standards of various harmful elements in chemical fertilizers should then be formulated and improved.

3.6.2. Coal Combustion

The coal usage decreased by 244.18, 316.96, and 281.05 million tons in 2008, 2013, and 2016, respectively. The coal usage rate has decreased via control in recent years. Industrial waste emissions have increased from 3755.8 billion cubic meters in 2008 to 7857 billion cubic meters in 2016 [12], i.e., almost doubling in eight years. Continuously increasing waste emissions will inevitably cause significant harm to the atmosphere and agricultural soil. Note that >70% of the coal was used for coal-fired power generation. The global Hg hazards associated with coal combustion and pollution caused by tens of millions of tons of coal gangue piles, spontaneous combustion, and rainwater leaching have considerably affected the society and environment. The evaluation of hazards associated with typical high-Hg coal was an urgent requirement with respect to environmental protection. The Hg content of coal in the study area was 0.14 mg/kg and that observed when burning coal was 0.15–0.16 mg/kg in various industries and processes such as power plants and coking [56]. In certain processes, the Hg content was considerably high. The environmental hazards caused by coal utilization cannot be ignored. Currently, the control technology of coal-fired mercury emissions is in the experimental development stage globally, and its commercial technology has not yet been completely developed [56,57]. Therefore, the importance to achieving inexpensive coal-fired Hg emission control is to develop a technique and provide technical and theoretical support for China's policy-making departments to propose reasonable and effective control measures for coal-fired Hg. The internal laws of coal-fired Hg, harmfulness of emissions, and mineral composition and morphological characteristics of high-Hg coal combustion should be explored [58–60]. Therefore, the selection of typical high-Hg coal for combustion emission test research and analysis of its precipitation rules are important for achieving environmental protection in China and the rational development and usage of coal resources.

3.6.3. Irrigation Water

The proportion of irrigation-water input was less from the perspective of the irrigation-water input flux. However, in certain regions, the level of HMs in sewage water irrigation should be considered. For example, in SJZ and BD, the Cd level in sewage irrigation was high [61,62]. The low surface concentration and high concentration of As were considerably related to groundwater [20]. Therefore, the level of HMs at both places should be considered before wastewater irrigation, and the water quality must meet the water quality standards for irrigation. In areas where wastewater irrigation has been applied for a long time, the content and accumulation rate of HMs in the soil, groundwater quality, residents' physical health, and surrounding ecological environment should be monitored.

El-Mageed (2021) reported that adding Si could reduce HM concentration and improve grain yield at the HMP level. We suggest that Si foliar can be used to enhance plant growth and productivity in irrigation with HMP [63]. Moreover, Edelstein (2018) and Zhan (2018) reported that arbuscular mycorrhizal fungi (AMF) could induce resistance to HMs. Hence, the direct strategy involves using AMF in regions irrigated with sewage irrigation in the study area [64,65].

3.6.4. Vehicular Emissions

Since 2008, the consumption of gasoline and diesel has increased on a yearly basis, from 2.11 million tons of gasoline to 4.949 million tons in 2016 and from 5.317 million tons of diesel in 2008 to 8.436 million tons in 2016 [12]. The increase in gasoline and diesel directly reflects the increase in car ownership. Pb and Cu were the primary factors responsible for air pollution [20,66–68].

Research results demonstrated that the HM concentration in roadside soil was zoned along the distance and decreased exponentially with an increase in the roadside distance [69–73]. For instance, MacKinnon (2011) reported that the Pb accumulation range caused by traffic activities was limited to 10 m of the expressway and 3 m of feeder road [67]. In future, we should focus on monitoring the content of HMs in arable soil within 10 m of the roadside. Moreover, green vegetation enrichment is an economical and effective technique to reduce HMP in the atmosphere [70]. Therefore, increasing the green belt or changing the planting structure of the green belt can improve the concentration of HMs in arable soil. Karmakar (2019) evaluated fifteen plants using the air pollution tolerance index, expected performance index, and metal accumulation index to determine their tolerance to air pollution, expected performance, and metal accumulation capacity [71]. Esfandiari (2020) and Mondal (2021) reported that common plant species as green belts could be accumulated to improve the HM concentration level [72,73].

Moreover, the relevance and effectiveness of remote sensing technology for the on-site identification of high-emission vehicles for inspection and maintenance plans should be considered in future [41].

## 4. Conclusions

The RI result of HMs demonstrated moderately potential ecological risk. In particular, the distribution map demonstrated that the concentration of HMs in the JZN plain is greater than that in the JD plain. The high values of As, Cu, Pb, Cd, and Hg were mainly distributed in the eastern BD. Meanwhile, the input of atmospheric deposition in the Hebei plain demonstrated significant regional and industrial characteristics. The input fluxes of atmospheric deposition, irrigation water and fertilizer are As (70.91 g/hm$^2$·a), Cr (275.81 g/hm2·a), Ni (138 g/hm$^2$·a), Cu (419.37 g/hm$^2$·a), Pb (475.11 g/hm$^2$·a), Cd (17.49 g/hm$^2$·a), Hg (1.33 g/hm$^2$·a), and Zn (2609.9 g/hm$^2$·a). The input of atmospheric deposition plays an important role in the Hebei plain. The input of irrigation water and fertilizer also showed a contribution rate for As (24%) and Hg (29%). The result of the source of HMs in the study area demonstrated traffic, fertilizer application, farming—Cd and Zn, and coal-burning—the Hg group represents the primary source of Cu, Pb, Cd, Zn, and Hg. In particular, diversified pollution can be observed based on the contribution rate of As. In general, the three input sources of atmospheric deposition, fertilizer, and irrigation water served as sinks for HMs in the Hebei plain, except for Ni and Cr. Hence, the relevant department must formulate strategies to control the input of HMs. In future, speciation in the atmosphere is important to the department of environment to perform the detailed analysis of the pollution characteristics (vehicular traffic and industrial emission), including mineral components.

**Supplementary Materials:** The following are available online at https://www.mdpi.com/article/10.3390/ijerph19042288/s1, Figure S1: The Cluster analysis combine map of heavy metals in study area by using Average Linkage (Between Groups), Table S1: Correlation analysis of heavy metals in the study area by the Pearson, Table S2: Rotated component matrix, Table S3: The mean values of Heavy metals in chemical fertilizers (mg/kg).

**Author Contributions:** Conceptualization, K.C.; methodology, C.L.; software, K.C. and C.L.; formal analysis, C.L.; investigation, K.C.; writing—original draft preparation, K.C. and C.L.; writing—review and editing, K.C. and C.L. All authors have read and agreed to the published version of the manuscript.

**Funding:** The work was supported by Natural Science Foundation of Hebei Province (CN) (Green Channel) under grant No. D2020403003. and in part by the Project of Introducing Overseas Students in Hebei Province under grant No. C20200307.

**Institutional Review Board Statement:** Not applicable.

**Informed Consent Statement:** Not applicable.

**Conflicts of Interest:** The authors declare that they have no known competing any financial and personal relationships with other people or organizations that could inappropriately influence (bias) their work in this paper.

## References

1. Kumar, S.S.; Kumar, A.; Singh, S.; Malyan, S.K.; Baram, S.; Sharma, J.; Pugazhendhi, A. Industrial wastes: Fly ash, steel slag and phosphogypsum-Potential candidates to mitigate greenhouse gas emissions from paddy fields. *Chemosphere* **2019**, *241*, 124824. [CrossRef] [PubMed]
2. Zhang, W.L.; Ji, H.J.; Kolbe, H.; Xu, A.G. Estimation of agricultural non-point source pollution in China and the alleviating strategies II. Status of agricultural non-point source pollution and the alleviating strategies in European and American countries. *Sci. Agric. Sinica* **2004**, *37*, 1018–1025.
3. Sun, B.; Zhang, L.; Yang, L.; Zhang, F.; Norse, D.; Zhu, Z. Agricultural non-point source pollution in China: Causes and mitigation measures. *Ambio* **2012**, *41*, 370–379. [CrossRef] [PubMed]
4. Quan, W.; Yan, L. Effects of agricultural non-point source pollution on eutrophica-tion of water body and its control measure. *Acta Ecol. Sinica.* **2002**, *22*, 291–299.

5. Guo, W.; Fu, Y.; Ruan, B.; Ge, H.; Zhao, N. Agricultural non-point source pollution in the Yongding River Basin. *Ecol. Indic.* **2014**, *36*, 254–261. [CrossRef]
6. MEP; MLR. Report on the National General Survey of Soil Contamination in China. 2014. Available online: http://english.www.gov.cn/policies/latest_releases/2014/08/23/content_281474983026954.htm (accessed on 13 January 2022).
7. Chinese Environmental Protection Administration (CEPA). *Environmental Quality Standard for Soils (GB15618-2018)*; Standards Press of China: Beijing, China, 2018. Available online: https://www.chinesestandard.net/PDF/English.aspx/GB15618-2018 (accessed on 13 January 2022).
8. Chinese Environmental Protection Administration (CEPA). *Measures for the Management of Soil Environment on Agricultural Land (Trial)*; Standards Press of China: Beijing, China, 2017. Available online: https://www.iisd.org/system/files/publications/financing-models-soil-remediation-china.pdf (accessed on 13 January 2022).
9. Wang, C.; Zhang, S.; Liu, J.; Xing, Y.; Li, M.; Liu, Q. Pollution level and risk assessment of heavy metals in a metal smelting area of Xiong'an New Area. *Geol. China* **2021**, *48*, 1697–1709. Available online: https://kns.cnki.net/kcms/detail/11.1167.P.20210301.1753.015.html (accessed on 13 January 2022).
10. Cai, K.; Li, C. Street dust heavy metal pollution source apportionment and sustainable management in a typical city—Shijiazhuang, China. *Int. J. Environ. Res. Public Health* **2019**, *16*, 2625. [CrossRef]
11. Shangguan, Y.; Cheng, B.; Zhao, L.; Hou, H.; Ma, J.; Sun, Z.J.; Huo, X.L. Distribution assessment and source identification using multivariate statistical analyses and artificial neutral networks for trace elements in agricultural soils in Xinzhou of Shanxi Province, China. *Pedosphere* **2018**, *28*, 542–554. [CrossRef]
12. CSY. National Bureau of Statistics of China. In *China Statistical Yearbook 2019*; China Statistical Press: Beijing, China, 2019.
13. Dong, T.X.; Yang, H.X.; Li, H.F.; Qiao, Y.H.; Su, D.C. Accumulation Characteristics of heavy metals in the soil with wheat-corn rotation system in north China. *J. Agric. Resour. Environ.* **2014**, *31*, 355–365.
14. Usery, E.L.; Finn, M.P.; Scheidt, D.J.; Ruhl, S.; Beard, T.; Bearden, M. Geospatial data resampling and resolution effects on watershed modeling: A case study using the agricultural non-point source pollution model. *J. Geogr. Syst.* **2004**, *6*, 289–306. [CrossRef]
15. Yang, S.; Dong, G.; Zheng, D.; Xiao, H.; Gao, Y.; Lang, Y. Coupling Xinanjiang model and SWAT to simulate agricultural non-point source pollution in Songtao watershed of Hainan, China. *Ecol. Model.* **2011**, *222*, 3701–3717. [CrossRef]
16. Mohammed, H.; Yohannes, F.; Zeleke, G. Validation of agricultural non-point source (AGNPS) pollution model in Kori watershed, South Wollo, Ethiopia. *Int. J. Appl. Earth Obs.* **2004**, *6*, 97–109. [CrossRef]
17. Cai, M.; Li, H.E.; Zhuang, Y.T.; Wang, Q.H. Application of modified export coefficient method in polluting load estimation of non-point source pollution. *J. Hydraul. Eng.* **2004**, *7*, 40–45.
18. Hou, Q.; Yang, Z.; Ji, J.; Yu, T.; Chen, G.; Li, J.; Yuan, X. Annual net input fluxes of heavy metals of the agro-ecosystem in the Yangtze River delta, China. *J. Geochem. Explor.* **2014**, *139*, 68–84. [CrossRef]
19. Guo, H.; Wang, T.; Louie, P.K.K. Source apportionment of ambient non-methane hydrocarbons in Hong Kong: Application of a principal component analysis/absolute principal component scores (PCA/APCS) receptor model. *Environ. Pollut.* **2004**, *129*, 489–498. [CrossRef]
20. Cai, K.; Li, C.; Na, S. Spatial distribution, pollution source, and health risk assessment of heavy metals in atmospheric depositions: A case study from the sustainable city of Shijiazhuang, China. *Atmosphere* **2019**, *10*, 222. [CrossRef]
21. Men, C.; Liu, R.; Xu, F.; Wang, Q.; Guo, L.; Shen, Z. Pollution characteristics, risk assessment, and source apportionment of heavy metals in road dust in Beijing, China. *Sci. Total Environ.* **2018**, *612*, 138–147. [CrossRef]
22. Othman, M.; Latif, M.T. Pollution characteristics, sources, and health risk assessments of urban road dust in Kuala Lumpur City. *Environ. Sci. Pollut. Res* **2020**, *27*, 11227–11245. [CrossRef]
23. Cai, K.; Song, Z. Cycling and total risks of multiple as fractions in the Beijing–Tianjin–Hebei area on the agricultural plain, China. *Ecotox. Environ. Safe* **2020**, *190*, 110097. [CrossRef]
24. Technical Standard of Geological Survey of China Geological Survey. *Specification for Multi-Objective Regional Geochemical Survey (1:250000) (DD2005-01)*; Standard Press of China: Beijing, China, 2005.
25. Standard Examination Methods for Drinking Water—Metal Parameters. GB/T5750.6-2006. National Health Commission of the People's Republic of China. Available online: http://www.nhc.gov.cn/wjw/pgw/201805/e3b8ea5eb9a345df8b74ee93087feccd.shtml (accessed on 13 January 2022).
26. Geological and Mineral Industry Standards of the People's Republic of China (DZ/T0279-2016). Analysis Methods for Regional Geochemical Sample. Ministry of Natural Resources of the People's Republic of China. Available online: http://g.mnr.gov.cn/201701/t20170123_1430106.html (accessed on 13 January 2022).
27. Muller, G. Index of geoaccumulation in sediments of the Rhine River. *Geojournal* **1969**, *2*, 108–118.
28. Hasan, A.B.; Kabir, S.; Reza, A.H.M.S.; Zaman, M.N.; Ahsan, A.; Rashid, M. Enrichment factor and geo-accumulation index of trace metals in sediments of the ship breaking area of Sitakund Upazilla (Bhatiary-Kumira), Chittagong, Bangladesh. *J. Geochem. Explor.* **2013**, *125*, 130–137. [CrossRef]
29. Chinese Environmental Protection Administration (CEPA). *Elemental Background Values of Soils in China*; Environmental Science Press of China: Beijing, China, 1990.
30. Hakanson, L. An ecological risk index for aquatic pollution control. A sedimentological approach. *Water Res.* **1980**, *14*, 975–1001. [CrossRef]

31. Mihankhah, T.; Saeedi, M.; Karbassi, A. A comparative study of elemental pollution and health risk assessment in urban dust of different land-uses in Tehran's urban area. *Chemosphere* **2020**, *241*, 124984. [CrossRef]
32. Yi, K.; Fan, W.; Chen, J.; Jiang, S.; Huang, S.; Peng, L.; Luo, S. Annual input and output fluxes of heavy metals to paddy fields in four types of contaminated areas in Hunan Province, China. *Sci. Total Environ.* **2018**, *634*, 67–76. [CrossRef]
33. Thurston, G.D.; Spengler, J.D. A quantitative assessment of source contributions to inhalable particulate matter pollution in metropolitan Boston. *Atmos. Pollut.* **1985**, *19*, 9–25. [CrossRef]
34. Wang, C.; Yang, Z.; Zhong, C.; Ji, J. Temporal–spatial variation and source apportionment of soil heavy metals in the representative river–alluviation depositional system. *Environ. Pollut.* **2016**, *216*, 18–26. [CrossRef]
35. Ministry of Ecology and Environment of the People's Republic of China (MEE). Soil Environmental Quality–Risk Control Standard for Soil Contamination of Agricultural Land (GB15618-2018). Beijing, China, 2018. Available online: http://www.mee.gov.cn/ywgz/fgbz/bz/bzwb/trhj/201807/t20180703_446029.shtml (accessed on 13 January 2022).
36. Zhou, Y.L.; Wang, Q.J.; Wang, C.W.; Liu, F.; Song, Y.T.; Guo, Z.J.; Yang, Z.B. Heavy metal pollution and cumulative effect of soil-crop system around typical enterprises in Xiong'an new district. *Environ. Sci.* **2021**. [CrossRef]
37. Shen, B.; Wang, X.; Zhang, Y.; Zhang, M.; Wang, K.; Xie, P.; Ji, H. The optimum pH and Eh for simultaneously minimizing bioavailable cadmium and arsenic contents in soils under the organic fertilizer application. *Sci. Total Environ.* **2020**, *711*, 135229. [CrossRef]
38. Guo, Z.J.; Zhou, Y.L.; Wang, Q.L.; Wang, C.W.; Song, Y.T.; Liu, F.; Yang ZKong, M. Characteristics of soil heavy metal pollution and health risk in Xiong'an New District. *China Environ. Sci.* **2021**, *41*, 431–441.
39. Guo, H.Q.; Yang, Z.H.; Li, H.L.; Ma, W.J.; Ren, J.F. Envrionmental quality and anthropogenic pollution assessment of heavy metals in topsoil of Hebei plain. *Geol. China* **2011**, *38*, 218–224.
40. Cui, X.T.; Qin, Z.Y.; Luan, W.L.; Song, Z.F. Assessment of the heavy metal pollution and the potential ecological hazard in soil and plain area of Baoding City of hebei province. *Geoscience* **2014**, *28*, 523–530.
41. Zhang, X.Z.; Wang, S.M.; Li, J.H. Research on the enrichment and origin of cadmium in soils in a coastal area of east Hebei Province. *Earth Envrion.* **2007**, *35*, 321–326.
42. Huang, R.J.; Cheng, R.; Jing, M.; Yang, L.; Li, Y.; Chen, Q.; Zhang, R. Source-specific health risk analysis on particulate trace elements: Coal combustion and traffic emission as major contributors in wintertime Beijing. *Environ. Sci. Technol.* **2018**, *52*, 10967–10974. [CrossRef] [PubMed]
43. Yang, S.; Yan, X.L.; Feng, Y.T. Spatial distribution and source identification of heavy metals in the farmland soil of the Caofeidian in Hebei Province. *Acta Sci. Circumstantiae* **2019**, *39*, 3064–3072.
44. Luo, L.; Ma, Y.; Zhang, S.; Wei, D.; Zhu, Y. An inventory of trace element inputs to agricultural soils in China. *J. Environ. Manag.* **2009**, *90*, 2524–2530. [CrossRef]
45. Zhang XZGuo HQLi, H.L. Distinguishing origins of elements in environmental geochemistry of Baiyangdian Billabong of Hebei Province, north China. *Earth Sci. Front.* **2008**, *15*, 90–96.
46. Jiang, W.; Hou, Q.; Yang, Z.; Yu, T.; Zhong, C.; Yang, Y.; Fu, Y. Annual input fluxes of heavy metals in agricultural soil of Hainan Island, China. *Environ. Sci. Pollut. Res.* **2014**, *21*, 7876–7885. [CrossRef]
47. Zeng, X.; Wang, Z.; Wang, J.; Guo, J.; Chen, X.; Zhuang, J. Health risk assessment of heavy metals via dietary intake of wheat grown in Tianjin sewage irrigation area. *Ecotoxicology* **2015**, *24*, 2115–2124. [CrossRef]
48. Zhou, Q.; Tang, Y. Coal combustion on environment pollution in China. In Proceedings of the 2011 International Conference on Electrical and Control Engineering, Yichang, China, 16–18 September 2011; IEEE: Manhattan, NY, USA, 2011; pp. 1482–1486. [CrossRef]
49. Zeng, T.; Sarofim, A.F.; Senior, C.L. Vaporization of arsenic, selenium and antimony during coal combustion. *Combust. Flame* **2001**, *126*, 1714–1724. [CrossRef]
50. Tian, H.Z.; Qu, Y.P. Inventories of atmospheric arsenic emissions from coal combustion in China, 2005. *Environ. Sci.* **2009**, *30*, 956–962.
51. Oyedele, D.J.; Asonugho, C.; Awotoye, O.O. Heavy metals in soil and accumulation by edible vegetables after phosphate fertilizer application. *Electron. J. Environ. Agric. Food Chem.* **2006**, *5*, 1446–1453.
52. Cheraghi, M.; Lorestani, B.; Merrikhpour, H. Investigation of the effects of phosphate fertilizer application on the heavy metal content in agricultural soils with different cultivation patterns. *Biol. Trace Elem. Res.* **2012**, *145*, 87–92. [CrossRef] [PubMed]
53. Atafar, Z.; Mesdaghinia, A.; Nouri, J.; Homaee, M.; Yunesian, M.; Ahmadimoghaddam, M.; Mahvi, A.H. Effect of fertilizer application on soil heavy metal concentration. *Environ. Monit. Assess.* **2010**, *160*, 83. [CrossRef] [PubMed]
54. Zuo, H.; Ma, L.; Wang, Z.; Liu, J.; Ma, C. Research on fertilizer application technology for seedlings and its development trends. *World For. Res.* **2010**, *23*, 39–43.
55. Jie, G.; Hua, G. Prospects on the technical innovation to increase fertilizer use efficiency. *Trans. Chinese Soc. Agric. Eng.* **2000**, *2*, 17–20.
56. Streets, D.G.; Hao, J.; Wu, Y.; Jiang, J.; Chan, M.; Tian, H.; Feng, X. Anthropogenic mercury emissions in China. *Atmos. Environ.* **2005**, *39*, 7789–7806. [CrossRef]
57. Hower, J.C.; Senior, C.L.; Suuberg, E.M.; Hurt, R.H.; Wilcox, J.L.; Olson, E.S. Mercury capture by native fly ash carbons in coal-fired power plants. *Prog. Energy Combust.* **2010**, *36*, 510–529. [CrossRef]

58. Pavlish, J.H.; Sondreal, E.A.; Mann, M.D.; Olson, E.S.; Galbreath, K.C.; Laudal, D.L.; Benson, S.A. Status review of mercury control options for coal-fired power plants. *Fuel Process Technol.* **2003**, *82*, 89–165. [CrossRef]
59. Galbreath, K.C.; Zygarlicke, C.J. Mercury transformations in coal combustion flue gas. *Fuel Process Technol.* **2000**, *65*, 289–310. [CrossRef]
60. Wang, S.X.; Zhang, L.; Li, G.H.; Wu, Y.; Hao, J.M.; Pirrone, N.; Ancora, M.P. Mercury emission and speciation of coal-fired power plants in China. Atmos. *Chem. Phys.* **2010**, *10*, 1183–1192.
61. Shao, J.Q.; Liu, C.C.; Yan, X.L. Cadmium distribution characteristics and environmental risk assessment in typical sewage irrigation area of Hebei Province. *Acta Sci. Circumstantiae* **2019**, *39*, 917–927.
62. Liu, Y.L.; Liu, S.Q.; Xun, Z.J.; Yan, Y.L.; Hou, D.L. Assessment of potential ecological risk of soll heavy metals in sewage irrigated area of baoding suburban. *J. Anhui Agri. Sci.* **2011**, *39*, 10330–10332.
63. El-Mageed, A.; Taia, A.; Shaaban, A.; El-Mageed, A.; Shimaa, A.; Semida, W.M.; Rady, M.O. Silicon defensive role in maize (*Zea mays* L.) against drought stress and metals-contaminated irrigation water. *Silicon* **2021**, *13*, 2165–2176. [CrossRef]
64. Edelstein, M.; Ben-Hur, M. Heavy metals and metalloids: Sources, risks and strategies to reduce their accumulation in horticultural crops. *Sci. Hortic.* **2018**, *234*, 431–444. [CrossRef]
65. Zhan, F.; Li, B.; Jiang, M.; Yue, X.; He, Y.; Xia, Y.; Wang, Y. Arbuscular mycorrhizal fungi enhance antioxidant defense in the leaves and the retention of heavy metals in the roots of maize. *Environ. Sci. Pollut. Res* **2018**, *25*, 24338–24347. [CrossRef] [PubMed]
66. Thorpe, A.; Harrison, R.M. Sources and properties of non-exhaust particulate matter from road traffic: A review. *Sci. Total Environ.* **2008**, *400*, 270–282. [CrossRef]
67. Habibi, K. Characterization of particulate lead in vehicle exhaust-experimental techniques. *Environ. Sci. Technol.* **1970**, *4*, 239–248. [CrossRef]
68. Osumi, K. Exhaust Gas Purification System and Exhaust Gas Purification Method. U.S. Patent 9,593,614, 14 March 2017.
69. MacKinnon, G.; MacKenzie, A.B.; Cook, G.T.; Pulford, I.D.; Duncan, H.J.; Scott, E.M. Spatial and temporal variations in Pb concentrations and isotopic composition in roaddust, farmland soil and vegetation in proximity to roads since cessation of use of leaded petrol in the UK. *Sci. Total Environ.* **2011**, *409*, 5010–5019. [CrossRef]
70. Saeedi, M.; Hosseinzadeh, M.; Jamshidi, A.; Pajooheshfar, S.P. Assessment of heavy metals contamination and leaching characteristics in highway side soils, Iran. *Environ. Monit. Assess.* **2009**, *151*, 231–241. [CrossRef]
71. Karmakar, D.; Padhy, P.K. Air pollution tolerance, anticipated performance, and metal accumulation indices of plant species for greenbelt development in urban industrial area. *Chemosphere* **2019**, *237*, 124522. [CrossRef]
72. Esfandiari, M.; Sodaiezadeh, H.; Ardakani, H. Assessment of heavy metals in Cypress (*Thuja orientalis* L.) in the Yazd Highway green belt. *Desert* **2020**, *25*, 15–23.
73. Mondal, S.; Singh, G. Air pollution tolerance, anticipated performance, and metal accumulation capacity of common plant species for green belt development. *Environ. Sci. Pollut. Res* **2021**, 1–12. [CrossRef] [PubMed]

Article

# Organochlorine Pesticides in Karst Soil: Levels, Distribution, and Source Diagnosis

Wei Chen [1,2,3,4,5,6,7,8,9], Faming Zeng [10], Wei Liu [8], Jianwei Bu [1,2,3,4], Guofeng Hu [11], Songshi Xie [12], Hongyan Yao [9], Hong Zhou [8], Shihua Qi [1,2,3,4,5,6,7] and Huanfang Huang [5,6,7,13,*]

1. State Key Laboratory of Biogeology and Environmental Geology, China University of Geosciences, Wuhan 430078, China; wei.chen@cug.edu.cn (W.C.); jwbu@cug.edu.cn (J.B.); shihuaqi@cug.edu.cn (S.Q.)
2. School of Environmental Studies, China University of Geosciences, Wuhan 430078, China
3. Hubei Key Laboratory of Environmental Water Science in the Yangtze River Basin, China University of Geosciences, Wuhan 430078, China
4. Hubei Provincial Engineering Research Center of Systematic Water Pollution Control, China University of Geosciences, Wuhan 430078, China
5. State Key Laboratory of Organic Geochemistry, Guangzhou Institute of Geochemistry, Chinese Academy of Sciences, Guangzhou 510640, China
6. Guangdong Province Key Laboratory of Environmental Protection, Chinese Academy of Sciences, Guangzhou 510640, China
7. Resources Utilization, and Guangdong-Hong Kong-Macao Joint Laboratory for Environmental Pollution and Control, Chinese Academy of Sciences, Guangzhou 510640, China
8. Institute of Geological Survey, China University of Geosciences, Wuhan 430074, China; wliu@cug.edu.cn (W.L.); zhouhong@cug.edu.cn (H.Z.)
9. Ecological Environment Monitoring Station, Ninth Division, Xinjiang Production and Construction Corps, Tacheng 834601, China; wowoyhy@163.com
10. School of Environmental and Chemical Engineering, Foshan University, Foshan 528000, China; famingzeng@fosu.edu.cn
11. China City Environment Protection Engineering Limited Company (CCEPC), Wuhan 430071, China; 54001@ccepc.com
12. Shandong Institute of Geological Survey, Jinan 250013, China; sdsddy@163.com
13. South China Institute of Environmental Sciences, MEE, Guangzhou 510535, China
* Correspondence: hhuanfang@outlook.com

**Abstract:** Excessive reclamation and improper use of agrochemicals in karst areas leads to serious non-point source pollution, which is of great concern and needs to be controlled, since contaminants can easily pollute groundwater due to the thin patchy soil and developed karst structures. The occurrences of organochlorine pesticides (OCPs) in karst soil were investigated by analyzing 25 OCPs in the karst soils near the Three Gorges Dam, China. The total concentrations of OCPs ranged 161–43,100 (6410 ± 9620) pg/g, with the most abundant compounds being $p,p'$-DDT and mirex. The concentration differences between the orchard and vegetable field and between upstream and downstream presented the influences of land-use type and water transport on the OCP spatial distributions. Composition analysis indicated the possible fresh inputs of lindane, technical DDT, aldrin, endrin, mirex, and methoxychlor. Their illegal uses implied an insufficient agrochemical management system in undeveloped karst areas. Principal component analysis with multiple linear regression analysis characterized the dominant sources from current agricultural use and current veterinary use in the study area. OCPs in the soils might not pose significant cancer risk for the residents, but they need to be controlled due to their illegal uses and bioaccumulation effect via the food chain.

**Keywords:** illegal use; non-point source pollution; agricultural use; veterinary use; Three Gorges

## 1. Introduction

Organochlorine pesticides (OCPs), mainly including dichlorodiphenyltrichloroethane (DDT), hexachlorocyclohexane (HCH), chlordane, endosulfan, aldrin, and mirex, are a class of synthetic chlorine-contained pesticides. They can effectively cause insect spasms and eventually kill insects by opening the sodium ion channel in the neurons or nerve cells of insects, causing them to fire spontaneously [1]. Because of the excellent insecticidal effects, OCPs were widely and largely used in agriculture during the 1950s–1980s worldwide [2,3]. With the disclosure of the high toxicity on humans and wildlife, including cancers, allergies, and neurologic, reproductive and immune dysfunctions [4,5], most OCPs were listed in the Stockholm Convention and banned in over 130 countries since the 1970s. Nevertheless, the OCP pollution is still of concern: (1) because of the persistence, the high OCP residues are still detected in the soil, water, sediment, atmosphere, and biota [6–8]; (2) OCPs can undergo long-range transport within the atmosphere, water, and migrant birds, even to the places without any pesticide applications, leading to global pollution [9,10]; and (3) due to the lipophilicity, OCPs in the soil or water would accumulate in plants and livestock and eventually threaten human health via the food chain [11]. In addition, illegal uses of OCPs are still found in some countries and regions due to poor pesticide management. Recently, Khuman et al. (2020) reported the ongoing usage of technical HCH contradicting the ban in the agriculture sector on India's southwest coast [12]. Fresh inputs were also observed for HCHs and heptachlor in soil and groundwater in the middle reaches of the Yangtze River Basin, China [13] and for DDTs in the soil from Mt. Shergyla, Tibetan Plateau, China [14]. These emphasize the need to continuously investigate the occurrence of OCPs in the environment and accordingly adjust policies for risk control.

With the area accounting for ca. 15% of the continental surface, karst is one of the most important landscapes on terrene and is home to a quarter of the global population [15]. Karst areas are mostly mountainous and dominated by agriculture economy. The soil resource is very precious in karst areas. On the one hand, as the carbonatite widely distributes, it is not easy to form soil in karst areas; the formation of one cm depth soil in karst areas might take 4000–8500 years [16]. On the other hand, as the transmissive network consisting of sinkholes, fissures, and conduits is well developed in the rainfall-dissolved carbonate bedrock [17], soil erosion is prevalent and severe in karst areas [18]. The soil in karst areas is generally thin, patchy, and fragile [19]. Nevertheless, farmers conduct agricultural activities, which is perhaps the most ubiquitous human activity on karst terranes, to feed themselves in this vulnerable soil layer. Agriculture has even been expanded to marginal soil on slopes and ridges due to the increase of population and the decline of land productivity [20], adversely affecting the ecology in karst areas, including the exacerbation of soil erosion, deforestation, and pollutions of fertilizers, pesticides, and agricultural wastes [21]. Among those, the non-point source pollution of agrochemicals in soil has raised great concerns because the soil contaminants pose adverse impacts on human health directly and via the food chain. To make matters worse, the thin patchy karst soil is not capable of buffering against pollutants; the soil contaminants can easily pollute the surface water and groundwater with rapid water runoff via highly permeable networks of fissures and conduits [22,23], leading to widespread pollution in the karst multimedia [24].

Many studies have focused on the OCP pollution in the karst water. The high OCP concentrations were reported in the surface river water (32.1–293, average 120 ng/L) [25], underground river water (2.58–320 ng/L) [25,26], spring water (0.30–32.2 ng/L) [27], and the sediment cores (0.85–63.1, average 8.11 ng/g) [28] in southwestern China, one of the largest karst areas in the world [29]. In the Yucatán karst area, México, severe OCP pollution (up to $1.36 \times 10^7$ ng/L for heptachlor) was also reported in groundwater [30]. By contrast, there are much fewer investigations of OCPs in karst soil [31,32]. Because OCPs in the water enter via the soil [33], the OCP investigation in the karst soil is fundamental and crucial for diagnosing the source, implementing effective management practice, and developing a regulatory system for risk control.

To study the occurrence of OCPs in karst soil, we collected soil samples from the Yichang karst area near the Three Gorges Dam, central China (a typical karst mountainous area) and analyzed 25 OCP compounds to (1) investigate the levels, compositions, and spatial distributions of OCPs in the karst soil; (2) diagnose and quantify the OCP sources in the karst area; and (3) assess the carcinogenic risk posed by OCPs in the karst soil to residents.

## 2. Materials and Methods

### 2.1. Study Area and Sample Collection

The karst region in southwestern China (ca. 780,000 km$^2$) [34], including Guizhou, western Guangxi, eastern Yunnan, southeastern Chongqing, southern Sichuan, western Hunan, and western Hubei, is the largest contiguous karst area with the most intense karst development in the world [35]. It is the most undeveloped remote mountainous area in China, and many counties therein are poverty-stricken. In these undeveloped areas, people rely on agricultural production but have relatively weak environmental awareness and low risk perception on handling agrochemicals.

The karst study area is in Yichang, western Hubei, with the area of approx. 2100 km$^2$ (Figure 1). It mainly includes Zigui County (the first county closest to the Three Gorges Dam), in addition to part of Changyang County and Yiling and Dianjun Districts of Yichang City. The subtropical monsoon climate prevails in Yichang, with the average annual precipitation of 1216 mm and temperatures of 2–33 °C. The karst study area belongs to a karst trough zone (a typical landscape in central and southern China) [36] and has complex karst landforms consisting of middle-low mountains and deep ravines (40–2057 m a.s.l.). Numerous sinkholes, dolines, and grooves are developed on the up-platform, and large karst springs emerge in the deep valley. The soil layer in the study area is loose and highly uneven (0–4 m). Yellow soil, lime soil, and purple soil were dominant in this region. The pH in the soil ranges between 4.8–6.5, and the total organic carbon concentrations vary between 15.1–30.0 g/kg [37].

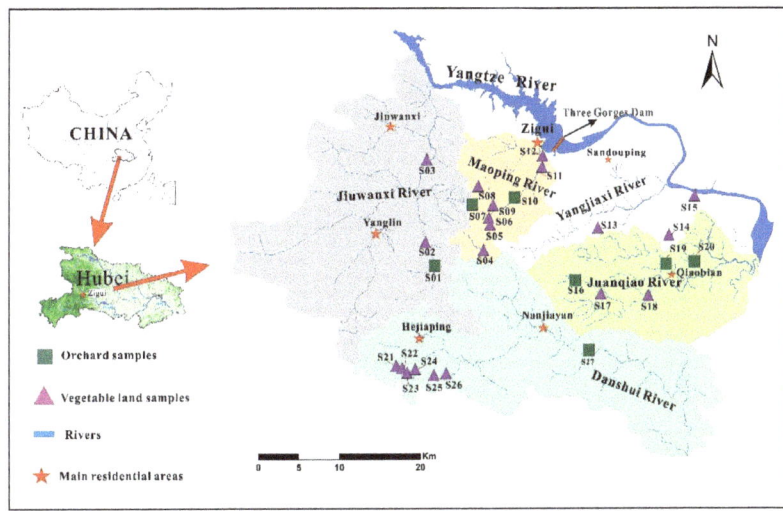

**Figure 1.** Location of the karst study area near Three Gorges Dam, China and the soil sampling sites.

Although the mountainous karst area is not suitable for farming, cultivation is the most predominant human activity in this undeveloped area (the urbanization rate of Yichang is 44.4% [38]). The arable lands are scattered in big depressions on the up-platforms, slopes, and at the bottom of valleys. Farmers forge a living from cultivation on this thin soil overlying carbonate rocks by growing vegetables, flue-cured tobacco, tea, oranges, chestnuts, and other cash crops. Under the excessive reclamation, the soil layer in Zigui

County had decreased by on about 3–5 cm depth per year [39] in the past and suffered severe agrochemical non-point source pollution [40].

The sampling campaign was conducted in October 2019 to avoid the impact of intensive agricultural activity. To collect soil samples that can represent the study area, sample sites were set in fields with relatively thick (>20 cm) and continuous soil layers (>20,000 m$^2$). Twenty-seven surface soil samples (0–20 cm, ca. 1 kg for each) were collected in agricultural fields by clean stainless-steel shovels, of which seven samples were collected from orchards and twenty samples were collected from vegetable fields (Figure 1). After collection, soil samples were wrapped with pre-baked aluminum foil, sealed in PE zip bags and stored in a car refrigerator (4 °C, in the dark) during the sampling and transportation. Once delivered to the laboratory, soil samples were freeze-dried, preserved in the freezer (−20 °C, in the dark), and pretreated within seven days.

*2.2. Sample Preparation and Analysis*

Each dry soil sample (10 g dry weight) was spiked with 20 ng of 2,4,5,6-tetrachloro-*m*-xylene (TC*m*X) and decachlorobiphenyl (PCB209) as recovery surrogates [41,42], and then Soxhlet-extracted with 150 mL of dichloromethane (DCM) for 24 h. Before Soxhlet extraction, clean activated copper granules were added to the collection flask to remove the elemental sulfur from the extract during the extraction. After extraction, each extract was concentrated, solvent-exchanged to *n*-hexane, and reduced to 3 mL by a rotary evaporator (Heidolph G3, Schwabach, Germany). A neutral alumina/silica gel (*v/v*, 1:2) column was then used to purify each concentrated extract. Target OCP compounds were eluted with DCM/*n*-hexane (2:3, 30 mL). The eluate was then concentrated to 0.2 mL under a high-purified gentle nitrogen stream (99.999%) and stored in the sample freezer (−20 °C, in the dark). Before performing the instrumental analysis, each sample was spiked with 20 ng of pentachloronitrobenzene as the internal standard [43]. More details of the sample preparation for OCPs could be found in previous studies [44,45].

Target OCPs were analyzed with an HP7890A gas chromatograph-$^{63}$Ni electron capture detector (GC-ECD, Agilent, Santa Clara, CA, USA) equipped with an HP-5MS column (30.0 m × 0.32 mm × 0.25 μm). According to a well-documented study in the same research group [44], the injector and detector temperatures were 290 and 300 °C, respectively, and the GC oven temperature program was set as: initially 100 °C for 1 min, 4 °C/min to 200 °C, 2 °C/min to 230 °C, and 8 °C/min to 280 °C for 15 min. The calibration curves used to quantify OCPs were built with six standards with increasing analytes concentrations (10, 20, 50, 100, 150, and 200 ng/mL) and 100 ng/mL pentachloronitrobenzene.

*2.3. Quality Assurance and Quality Control (QA/QC)*

Method blanks, parallel samples, blank solvent, QC standard samples, recovery surrogates, and internal standards were used to perform the QA and QC during the sample pretreatment and instrumental analysis. A method blank sample and a parallel sample were pretreated in each batch (16 samples) with the same procedures for the sample extraction and purification. For the instrumental QC, a blank solvent and a standard solution of OCPs were analyzed between every ten-sample analysis to check for interference/cross-contamination and instrument stability. Target compounds were undetectable in all blank samples and solvents. Relative standard deviation values were within 20% for the parallel samples and 10% for the QC standards. The recoveries of TC*m*X and PCB209 were 74.8 ± 17.3% and 86.9 ± 24.1%, respectively. Three times the signal-to-noise levels were used as the detection limits for target compounds. The method detection limits (MDLs) of OCPs were in the range of 1–20 pg/g (Table S1).

All data reported in this study were based on the GC-ECD analysis, but samples with high OCP concentrations were also confirmed with a gas chromatography–mass spectrometer (GC-MS, 6890N GC-5975MSD, Agilent, Santa Clara, CA, USA), and results showed the low relative percentile differences of OCP concentrations between GC-ECD and

GC-MS analysis (<10%). Because of the higher MDLs of OCPs by GC-MS compared with GC-ECD, samples with relatively low OCP concentrations were not analyzed with GC-MS.

### 2.4. Data Analysis

According to the parent–daughter relationship and commercial formulas, twenty-five OCPs were divided into eight groups: HCB, HCHs ($\alpha$-HCH, $\beta$-HCH, $\gamma$-HCH, and $\delta$-HCH), DDTs ($p,p'$-DDT, $o,p'$-DDT, $p,p'$-DDE, $o,p'$-DDE, $p,p'$-DDD, $o,p'$-DDD), CHLs (*trans*-chlordane, *cis*-chlordane, heptachlor, and heptachlor epoxide), ENDOs ($\alpha$-endosulfan, $\beta$-endosulfan, and endosulfan sulfate), DRINs (aldrin, dieldrin, endrin, endrin aldehyde, and endrin ketone), mirex, and methoxychlor. All concentrations in the soil were reported on a pg/g dry weight (dw) basis.

The Kruskal–Wallis test and Spearman correlation analysis were conducted to investigate the difference and correlation between OCP groups. The principal component analysis (PCA) with multiple linear regression analysis (MLRA) was used to identify and quantify the OCP sources in the karst soil. The methodology details were presented in Text S1. These statistical analyses were all performed by SPSS Statistics 25 (IBM, Chicago, IL, USA).

Human exposure to OCPs in soil is mainly via ingestion, dermal contact, and inhalation. In this study, the incremental lifetime risk of cancer ($ICLR_{total}$) exposure to 25 OCPs in the soil via ingestion, dermal contact, and inhalation was assessed according to the US EPA Exposure Factors Handbook–1997 [46]. The $ICLR_{total}$ was assessed for three population groups: children (3–10 years old), adolescents (11–18 years old), and adults (19–64 years old). Furthermore, risks of males and females were estimated separately. The parameters and calculation methods were analytically presented in Text S2.

## 3. Results and Discussion

### 3.1. General Comments on OCP Concentrations

The total concentrations of OCPs ($\sum_{25}$OCPs) were in the range (mean ± standard deviation) of 161–43,100 (6410 ± 9620) pg/g (Table S1). The most abundant compounds were $p,p'$-DDT (1640 ± 5560 pg/g) and mirex (1410 ± 1720 pg/g) (Figure 2), accounting for the average 16.0% and 34.7% of the $\sum_{25}$OCPs, respectively. With the detection rates of >85%, HCB, $\alpha$-HCH, $\beta$-HCH, $\gamma$-HCH, $p,p'$-DDT, $p,p'$-DDD, aldrin, and mirex were prevalent in the study karst soil (Figure 2).

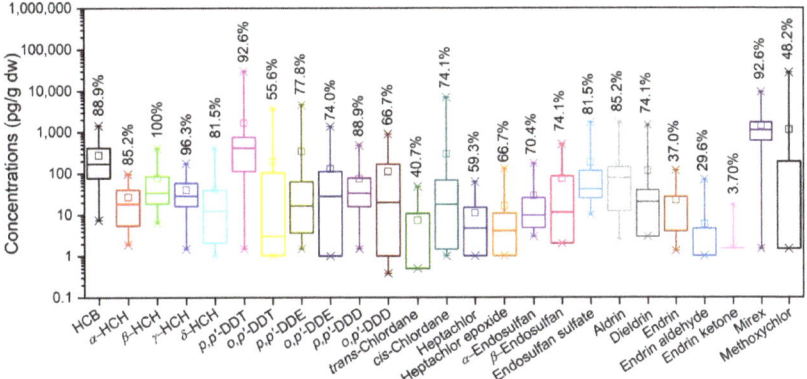

**Figure 2.** Concentrations and detection rates of individual OCPs in the study karst soil.

Compared with OCP concentrations in agricultural soils in other areas (Table S2), the HCB concentrations herein were within the ranges found in the Indus River Basin, Pakistan (400–1900 pg/g) [47] and Central Germany (570–3750 pg/g) [48], while these areas had higher $\sum$HCHs and $\sum$DDTs concentrations than ours. The $\sum$HCHs and $\sum$DDTs concentrations herein were also lower than those observed in the Pearl River Delta, southern

China (<MDL–24,100 and 520–414,000 pg/g for $\sum$HCHs and $\sum$DDTs, respectively) [49], and the Sichuan Basin, southwestern China (69–3190 (avg. 1780) and 1870–25,200 (avg. 13,500) pg/g for $\sum$HCHs and $\sum$DDTs, respectively) [45].

*3.2. Influence of Land-Use Type and Water Transport on the OCP Spatial Variation*

Due to the hilly terrain and small farmland area, mechanized farming is not widespread in the study area. Individuals cultivated farmlands without unified management, which resulted in the high coefficients of spatial variations (CV) for $\sum_{25}$OCPs (CV: 150%, Figure 3) and individual OCP compounds (CV: 94.8–520%, Figures S1–S3). The highest $\sum_{25}$OCPs concentrations were found in Sites S15 and S19 from a vegetable field and an orange orchard, respectively (Figure 3), which might be attributed to the improper use of pesticides by farmers, or there might be agrochemical dumps in these sites. The Spearman correlation analysis (Table S3) showed that (1) HCB, HCHs, DDTs, CHLs, ENDOs, and DRINs were significantly correlated with each other; and (2) mirex and methoxychlor were not significantly correlated with any other OCP groups. These results indicated similar spatial distributions for OCP groups except for mirex and methoxychlor (Figures S1–S3).

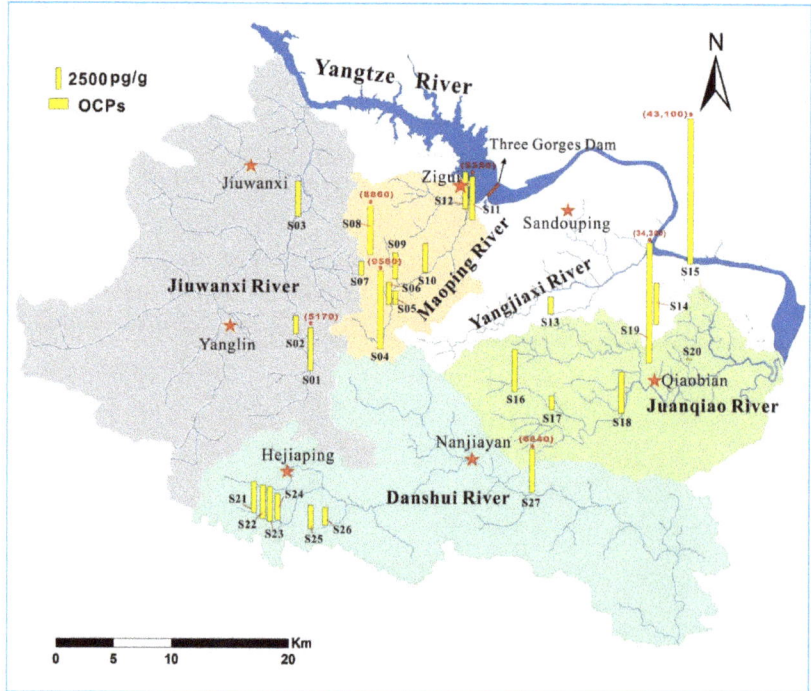

**Figure 3.** The spatial distribution of soil OCPs in different river basins.

The land-use type might affect the spatial distributions of OCPs in the karst area. The $\sum_{25}$OCPs concentrations in the orchard (9000 ± 11,500 pg/g) were higher than those in the vegetable field (5510 ± 9020 pg/g). Specifically, the vegetable field had higher concentrations of HCB, HCHs, DDTs, CHLs, ENDOs, and DRINs compared with the orchard, while the concentrations of mirex and methoxychlor in the orchard were higher, although these comparisons were not significant (Kruskal–Wallis test, $p > 0.05$, Table S4). The higher concentrations of mirex and methoxychlor in the orchard soil might be due to their uses to treat tree mites, poultry, and livestock (and their sheds), as farmers keep poultry and livestock in orchards in the study area. Aside from the land-use type, the hydrogeological condition might also affect the spatial distributions of OCPs. Compared with the concentrations in upstream areas, higher concentrations were generally found in corresponding downstream sites for HCB, HCHs, DDTs, ENDOs, and DRINs in surface river basins (Figures S1 and S2). For example, Site S27 had higher $\sum$DDTs concentration than Sites S21–S26 in the Danshui River basin (Figure S1). The higher DDT concentrations in the downstream soil were also observed in the Minjiang River, Fujian, China [50]. This might be attributed to the collection of contaminants via tributaries and surface runoff, and the more extensive cultivations in the downstream areas. Mirex and methoxychlor had opposite spatial distributions with other OCP groups, i.e., they had higher concentrations in upstream areas (Figure S3), suggesting that compared with the water transport, the land-use type affected the spatial distributions of mirex and methoxychlor more significantly.

*3.3. Source Diagnosis for OCPs by Composition Analysis*
3.3.1. HCB

HCB accounted for avg. 7.84% of $\sum_{25}$OCPs in the soil (Figure S4a). HCB was a pesticide used to treat seeds and control wheat bunt [51], and was banned in 2009 in China [52]. However, HCB may still be emitted during industrial manufacturing, as HCB is a material of fireworks, ammunition, and synthetic rubbers [51,53]. In addition, coal combustion, waste incineration, and fuel combustion may also release HCB [51,54]. With the high atmosphere transport potential, HCB might also come from the long-range atmosphere transport from other areas. The study area is a remote mountainous area without industries around; thus, the industrial emission was not the main source of HCB herein. The contribution of long-range transport might also be only marginally indicated due to the high spatial variation of HCB (CV: 114%). Therefore, the locatable agricultural use was deemed the primary source for HCB in the soil.

3.3.2. HCHs

The $\sum$HCHs concentrations accounted for avg. 5.43% of $\sum_{25}$OCPs in the soil (Figure S4a). As shown in Figure S4b, $\beta$-HCH was the most abundant compound among four HCH isomers (accounting for avg. 38.9% of $\sum$HCHs), followed by $\gamma$-HCH. HCHs were generally introduced into the environment via the agricultural uses of technical HCH and lindane. Sources of HCHs could be distinguished as technical HCH and lindane have different formulas: technical HCH generally contains $\alpha$-HCH (60–70%), $\beta$-HCH (5–12%), $\gamma$-HCH (10–15%), $\delta$-HCH (6–10%), and other isomers (3–4%), while lindane contains a high content of $\gamma$-HCH (> 90%) [44,45]. In the environment, both $\alpha$-HCH and $\gamma$-HCH can degrade ($\gamma$-HCH is more easily degraded) to $\beta$-HCH, which is more stable than its parent HCH compounds [55,56]. Furthermore, $\gamma$-HCH might be converted or biodegraded to $\alpha$-HCH via photoisomerization and biodegradation [57,58]. $\alpha$-/$\gamma$-HCH values of <4, 4–7, and >7 could therefore indicate the current-use of lindane, the current-use of technical HCH, and historical use of technical HCH, respectively.

The ratios of $\beta$-HCH/($\alpha$-HCH + $\gamma$-HCH) ranged from 0.09 to 5.41 (median: 0.87, Figure 4a) in the study karst soil. Only 40.7% of samples had ratios of >1, indicating that HCHs in the soil had not been highly degraded, i.e., there might be fresh input of HCHs herein. This was also supported by the low ratios of $\alpha$-HCH/$\gamma$-HCH in the soil; the ratios of $\alpha$-HCH/$\gamma$-HCH ranged from 0.04 to 11.1 (median: 0.34), with low ratios (<4) found in 92.6% samples, indicating the possible current use of lindane in the karst study area.

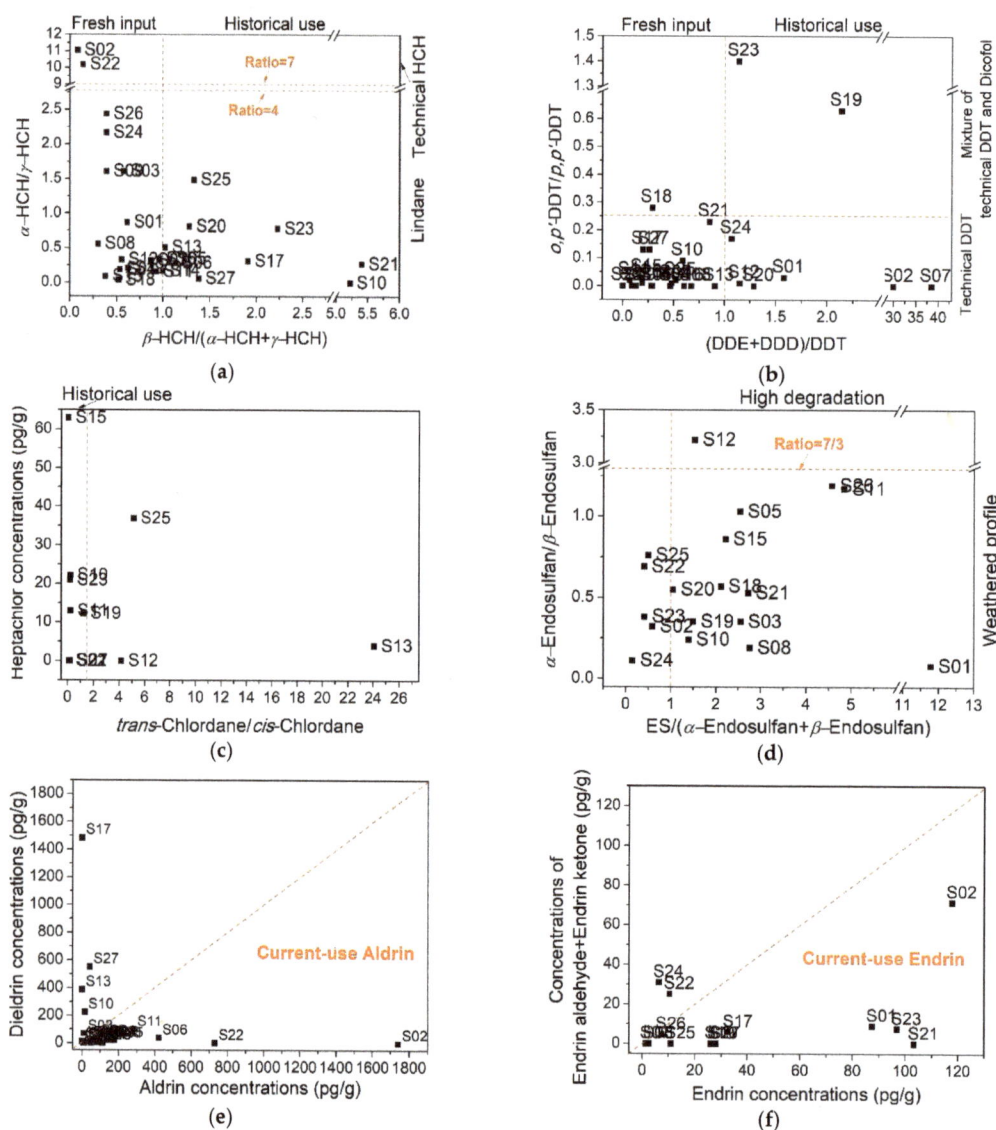

**Figure 4.** Isomeric and metabolic ratios for identifying the sources of HCH (**a**), DDT (**b**), chlordane (**c**), endosulfan (**d**), aldrin (**e**), and endrin (**f**) in the study karst soil. Some samples were not plotted because of undetectable target OCP compounds. Results showed the possible current uses of lindane, technical DDT, aldrin, and endrin in the soil.

### 3.3.3. DDTs

DDTs were one of the most important groups in the soil, accounting for avg. 28.2% of $\sum_{25}$OCPs (Figure S4a). The most abundant compound was $p,p'$-DDT, accounting for avg. 54.9% of $\sum$DDTs (Figure S4c). Parent DDTs mainly degrade to DDE and DDD ($p,p'$- and $o,p'$-isomers included) under aerobic and anaerobic conditions, respectively. In this study, the values of DDE/DDD were in the range of 0–50.3, with low values (<1) found in 45.8% samples. This indicated the existence of anaerobic degradation of parent-DDTs in surface soil. Ratios of (DDE + DDD)/DDT ranged from 0.07 to 38.5, with low ratios (<1) found in 69.2% samples (Figure 4b), suggesting that DDTs in most sites were not highly degraded. Further analysis showed the low ratios (<0.25) of $o,p'$-DDT/$p,p'$-DDT in 88.0% of sites (Figure 4b), indicating the possible input of technical DDT (rather than dicofol) in our study area. This is unexpected because the agricultural use of technical DDT was banned in 1983, and its exception use for vector-control was also banned in 2009 in China [52]. The fresh input of technical DDT was also found in the air from a karst cave in Guilin [59], and in the soil from Chongqing, southwestern China [32]. Results herein implied the illegal use of technical DDT, and therefore insufficient pesticide management in the study area [60].

### 3.3.4. CHLs

As a broad-spectrum insecticide on a range of crops, chlordane was used extensively to control termites. Technical chlordane in the international market contains 13% *trans*-chlordane, 11% *cis*-chlordane, and 5% heptachlor [61]. Of note, *trans*-chlordane is more prone to be photodegraded than *cis*-chlordane [62]. Thus, the *trans*-chlordane/*cis*-chlordane ratio is expected to be lower than 1.56 in the environment. In the study karst soil, the $\sum$CHLs concentrations accounted for avg. 5.04% of $\sum_{25}$OCPs (Figure S4a). Here, *trans*-chlordane and *cis*-chlordane were rarely detected, with detection rates of 40.7% and 74.1%, respectively (Figure 2). Among the soil samples with both detectable *trans*-chlordane and *cis*-chlordane, *trans*-chlordane/*cis*-chlordane ratios varied from 0.06 to 24.1, with low ratios (<1) observed in 72.2% samples. This indicated the weathered chlordane profile in most soils. A very high *trans*-chlordane/*cis*-chlordane ratio (24.1) was found in Site S13. This might be attributed to the possible high aerobic degradation of *cis*-chlordane [63], rather than the use of heptachlor (commercial heptachlor contains 20–22% *trans*-chlordane), since the heptachlor concentration in Site S13 was low (4.02 pg/g, Figure 4c).

### 3.3.5. ENDOs

Endosulfan was mainly used in cotton cultivation and was banned in China since 26 March 2019 [64]. The commercial endosulfan contains α-endosulfan and β-endosulfan in a ratio of 7:3 [65]. Both isomers can be degraded to endosulfan sulfate in the environment. In the study karst soil, endosulfan sulfate accounted for avg. 58.2% of $\sum$ENDOs (Figure S4e). The ratios of endosulfan sulfate/(α-endosulfan + β-endosulfan) ranged 0–34.3, with high values (>1) found in 73.9% samples (Figure 4d), indicating the high degradation of endosulfan in the soil. In samples with low ratio values (<1) of endosulfan sulfate/(α-endosulfan + β-endosulfan), the ratios of α-endosulfan/β-endosulfan ranged between 0.11–0.76 (Figure 4d), showing the weathered profile of endosulfan in the soil, as α-endosulfan was more prone to volatilize from the surface than β-endosulfan [66]. Considering the above, there might be no fresh input of endosulfan in the karst study area.

### 3.3.6. DRINs

The concentrations of $\sum$DRINs merely accounted for avg. 6.88% of $\sum_{25}$OCPs (Figure S4a), with the most abundant compound being aldrin (Figure S4f). Aldrin was used to kill termites, grasshoppers, and other insect pests. Its use had been banned since 2002 in China [67]. The aldrin concentration in the environment is generally low because aldrin can rapidly convert to dieldrin [68]. However, high ratios of aldrin/dieldrin (>1) were observed in 71.4% samples in this study, indicating the fresh input of aldrin. Endrin and its degradation products endrin aldehyde and endrin ketone were rarely detected in the soil (detection rates: <37.0%, Table S1). Among seven soil samples detected with at least one of these compounds, five samples have low ratios (<1) of (endrin aldehyde + endrin ketone)/endrin, indicating the possible current-use of endrin on a small scale.

### 3.3.7. Mirex and Methoxychlor

Mirex is mainly used to combat ants and termites. It has also been used as a fire retardant in plastics, rubber, and electrical goods [68]. In China, the production, circulation, use, import, and export of Mirex had been banned since 2009 [52]. In the study karst soil, mirex was highly and frequently detected; it accounted for avg. 34.7% of $\sum_{25}$OCPs (Figure S4a) with a detection rate of 92.6% (Table S1). In addition, mirex had high spatial variation (CV: 123%) with concentrations up to 9300 pg/g. The study area has suffered serious termite hazards. Based on the investigation and field survey of termite hazards in Zigui County conducted in January 2015, twenty-three out of thirty towns had serious hazards [69]. Using toxic pesticides is one of the methods used by residents to control termites. The high abundance and prevalence of mirex in the soil, accompanied by the serious termite hazards and termite control methods, might indicate the current use of mirex in the study area.

Methoxychlor was initially developed as a DDT replacement [70]. It was widely used in both agriculture (to treat field crops, vegetables, fruits, stored grains) and veterinary practices (to treat livestock, pets, homes, gardens) to combat biting flies, houseflies, and mosquito larvae [68]. Due to the acute toxicity, bioaccumulation, and endocrine disruption activity, methoxychlor was banned in 2003 in the USA [71] and is proposed for listing under the Stockholm Convention [70]. However, it is still used in some areas in China [72]. In the study karst soil, methoxychlor concentrations accounted for avg. 6.15% of $\sum_{25}$OCPs (Figure S4a). Although methoxychlor was not prevalent in the soil (detection rate: 48.2%, Table S1), it had high spatial variation (CV: 470%) with concentrations up to 27,700 pg/g. This partly indicated the possible current input of methoxychlor herein.

### 3.4. Characteristics and Contributions of Sources

The OCP sources in the study karst soil were characterized by the PCA with MLRA, which were widely used in previous studies. For example, the sources of dicofol-type DDT, historical residues, and fresh technical DDT were drawn by the PCA + MLRA to explain 55%, 21%, and 17% of DDTs in the Pearl River Delta soil, southern China [49]. The PCA + MLRA indicated a greater contribution of the forest filter effect than the mountain cold trapping effect for the occurrence of polychlorinated biphenyls in the forest soil of Mt. Gongga, eastern Tibet [73].

The log-transformed concentrations of HCB, HCHs, DDTs, CHLs, ENDOs, DRINs, mirex, and methoxychlor were used to perform the PCA (methodology details could be found in Text S1). Three PCs with eigenvalues greater than one were extracted, explaining 68.4% of the total variability. PC1 explained 31.2% of the total variability and had high loadings of HCHs, DDTs and DRINs, and medium loading of HCB; PC2 explained 22.2% and had high loadings of ENDOs and medium loading of CHLs; PC3 explained 15.0% and had high loading of methoxychlor and medium loading of mirex (Table S5 and Figure 5a). According to the source diagnosis results for different OCP groups in Section 3.3, PC1 and PC2 mainly contained compounds with possible fresh inputs and with weathered profiles, respectively. Therefore, PC1 and PC2 were identified as the current-use source and historical source, respectively. PC3 contained mirex and methoxychlor. There might also be current uses for these two compounds. However, they were clearly distinguished from other current-use pesticides (HCHs, DDTs, and DRINs in our study) (Figure 5a), indicating the different usages of mirex and methoxychlor. Mirex is mainly used against ants and termites. Methoxychlor was also widely used in veterinary practices to combat biting flies, houseflies, mosquito larvae, and cockroaches, aside from the agricultural use [68]. Therefore, PC3 might indicate the veterinary source of pesticides.

Soil samples were not regularly grouped (even for those located nearby) based on their PC scores (Figure 5b), showing the mixture OCP sources for most soils. For example, Sites S21–S26 were located nearby (Figure 1), but they were plotted separately in Figure 5b. This was attributed to the different pesticide uses due to individuals' cultivation without unified management in the karst study area. Sites S15 and S19 were separated with predominant proportions of $\sum$DDTs (91.6%) and methoxychlor (80.8%), respectively, which were different from other samples.

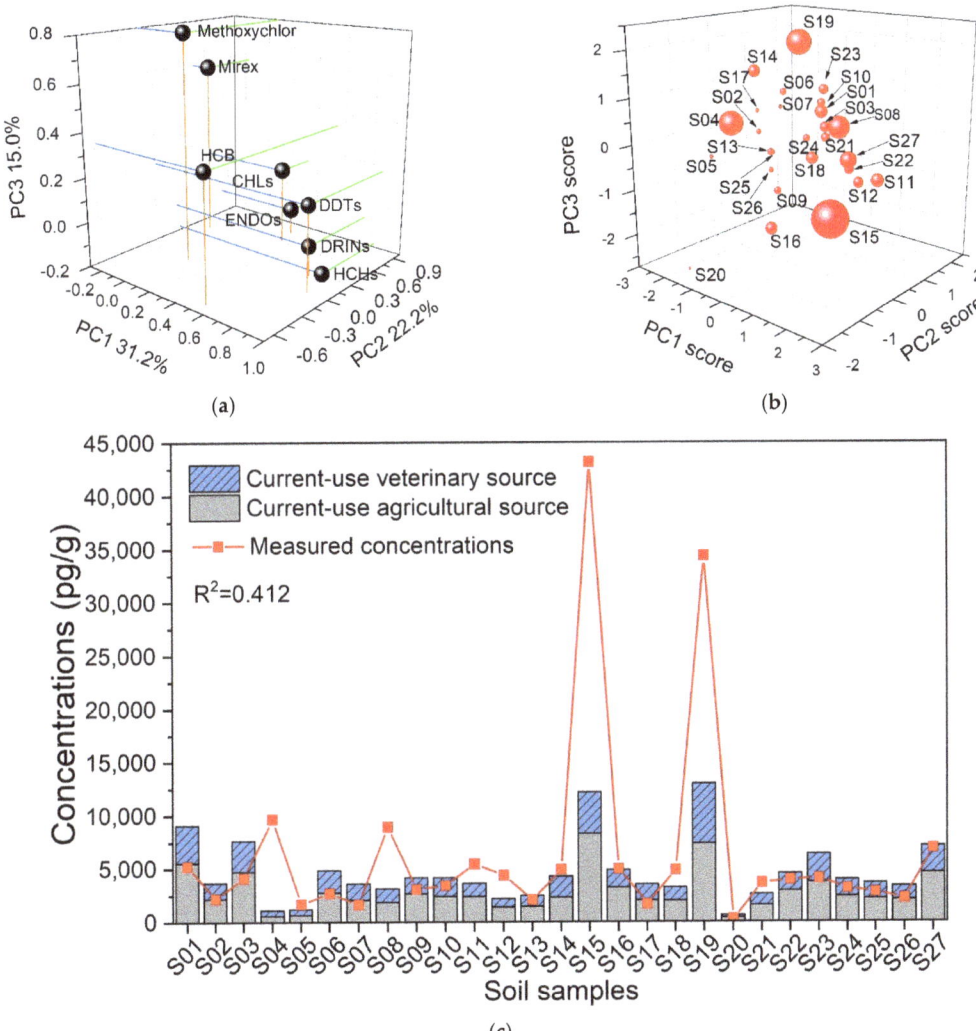

**Figure 5.** Loading profiles of PCs (**a**), factor scores of each soil samples (**b**), and contributions of current-use agriculture source and current-use veterinary source to the $\sum_{25}$OCP concentrations in each soil samples (**c**), based on the PCA + MLRA analysis. PC1, PC2, and PC3 in (**a**) indicate the current agricultural use, historical agricultural use, and current veterinary use, respectively. The point sizes in (**b**) represent the concentration levels of $\sum$OCPs. The poor fits between modeled concentrations and measured concentrations in Sites S15 and S19 in (**c**) indicate the existence of other dominant pesticide sources (e.g., agrochemical waste dumps).

Subsequently, the MLRA was performed to apportion the contributions of each source to the $\sum_{25}$OCPs concentrations (see Text S1 for methodology details). The factor scores of PC1 and PC3 entered the regression equation, while PC2 was removed, suggesting the minor historical agricultural use for OCPs in the soil. The regression of PC1 and PC3 could merely explain 41.2% of the variation of log$\sum_{25}$OCPs (the dependent variable), indicated by the adjusted $R^2$ value of 0.412 (Table S6, Figure S5). The fitted MLRA equation had statistical significance (ANOVA, $p < 0.05$, Table S7). The regression coefficients are shown in Table S8, and accordingly, the regression equation was:

$$\log \sum\nolimits_{25} \text{OCPs} = 0.566 \times \text{PCS}_1 + 0.368 \times \text{PCS}_3 \quad (1)$$

where $\text{PCS}_1$ and $\text{PCS}_3$ were the factor scores of PC1 and PC3, respectively.

The contributions of PC1 and PC3 to the $\sum_{25}$OCPs concentrations for each soil sample were then calculated, and are shown in Figure 5c. The MLRA could not exactly model the high $\sum_{25}$OCPs concentrations in Sites S15 and S19, indicating that there might be other dominant sources in these sites, such as agrochemical waste dumps. The current agricultural use was dominant for the occurrence of OCPs herein, explaining 53.0–68.2% (average 60.5%) of $\sum_{25}$OCPs. With the percent contributions of 31.8–47.0%, the current use of pesticides in veterinary practices should also be of concern.

### 3.5. Risk Assessment

The Soil Environmental Quality–Risk Control Standard for Soil Contamination of Agricultural Land (GB 15618-2018) from the China and Soil Remediation Circular 2009, from the Netherlands, was consulted to indicate the OCP pollution level in the soil. In the karst study area, the concentrations of $\sum$HCHs and $\sum$DDTs were lower than the Chinese risk screening values (both $1.0 \times 10^5$ pg/g) for soil contamination of agricultural land [74]. Besides, all OCP concentrations were lower than the soil remediation intervention values for HCHs ($1.2 \times 10^6$–$1.70 \times 10^7$ pg/g), DDTs ($1.7 \times 10^6$–$3.40 \times 10^6$ pg/g), aldrin ($3.2 \times 10^5$ pg/g), and chlordane, endosulfan, and heptachlor epoxide ($4.0 \times 10^6$ pg/g) (Table S1) [75], indicating that the functional properties of soil in the karst study area for human, plant and animal life are not seriously impaired or threatened.

The incremental cancer risks calculated for children, adolescence, and adults were in the ranges of $9.52 \times 10^{-11}$–$1.46 \times 10^{-8}$, $8.76 \times 10^{-11}$–$1.14 \times 10^{-8}$, and $1.31 \times 10^{-10}$–$2.01 \times 10^{-8}$, respectively (calculation details were presented in Text S2). The risk for adults was the highest among the three groups. In each group, males' risk was slightly lower than that for females because of males' higher body weight (Table S9). A risk lower than $10^{-6}$ was considered acceptable [76,77]. Therefore, OCPs in the soil would not pose a significant risk to residents. Nevertheless, the persistence and lipophilic affinity of OCPs would result in the bioaccumulation and biomagnification of these substances in crops and livestock, and they might eventually threaten human health via the food chain. Besides, as discussed in Section 3.3, fresh inputs of OCPs were found in the karst study area. Therefore, the risk of OCPs in soil should not be ignored.

## 4. Conclusions

The soil in karst areas has suffered severe non-point source pollution of agrochemicals due to the excessive reclamation and improper or illegal use of agrochemicals, which is especially crucial to control in karst areas since soil contaminants can easily enter surface water and groundwater owing to the thin patchy soil, fast water runoff, and developed karst fissures and caves. This study elaborated the occurrences of OCPs in karst soil by analyzing 25 OCPs in the soil from the Yichang karst area near the Three Gorges Dam, China. Results showed the total OCP concentrations of 161–43,100 pg/g dw. HCB, $\alpha$-HCH, $\beta$-HCH, $\gamma$-HCH, $p,p'$-DDT, $p,p'$-DDD, aldrin, and mirex were frequently detected, of which $p,p'$-DDT and mirex were the most abundant compounds. The OCP spatial distributions were affected by the land-use type and water transport. The isomeric and metabolic ratios indicated the possible fresh inputs of lindane, technical DDT, aldrin, endrin, mirex, and methoxychlor. The PCA with MLRA analysis characterized the dominant sources of pesticides from current agricultural use and current veterinary use in the study karst soil.

The illegal uses and prevalence of OCPs implied the poor agrochemical management system, and farmers' relatively weak environmental awareness and low-risk perception of handling agrochemicals, which might be attributed to poverty, low level of education, and lack of regulation in the agricultural sector. This social condition is a problem in the study area and in many remote karst areas worldwide, which is of great concern. Strict market regulation and professional training are urgently needed to prevent the illegal production, sale, and use of prohibited agrochemicals. Government and the public should recognize the ecological vulnerability in karst areas and take mitigation measures for sustainable development.

**Supplementary Materials:** The following are available online at https://www.mdpi.com/article/10.3390/ijerph182111589/s1, Text S1: Principle component analysis with multiple linear regression analysis; Text S2: Carcinogenic risk calculation; Figure S1: Spatial distributions of HCB, HCHs, DDTs, and DRINs in different river basins; Figure S2: Spatial distributions of CHLs and ENDOs in different river basins; Figure S3: Spatial distributions of mirex and methoxychlor in different river basins; Figure S4: The OCP compositions in the soil from the study karst area; Figure S5: The linear fit between the measured log$\sum$25OCPs and modeled log$\sum$25OCPs by the MLRA; Table S1: The OCP concentrations in the study karst soil; Table S2: Comparisons of OCPs in the agricultural soil between the study karst area and other areas around the world; Table S3: Spearman correlation coefficients between OCP groups; Table S4: OCP concentrations in different land use types; Table S5: Rotated component matrix; Table S6: The summary of the MLRA model for the OCP data in the study karst soil; Table S7: The ANOVA result for the MLRA; Table S8: The coefficients for the regression equation by MLRA; Table S9: The incremental lifetime cancer risk exposure to OCPs in soil from the study karst area.

**Author Contributions:** Formal analysis, Data curation, Visualization, Writing—original draft, W.C.; Investigation, Visualization, Writing—review and editing, F.Z.; Investigation, Resources, Writing—review and editing, Visualization, Supervision, W.L.; Methodology, Data curation, Writing—review and editing, J.B.; Formal analysis, Writing—review and editing, G.H.; Data curation, Writing—review and editing, S.X.; Methodology, Writing—review and editing, H.Y.; Supervision, Funding acquisition, H.Z.; Resources, Supervision, S.Q.; Conceptualization, Methodology, Investigation, Resources, Writing—review and editing, H.H. All authors have read and agreed to the published version of the manuscript.

**Funding:** This research was funded by the National Key Research and Development Program (2019YFC1805502), the National Natural Science Foundation of China (41907327 and 42007178), the Natural Science Foundation of Hubei (2019CFB372 and 2020CFB463), Open Funds from State Key Laboratory of Organic Geochemistry (SKLOG202008), the Guangxi Key Science and Technology Innovation Base on Karst Dynamics (KDL and Guangxi202002), the China Geological Survey (DD20190824), Fundamental Research Funds for the Central Universities (CUG 190644 and CUGL180817), the special fund from the Hubei Provincial Engineering Research Center of Systematic Water Pollution Control (20190814), and Open Funds from Hubei Key Laboratory of Environmental Water Science in the Yangtze River Basin.

**Institutional Review Board Statement:** Not applicable.

**Informed Consent Statement:** Not applicable.

**Data Availability Statement:** The data that support the findings of this study are available from the corresponding author upon reasonable request.

**Acknowledgments:** The authors would like to express grateful thanks to Longliang Wang and Bo Peng for sampling assistance; to Zhaoyang Wei, Zhe Qian, Chang Pu and Ziqiong Zhang for sample preparation; to the anonymous reviewers for comments and suggestions in improving the manuscript; and special thanks to Julia Ellis Burnet for always helping with the English proofing.

**Conflicts of Interest:** The authors declare no conflict of interest.

# References

1. Ganzel, B. How Insecticides Work. Available online: https://livinghistoryfarm.org/farminginthe70s/pests_06.html (accessed on 15 March 2021).
2. Li, Y.; Cai, D.; Singh, A. Technical Hexachlorocyclohexane Use Trends in China and Their Impact on the Environment. *Arch. Environ. Contam. Toxicol.* **1998**, *35*, 688–697. [CrossRef]
3. Pacyna, J.M.; Breivik, K.; Münch, J.; Fudala, J. European Atmospheric Emissions of Selected Persistent Organic Pollutants, 1970–1995. *Atmos. Environ.* **2003**, *37*, 119–131. [CrossRef]

4. WHO. *Public Health Impact of Pesticides Used in Agriculture*; World Health Organization: Geneva, Switzerland, 1990.
5. Ledirac, N.; Antherieu, S.; d'Uby, A.D.; Caron, J.C.; Rahmani, R. Effects of Organochlorine Insecticides on MAP Kinase Pathways in Human HaCaT Keratinocytes: Key Role of Reactive Oxygen Species. *Toxicol. Sci.* **2005**, *86*, 444–452. [CrossRef]
6. Polanco-Rodríguez, A.G.; López, M.I.R.; Casillas, Á.D.; León, J.A.A.; Banik, S.D. Impact of Pesticides in Karst Groundwater. Review of Recent Trends in Yucatan, Mexico. *Groundw. Sustain. Dev.* **2018**, *7*, 20–29. [CrossRef]
7. Tsygankov, V.Y. Organochlorine Pesticides in Marine Ecosystems of the Far Eastern Seas of Russia (2000–2017). *Water Res.* **2019**, *161*, 43–53. [CrossRef] [PubMed]
8. Zhang, J.; Xing, X.; Qi, S.; Tan, L.; Yang, D.; Chen, W.; Yang, J.; Xu, M. Organochlorine Pesticides (OCPs) in Soils of the Coastal Areas along Sanduao Bay and Xinghua Bay, Southeast China. *J. Geochem. Explor.* **2013**, *125*, 153–158. [CrossRef]
9. Van den Brink, N.W. Directed Transport of Volatile Organochlorine Pollutants to Polar Regions: The Effect on the Contamination Pattern of Antarctic Seabirds. *Sci. Total Environ.* **1997**, *198*, 43–50. [CrossRef]
10. Wania, F.; Mackay, D. A Global Distribution Model for Persistent Organic Chemicals. *Sci. Total Environ.* **1995**, *160–161*, 211–232. [CrossRef]
11. Gerber, R.; Smit, N.J.; Van Vuren, J.H.J.; Nakayama, S.M.M.; Yohannes, Y.B.; Ikenaka, Y.; Ishizuka, M.; Wepener, V. Bioaccumulation and Human Health Risk Assessment of DDT and Other Organochlorine Pesticides in an Apex Aquatic Predator from a Premier Conservation Area. *Sci. Total Environ.* **2016**, *550*, 522–533. [CrossRef] [PubMed]
12. Khuman, S.N.; Vinod, P.G.; Bharat, G.; Kumar, Y.S.M.; Chakraborty, P. Spatial Distribution and Compositional Profiles of Organochlorine Pesticides in the Surface Soil from the Agricultural, Coastal and Backwater Transects along the South-West Coast of India. *Chemosphere* **2020**, *254*, 126699. [CrossRef]
13. Pan, H.; Lei, H.; He, X.; Xi, B.; Xu, Q. Spatial Distribution of Organochlorine and Organophosphorus Pesticides in Soil-Groundwater Systems and Their Associated Risks in the Middle Reaches of the Yangtze River Basin. *Environ. Geochem. Health* **2019**, *41*, 1833–1845. [CrossRef]
14. Luo, Y.; Yang, R.; Li, Y.; Wang, P.; Zhu, Y.; Yuan, G.; Zhang, Q.; Jiang, G. Accumulation and Fate Processes of Organochlorine Pesticides (OCPs) in Soil Profiles in Mt. Shergyla, Tibetan Plateau: A Comparison on Different Forest Types. *Chemosphere* **2019**, *231*, 571–578. [CrossRef] [PubMed]
15. Hartmann, A.; Goldscheider, N.; Wagener, T.; Lange, J.; Weiler, M. Karst Water Resources in a Changing World: Review of Hydrological Modeling Approaches. *Rev. Geophys.* **2014**, *52*, 218–242. [CrossRef]
16. Dan, X.; He, D.; Wu, X.; Wu, Z.; Li, M.; Tu, Z.; Dan, W. Ecological characteristics of karst areas in China and the hazard of rocky desertification. *Cent. South For. Inventory Plan.* **2018**, *37*, 62–66. (In Chinese)
17. Zhu, J.; Nolte, A.M.; Jacobs, N.; Ye, M. Using Machine Learning to Identify Karst Sinkholes from LiDAR-Derived Topographic Depressions in the Bluegrass Region of Kentucky. *J. Hydrol.* **2020**, 125049. [CrossRef]
18. Zeng, F.; Jiang, Z.; Shen, L.; Chen, W.; Yang, Q.; Zhang, C. Assessment of Multiple and Interacting Modes of Soil Loss in the Karst Critical Zone, Southwest China (SWC). *Geomorphology* **2018**, *322*, 97–106. [CrossRef]
19. Fenton, O.; Mellander, P.E.; Daly, K.; Wall, D.P.; Jahangir, M.M.R.; Jordan, P.; Hennessey, D.; Huebsch, M.; Blum, P.; Vero, S.; et al. Integrated Assessment of Agricultural Nutrient Pressures and Legacies in Karst Landscapes. *Agric. Ecosyst. Environ.* **2017**, *239*, 246–256. [CrossRef]
20. Li, S.; Xu, S.; Wang, T.; Yue, F.; Peng, T.; Zhong, J.; Wang, L.; Chen, J.; Wang, S.; Chen, X.; et al. Effects of Agricultural Activities Coupled with Karst Structures on Riverine Biogeochemical Cycles and Environmental Quality in the Karst Region. *Agric. Ecosyst. Environ.* **2020**, *303*, 107120. [CrossRef]
21. Coxon, C. Agriculture and Karst. In *Karst Management*; van Beynen, P.E., Ed.; Springer: Berlin/Heidelberg, Germany, 2011; pp. 103–138, ISBN 978-94-007-1207-2.
22. Wang, Z.; Li, S.; Yue, F.; Qin, C.; Buckerfield, S.; Zeng, J. Rainfall Driven Nitrate Transport in Agricultural Karst Surface River System: Insight from High Resolution Hydrochemistry and Nitrate Isotopes. *Agric. Ecosyst. Environ.* **2020**, *291*, 106787. [CrossRef]
23. Xu, S.; Lang, Y.; Zhong, J.; Xiao, M.; Ding, H. Coupled Controls of Climate, Lithology and Land Use on Dissolved Trace Elements in a Karst River System. *J. Hydrol.* **2020**, *591*, 125328. [CrossRef]
24. Gillieson, D.S. Management of Caves. In *Karst Management*; van Beynen, P.E., Ed.; Springer: Berlin/Heidelberg, Germany, 2011; pp. 141–158, ISBN 978-94-007-1206-5.
25. Xu, X.; Sun, Y.; Wang, P.; Alam, M. The Comparison of Organochlorine Pesticides between Underground Water and Surface Water in Karst Area. *China Environ. Sci.* **2013**, *33*, 1630–1637. (In Chinese)
26. Guo, F.; Yuan, D.; Qin, Z. Groundwater Contamination in Karst Areas of Southwestern China and Recommended Countermeasures. *Acta Carsologica* **2010**, *39*. [CrossRef]
27. Huang, H.; Liu, H.; Xiong, S.; Zeng, F.; Bu, J.; Zhang, B.; Liu, W.; Zhou, H.; Qi, S.; Xu, L.; et al. Rapid Transport of Organochlorine Pesticides (OCPs) in Multimedia Environment from Karst Area. *Sci. Total Environ.* **2021**, *775*, 145698. [CrossRef] [PubMed]
28. Qian, Z.; Mao, Y.; Xiong, S.; Peng, B.; Liu, W.; Liu, H.; Zhang, Y.; Chen, W.; Zhou, H.; Qi, S. Historical Residues of Organochlorine Pesticides (OCPs) and Polycyclic Aromatic Hydrocarbons (PAHs) in a Flood Sediment Profile from the Longwang Cave in Yichang, China. *Ecotoxicol. Environ. Saf.* **2020**, *196*, 110542. [CrossRef] [PubMed]
29. Green, S.M.; Dungait, J.A.J.; Tu, C.; Buss, H.L.; Sanderson, N.; Hawkes, S.J.; Xing, K.; Yue, F.; Hussey, V.L.; Peng, J.; et al. Soil Functions and Ecosystem Services Research in the Chinese Karst Critical Zone. *Chem. Geol.* **2019**, *527*, 119107. [CrossRef]

30. Polanco-Rodríguez, A.G.; Alberto, J.A.N.; Sánchez, J.S.; Rejón, G.J.M.; Gómez, J.M.; Del Valls Casillas, T.A. Contamination by Organochlorine Pesticides in the Aquifer of the Ring of Cenotes in Yucatán, México: Contamination by Organochlorine Pesticides. *Water Environ. J.* **2015**, *29*, 140–150. [CrossRef]
31. Wang, Y.; Guo, S.; Xue, R.; Qi, S.; Xu, Y.; Xue, B.; Yuan, D. Organochlorine Pesticides in the Soil of a Karst Cave in Guilin, China. *Environ. Monit. Assess.* **2011**, *180*, 489–500. [CrossRef]
32. Sun, Y. Study on Migration and Transformation Characteristics of OCPs and PAHs in Epikarst System. Ph.D. Thesis, Southwest University, Chongqing, China, 2012.
33. Arias-Estévez, M.; López-Periago, E.; Martínez-Carballo, E.; Simal-Gándara, J.; Mejuto, J.C.; García-Río, L. The Mobility and Degradation of Pesticides in Soils and the Pollution of Groundwater Resources. *Agric. Ecosyst. Environ.* **2008**, *123*, 247–260. [CrossRef]
34. Teng, Y.; Su, L.; Li, W. Searching Resolutions for the Water Resources Dilemma in the Karst Area of Southwest China. Available online: https://www.cgs.gov.cn/xwl/ddyw/201610/t20161018_409605.html (accessed on 16 February 2021).
35. Yu, D.; Su, L. Revealing the Secrets of Karst in Southwest China: Perspective of Karst Dynamics and Global Change. Available online: https://www.cgs.gov.cn/ddztt/jdqr/49dqr/kpzs/201804/t20180426_456338.html (accessed on 16 February 2021).
36. Liu, W.; Wang, Z.; Chen, Q.; Yan, Z.; Zhang, T.; Han, Z.; Chen, W.; Zhou, H. An Interpretation of Water Recharge in Karst Trough Zone as Determined by High-Resolution Tracer Experiments in Western Hubei, China. *Environ. Earth Sci.* **2020**, *79*, 357. [CrossRef]
37. Zhou, L.; Duan, Z.; Han, Q.; Xia, L.; Zhang, G. Comprehensive evaluation of soil fertility in citrus orchards in Zigui County. *Jiangsu J. Agric. Sci.* **2019**, *35*, 1346–1353. (In Chinese)
38. Yichang Bureau of Statistics. *Yichang Statistical Yearbook (2020)*; China Statistics Press: Beijing, China, 2020.
39. Wu, Q. The Benefit Evaluation of Converting Farmland to Forest in Zigui. Master's Thesis, Beijing Forestry University, Beijing, China, 2011.
40. Zheng, J.; Xiang, C.; Hu, D.; Yi, J.; Li, L. Current situation and countermeasures of agricultural non-point source pollution in Zigui County. *Hubei Plant Prot.* **2013**, *6*, 3–5. (In Chinese)
41. Huang, Y.; Zhang, R.; Li, K.; Cheng, Z.; Zhong, G.; Zhang, G.; Li, J. Experimental Study on the Role of Sedimentation and Degradation Processes on Atmospheric Deposition of Persistent Organic Pollutants in a Subtropical Water Column. *Environ. Sci. Technol.* **2017**, *51*, 4424–4433. [CrossRef] [PubMed]
42. Ali, U.; Riaz, R.; Sweetman, A.J.; Jones, K.C.; Li, J.; Zhang, G.; Malik, R.N. Role of Black Carbon in Soil Distribution of Organochlorines in Lesser Himalayan Region of Pakistan. *Environ. Pollut.* **2018**, *236*, 971–982. [CrossRef]
43. Huang, H.; Li, J.; Zhang, Y.; Chen, W.; Ding, Y.; Chen, W.; Qi, S. How Persistent Are POPs in Remote Areas? A Case Study of DDT Degradation in the Qinghai-Tibet Plateau, China. *Environ. Pollut.* **2020**, *263*, 114574. [CrossRef] [PubMed]
44. Chen, W.; Jing, M.; Bu, J.; Burnet, J.E.; Qi, S.; Song, Q.; Ke, Y.; Miao, J.; Liu, M.; Yang, C. Organochlorine Pesticides in the Surface Water and Sediments from the Peacock River Drainage Basin in Xinjiang, China: A Study of an Arid Zone in Central Asia. *Environ. Monit. Assess.* **2011**, *177*, 1–21. [CrossRef] [PubMed]
45. Huang, H.; Ding, Y.; Chen, W.; Zhang, Y.; Chen, W.; Chen, Y.; Mao, Y.; Qi, S. Two-Way Long-Range Atmospheric Transport of Organochlorine Pesticides (OCPs) between the Yellow River Source and the Sichuan Basin, Western China. *Sci. Total Environ.* **2019**, *651*, 3230–3240. [CrossRef]
46. U.S. EPA. *Exposure Factors Handbook (1997, Final Report)*; EPA/600/P-95/002F a-c; U.S. Environmental Protection Agency: Washington, DC, USA, 1997.
47. Sultana, J.; Syed, J.H.; Mahmood, A.; Ali, U.; Rehman, M.Y.A.; Malik, R.N.; Li, J.; Zhang, G. Investigation of Organochlorine Pesticides from the Indus Basin, Pakistan: Sources, Air–Soil Exchange Fluxes and Risk Assessment. *Sci. Total Environ.* **2014**, *497–498*, 113–122. [CrossRef]
48. Manz, M.; Wenzel, K.D.; Dietze, U.; Schüürmann, G. Persistent Organic Pollutants in Agricultural Soils of Central Germany. *Sci. Total Environ.* **2001**, *277*, 187–198. [CrossRef]
49. Li, J.; Zhang, G.; Qi, S.; Li, X.; Peng, X. Concentrations, Enantiomeric Compositions, and Sources of HCH, DDT and Chlordane in Soils from the Pearl River Delta, South China. *Sci. Total Environ.* **2006**, *372*, 215–224. [CrossRef]
50. Huang, H.; Zhang, Y.; Chen, W.; Chen, W.; Yuen, D.A.; Ding, Y.; Chen, Y.; Mao, Y.; Qi, S. Sources and Transformation Pathways for Dichlorodiphenyltrichloroethane (DDT) and Metabolites in Soils from Northwest Fujian, China. *Environ. Pollut.* **2018**, *235*, 560–570. [CrossRef]
51. Taylor, J.; Wilson, J.D. *Toxicological Profile for Hexachlorobenzene*; U.S. Department of Health and Human Services: Washington, DC, USA, 2002.
52. Ministry of Ecology and Environment. *PRC Announcement on the Prohibition of the Production, Circulation, Use and Import and Export of DDT, Chlordane, Mirex and Hexachlorobenzene*; Ministry of Ecology and Environment: Beijing, China, 2009.
53. Tong, M.; Yuan, S. Physiochemical Technologies for HCB Remediation and Disposal: A Review. *J. Hazard. Mater.* **2012**, *229–230*, 1–14. [CrossRef]
54. Zhang, L.; Huang, Y.; Dong, L.; Shi, S.; Zhou, L.; Zhang, T.; Mi, F.; Zeng, L.; Shao, D. Levels, Seasonal Patterns, and Potential Sources of Organochlorine Pesticides in the Urban Atmosphere of Beijing, China. *Arch. Environ. Contam. Toxicol.* **2011**, *61*, 159–165. [CrossRef]
55. Buser, H.R.; Mueller, M.D. Isomer and Enantioselective Degradation of Hexachlorocyclohexane Isomers in Sewage Sludge under Anaerobic Conditions. *Environ. Sci. Technol.* **1995**, *29*, 664–672. [CrossRef]

56. Li, Y. Global Technical Hexachlorocyclohexane Usage and Its Contamination Consequences in the Environment: From 1948 to 1997. *Sci. Total Environ.* **1999**, *232*, 121–158. [CrossRef]
57. Benezet, H.J.; Matsumura, F. Isomerization of γ-BHC to α-BHC in the Environment. *Nature* **1973**, *243*, 480–481. [CrossRef]
58. Malaiyandi, M.; Shah, S.M. Evidence of Photoisomerization of Hexachlorocyclohexane Isomers in the Ecosphere. *J. Environ. Sci. Health* **1984**, *19*, 887–910. [CrossRef]
59. Wang, Y.; Guo, S.; Xu, Y.; Wang, W.; Qi, S.; Xing, X.; Yuan, D. The Concentration and Distribution of Organochlorine Pesticides in the Air from the Karst Cave, South China. *Environ. Geochem. Health* **2012**, *34*, 493–502. [CrossRef]
60. Tan, B.; Zhang, B.; Mei, P.; Fu, L.; Tan, C. Problems and countermeasures of pesticide supervision in Zigui County. *Hubei Plant Prot.* **2016**, *53–54*, 59.
61. Bidleman, T.F.; Jantunen, L.M.M.; Helm, P.A.; Brorström-Lundén, E.; Juntto, S. Chlordane Enantiomers and Temporal Trends of Chlordane Isomers in Arctic Air. *Environ. Sci. Technol.* **2002**, *36*, 539–544. [CrossRef] [PubMed]
62. Liu, X.; Zhang, G.; Li, J.; Yu, L.; Xu, Y.; Li, X.; Kobara, Y.; Jones, K.C. Seasonal Patterns and Current Sources of DDTs, Chlordanes, Hexachlorobenzene, and Endosulfan in the Atmosphere of 37 Chinese Cities. *Environ. Sci. Technol.* **2009**, *43*, 1316–1321. [CrossRef]
63. Cuozzo, S.A.; Fuentes, M.S.; Bourguignon, N.; Benimeli, C.S.; Amoroso, M.J. Chlordane Biodegradation under Aerobic Conditions by Indigenous Streptomyces Strains. *Int. Biodeterior. Biodegrad.* **2012**, *66*, 19–24. [CrossRef]
64. Ministry of Ecology and Environment. PRC Announcement on Banning the Production, Circulation, Use, Import and Export of Lindane and Other POPs (000014672/2019-00287). Available online: http://www.mee.gov.cn/xxgk2018/xxgk/xxgk01/201903/t20190312_695462.html (accessed on 16 May 2020).
65. Chakraborty, P.; Zhang, G.; Li, J.; Xu, Y.; Liu, X.; Tanabe, S.; Jones, K.C. Selected Organochlorine Pesticides in the Atmosphere of Major Indian Cities: Levels, Regional versus Local Variations, and Sources. *Environ. Sci. Technol.* **2010**, *44*, 8038–8043. [CrossRef]
66. Jia, H.; Liu, L.; Sun, Y.; Sun, B.; Wang, D.; Su, Y.; Kannan, K.; Li, Y. Monitoring and Modeling Endosulfan in Chinese Surface Soil. *Environ. Sci. Technol.* **2010**, *44*, 9279–9284. [CrossRef] [PubMed]
67. Ministry of Agriculture and Rural Affairs. *PRC Announcement of the Ministry of Agriculture of the People's Republic of China No.199*; Ministry of Agriculture and Rural Affairs: Beijing, China, 2002.
68. Stockholm Convention. All POPs Listed in the Stockholm Convention. Available online: http://www.pops.int/TheConvention/ThePOPs/AllPOPs/tabid/2509/Default.aspx (accessed on 12 May 2020).
69. Dong, Y. Investigation of termite damage in the Three Gorges Reservoir Area (Zigui County). *Hubei Plant Prot.* **2017**, 29–30, 40. (In Chinese)
70. Stockholm Convention. UN Environment Chemicals Proposed for Listing under the Convention. Available online: http://www.pops.int/TheConvention/ThePOPs/ChemicalsProposedforListing/tabid/2510/Default.aspx (accessed on 1 February 2021).
71. U.S. EPA. Methoxychlor Reregistration Eligibility Decision (RED) (No. EPA 738-R-04-010). Available online: https://archive.epa.gov/pesticides/reregistration/web/html/methoxychlor_red.html (accessed on 1 February 2021).
72. Cheng, L.; Song, W.; Rao, Q.; Zhou, J.; Zhao, Z. Bioaccumulation and Toxicity of Methoxychlor on Chinese Mitten Crab (Eriocheir Sinensis). *Comp. Biochem. Physiol. Part C Toxicol. Pharmacol.* **2019**, *221*, 89–95. [CrossRef] [PubMed]
73. Liu, X.; Li, J.; Zheng, Q.; Bing, H.; Zhang, R.; Wang, Y.; Luo, C.; Liu, X.; Wu, Y.; Pan, S.; et al. Forest Filter Effect versus Cold Trapping Effect on the Altitudinal Distribution of PCBs: A Case Study of Mt. Gongga, Eastern Tibetan Plateau. *Environ. Sci. Technol.* **2014**, *48*, 14377–14385. [CrossRef] [PubMed]
74. State Administration for Market Regulation; Ministry of Ecology and Environment. *PRC Soil Environmental Quality–Risk Control Standard for Soil Contamination of Agricultural Land (GB 15618-2018)*; State Administration for Market Regulation: Beijing, China; Ministry of Ecology and Environment: Beijing, China, 2018.
75. Ministry of Housing, Spatial Planning and Environmental Management. *Soil Remediation Circular 2009*; Ministry of Housing, Spatial Planning and Environmental Management: Amsterdam, The Netherlands, 2009.
76. U.S. EPA. Guidelines for Carcinogen Risk Assessment. Available online: https://www.epa.gov/risk/guidelines-carcinogen-risk-assessment (accessed on 15 January 2021).
77. Zhou, P.; Zhao, Y.; Li, J.; Wu, G.; Zhang, L.; Liu, Q.; Fan, S.; Yang, X.; Li, X.; Wu, Y. Dietary Exposure to Persistent Organochlorine Pesticides in 2007 Chinese Total Diet Study. *Environ. Int.* **2012**, *42*, 152–159. [CrossRef]

Article

# Arsenic Release from Soil Induced by Microorganisms and Environmental Factors

Yitong Yin [1], Ximing Luo [1,2,*], Xiangyu Guan [1,2], Jiawei Zhao [1], Yuan Tan [1], Xiaonan Shi [1], Mingtao Luo [1] and Xiangcai Han [1,3]

[1] School of Ocean Sciences, China University of Geosciences (Beijing), Beijing 100083, China; kaqichuan@163.com (Y.Y.); guanxy@cugb.edu.cn (X.G.); 2011200021@cugb.edu.cn (J.Z.); tanyuan0228@163.com (Y.T.); 2005200052@cugb.edu.cn (X.S.); luomingtao1996@163.com (M.L.); 2001200152@cugb.edu.cn (X.H.)
[2] Beijing Key Laboratory of Water Resources and Environmental Engineering, China University of Geosciences (Beijing), Beijing 100083, China
[3] Yantai Coastal Zone China Geological Survey, Yantai 264000, China
* Correspondence: luoxm@cugb.edu.cn

**Abstract:** In rhizospheric soil, arsenic can be activated by both biological and abiotic reactions with plant exudates or phosphates, but little is known about the relative contributions of these two pathways. The effects of microorganisms, low-molecular-weight organic acid salts (LMWOASs), and phosphates on the migration of As in unrestored and nano zero-valent iron (nZVI)-restored soil were studied in batch experiments. The results show that As released by microbial action accounted for 17.73%, 7.04%, 92.40%, 92.55%, and 96.68% of the total As released in unrestored soil with citrate, phytate, malate, lactate, and acetate, respectively. It was only suppressed in unrestored soil with oxalate. In restored soil, As was still released in the presence of oxalate, citrate, and phytate, but the magnitude of As release was inhibited by microorganisms. The application of excess nZVI can completely inhibited As release processes induced by phosphate in the presence of microorganisms. Microbial iron reduction is a possible mechanism of arsenic release induced by microorganisms. Microorganisms and most environmental factors promoted As release in unrestored soil, but the phenomenon was suppressed in restored soil. This study helps to provide an effective strategy for reducing the secondary release of As from soils due to replanting after restoration.

**Keywords:** low-molecular-weight organic acid salts; phosphate; arsenic-contaminated soil; microorganisms; nano zero-valent iron (nZVI)

## 1. Introduction

Arsenic (As) is one of the most harmful and widespread pollutants in the natural environment, and As-contaminated soils are widespread globally [1]. Hence, the remediation of soils contaminated with arsenic has become a focus of global concern. Recently, Fe-based materials, biochar, and composites have been studied as amendments for stabilizing As in soil [2]. However, further studies have indicated that, among composite materials, Fe-based materials play a significant role in stabilizing As [3–7]. Among the Fe-based materials used for remediation in recent years, nano zero-valent iron (nZVI) has received increasing attention due to its large specific surface area and high reactivity [8].

NZVI is used to immobilize As by promoting the transformation of more mobile As fractions into less mobile fractions. Hou et al. (2020) found that the proportion of amorphous hydrous oxide–bound As and residual As was increased after using a sponge iron filter containing large amounts of zero-valent iron to restore soil [9]. In addition, Li et al. (2020) found that the percentage of the acid-soluble As decreased, while the reducible As increased by 25.4%. Zeolite-supported nZVI has been used to immobilize As in alkaline soils [10]. These studies on the transformation of soil As fractions indicate that soil restoration efforts can be effective, because the application of nZVI enhances the conversion of soluble As to the insoluble fraction. However, several studies have found that the amorphous hydrous oxide–bound As and the reducible As fraction are potentially bioavailable [11]. Additionally, An et al. (2019) suggested that chemical analysis alone is insufficient to assess the ecotoxicological responses of As in soil. Therefore, further assessment of the biological responses of restored soil is needed to test the stability of the in situ immobilization of As under actual replanting conditions [12].

The biogeochemical cycle of As involves several physical and chemical processes (precipitation/solubilization, adsorption/desorption, and redox processes), as well as biological processes, especially those involving microorganism reactions [13]. Many studies have verified that microorganisms are a key mediator of the biogeochemical release and activation of As [14–16]. Several of the potential mechanisms of microbial involvement in As release have been verified and summarized, including As desorption from adsorption sites and As release by the reductive dissolution of iron minerals [17,18]. In recent years, it has been accepted that the development of anoxic conditions in soils leads to increased As mobility, mainly through direct As(V) reduction to As(III) and reductive dissolution of Fe(III) minerals [19]. This poses a challenge to practical applications in contaminated sites, because with either direct plant cultivation or replanting after remediation, processes such as rhizospheric interaction and fertilization may cause changes in the soil environment. These changes may also affect the re-release of As via adsorption/desorption processes or the dissolution of iron minerals.

Plants exude large amounts of photosynthesis-derived carbon (11–40%) via root exudates [20]. Among them, low-molecular-weight organic acids (LMWOAs), as one of the main exudates, are usually in a dissociated mildly acidic state [21]. Acetic acid, oxalic acid, malic acid, and citric acid are typical LMWOAs present in plant root exudates [22]. However, unlike typical plants, phytate (inositol hexaphosphate) has been detected in the root exudates of ferns such as *Pteris vittata* L. (Chinese Brake fern) [23]. As well as being active components of root exudates that can mobilize nutrients such as Fe and P, LMWOAs are also important sources of soil organic carbon [24]. Compared to complex organic matter (e.g., humic acid), microorganisms often utilize LMWOAs as available labile carbon sources [21]. The diversity of microbial community structures and the dynamics of phylogenetic composition can also be regulated by LMWOAs [25]. In addition to being utilized as organic carbon by microorganisms, LMWOAs can impact the soil environment via their functional groups. LMWOAs are often used as soil leaching reagents for soil washing because of their functional groups and organic ligands [2,26,27]. Several experiments have demonstrated the ability of LMWOAs to extract Fe-bound As and residual As from soil [28]. Therefore, replanting on stabilized soils is likely to induce the release of As from the rhizosphere environment.

Phosphorus (P) fertilizers are usually applied to promote plant growth. Phosphate ($PO_4^{3-}$), the main component of phosphate fertilizers, has multiple influences on As bioavailability. Phosphate can significantly suppress the adsorption of As to iron (hydr)oxides and to soils because of the structural similarity between phosphate ($PO_4^{3-}$) and arsenate ($AsO_4^{3-}$). Ji et al. (2019) and Deng et al. (2020) described in detail the effects of phosphate on As behavior in paddy soils [29,30]. Phosphorus is also an essential nutrient for both crops and microorganisms; however, both arsenate and phosphate can be taken up by bacteria via the same phosphate transporters, such as Pst and Pit [31]. Wang et al. (2020) demonstrated the occurrence of phosphate-stimulated As(V) reduction via faster bacterial reproduction and accelerated As desorption/sorption mediated by Bacillus XZM and suggested that phosphate may regulate the biogeochemical behavior of As [32].

This poses a challenge to practical applications in contaminated sites, because with either direct plant cultivation or replanting after remediation, processes such as rhizospheric interaction and fertilization may cause changes in the soil environment. The presence of LMWOAs and phosphate influences As re-release in the soil via both physicochemical and microbial processes. There are many studies of the effects of phosphate on As behavior with the participation of microorganisms in soils. More consideration has been given to the physicochemical effects of LMWOAs on As migration in soils [33–38]. However, there has been comparatively little consideration of the role of microorganisms in the action of LMWOAs [39,40]. Many studies have demonstrated that As release is closely related to Fe during the reaction process between As and LMWOAs or phosphate [29,30,35–37]. Therefore, we need to monitor the changes to Fe throughout the whole reaction process, including Fe(II) and Fe(III). In addition, the effects of these influencing factors on the re-release of As in soils remediated with nZVI have not been investigated in detail.

In this study, we evaluated the effectiveness of excess nZVI for the remediation of As-contaminated soils. The study area comprises a transition zone between agricultural soils and river sediments and has not previously been cultivated. The objectives of the study were as follows: (1) to determine the influence of low-molecular-weight organic acid salts (LMWOASs) and phosphates in restored or unrestored soils and (2) to assess the potential importance of LMWOASs and phosphates for As mobilization with natural soil microbial communities in restored and unrestored soils.

## 2. Materials and Methods

### 2.1. Soil Sampling and Pretreatment

Soil samples were collected in July 2021 from the surface layer (0–20 cm) of As-contaminated soil in the transition zone between agricultural land and a river in the vicinity of a gold mine in Dandong, Liaoning Province, China (123°42′ E, 40°44′ N). After collection, the soils were air-dried and then passed through a 2 mm sieve. The characterization methods for each parameter are summarized in the Supplementary Materials [41,42].

Then, part of air-dried and sieved soil underwent sterilization (121 °C for 1 h) using steam sterilization pot GI54T (Zealway, Xiamen, China) to obtain the sterilized soil. Five percent (by weight) of nZVI was mixed thoroughly with 20 g of sterile soil in a 100 mL beaker (group of 40 samples, consisting of restored soil). Nano zero-valent iron (nZVI) was supplied by Xindun Co., Ltd. (Nangong, China). The remaining sterile soil samples (group of 40 samples, consisting of unrestored soil) without additions were used as a control. The mixture was homogenized with a water-holding capacity of ~70% at room temperature for 7 days. The subsamples were then freeze-dried, and the soil-available As was extracted with 0.5 M $NaHCO_3$ [43].

### 2.2. Microcosm Experiments

Ten grams of original sieved soil was suspended in 50 mL of sterile 10 mM Tris-HCl buffer (pH 7.2). These soil suspensions were then used as a source of soil microbial communities.

#### 2.2.1. Experiment I (LMWOASs)

Two grams of dried soil subsample was weighed in 20 mL of brown serum vials and autoclaved at 121 °C for 1 h. Then, sterilized oxalate, citrate, malate, lactate, acetate, and phytate solutions (10 mL) were added separately to the sterilized restored or unrestored soil as a carbon source. Specifically, sodium oxalate, sodium citrate, sodium malate, sodium lactate, sodium acetate, and sodium phytate were added at concentrations of 11.17, 7.17, 7.42, 6.23, 6.84, and 25.66 g/L, respectively, to obtain an initial concentration of total organic carbon (TOC) of LMWOASs of 2 g/L. Sodium hydroxide and hydrochloric acid were used to regulate the pH of the LMWOAS solutions to 7.2. The incubation samples were $N_2$-bubbled for 5 min. Each soil suspension (1 mL) was then inoculated into the serum vials in biotic groups, and the remaining soil suspensions were autoclaved (121 °C, 1 h) and then inoculated in abiotic controls. Finally, the samples were incubated under anaerobic conditions at 30 °C in the dark on a reciprocal shaker (120 rpm). All treatments were carried out in triplicate. The soil was sampled after 3, 10, 17, 24, and 38 days of incubation.

### 2.2.2. Experiment II (Phosphates)

Microcosm incubations were constructed in 20 mL of brown serum vials containing 2 g of sterilized restored or unrestored soil and 10 mL of sterilized medium. The salt medium supplement used in this study was modified from Yamamura et al. (2003) and contained (per L): $(NH_4)_2SO_4$ (2.27 mM), $MgSO_4 \cdot 7 H_2O$ (0.57 mM), NaCl (1.71 mM), $KH_2PO_4$ (0.0, 0.07, 0.35 mM), $Na_2HPO_4 \cdot 12 H_2O$ (0.0, 0.06, 0.30 mM), $C_3H_5O_3Na$ (20 mM), and 1 mL of trace element solution [44]. The final concentrations of P in the medium were 0 mM, 0.13 mM, and 0.65 mM. The pH of the medium was adjusted to 7.2. The incubation samples were $N_2$-bubbled for 5 min, and each soil suspension (1 mL) was then inoculated into the serum vials. Finally, the samples were incubated under anaerobic conditions at 30 °C in the dark on a reciprocal shaker (120 rpm). All treatments were carried out in triplicate. The soil was sampled after 1, 2, 3, 4, 5, 7, 9, and 11 days of incubation.

### 2.3. Geochemical Analysis

The pH and oxidation–reduction potential (ORP) of the microcosm slurries were measured using a PB-10 device (Sartorius, Germany) and a portable ORP meter (HACH, Shanghai, China), immediately after sampling. When HCl-extractable Fe(II) was analyzed, 1 mL of 2 M HCl was mixed with the same volume of slurry and then mixed vigorously before filtration [45]. The supernatants in the brown serum vials were collected and filtered through a 0.22 μm polytetrafluoroethylene (PTFE) membrane after centrifugation (1100× $g$ for 10 min) with a 3K15 high-speed centrifuge (SIGMA, Germany). Slurries were also filtered through a 0.22 μm filter. Fe(II) was quantified using the 1,10-phenanthroline colorimetric method using a UV1800 spectrophotometer (Shimadzu, Japan) [46], which was conducted immediately. Part of the filtrate was acidified with 0.1 M $HNO_3$, and then total As, Fe, and P were analyzed by inductively coupled plasma–optical emission spectroscopy (ICP–OES) (Spectro Blue Sop, Germany). The ferric iron was calculated from the concentrations of total iron and ferrous iron. TOC levels were determined using TOC-L CPH equipment (Shimadzu, Japan). The instrument parameters for ICP–OES and TOC-L CPH equipment are summarized in the Supplementary Materials [41,42].

### 2.4. Quality Assurance/Quality Control (QA/QC) and Statistical Analysis

For quality assurance and quality control, measures for quality assurance and quality control were taken in the experimental design and laboratory analyses. All control and experimental treatments were carried out in triplicate throughout the whole experiment. In routine analysis, for every 10 samples, 1 laboratory blank, 1 parallel sample, and 1 spiked blank sample were added to the measuring sequence. If the sample number was less than 10, then 1 laboratory blank, 1 parallel sample, and 1 spiked blank sample were also taken.

In our analysis, the concentrations of target substances in all blanks were less than the corresponding MDLs. The standard solutions were provided by Beijing Wanjia Shouhua Biotechnology Co. (Beijing, China). The Chinese national standard was provided by the National Research Center for certified Reference Materials of China. The recovery range of spiked blank was 94–103%. The relative standard deviation (RSD) was lower than 5% for all of the tests. All statistical analyses were performed using IBM SPSS Statistics version 22.0.

## 3. Results and Discussion

### 3.1. Arsenic Release under LMWOAS Treatments in Unrestored Soil

The basic physical and chemical properties of the soils are given in Table 1. The results of other characterizations are shown in the Supplementary Materials [41,42].

A comparative analysis under different LMWOAS treatments revealed significant differences ($p < 0.05$) in As release from in situ soil (Figure 1). The effectiveness of As release under the LMWOAS treatments in abiotic control assays are ordered as follows: oxalate > phytate > citrate > malate > lactate > acetate. However, for the biotic microcosms, the dissolved As released from each group changed over time. The final rank order of the treatments after 38 days was as follows: malate > acetate ≈ lactate ≈ oxalate ≈ phytate > citrate.

The maximum release of As in the biotic control assay was observed with the addition of malate, which released 18.47% of the arsenic in the soil.

Table 1. Basic physical and chemical properties of the studied soil.

| Property | Value/Content |
| --- | --- |
| pH | 7.61 |
| Organic matter/(mg/kg) | 10,500 |
| Total C/(mg/kg) | 14,500 |
| Total P/(mg/kg) | 549 |
| Total S/(mg/kg) | 3200 |
| Total N/(mg/kg) | 636 |
| Total Fe/(mg/kg) | 33,400 |
| Total As/(mg/kg) | 1944 |
| Available As/(mg/kg) | 35 |

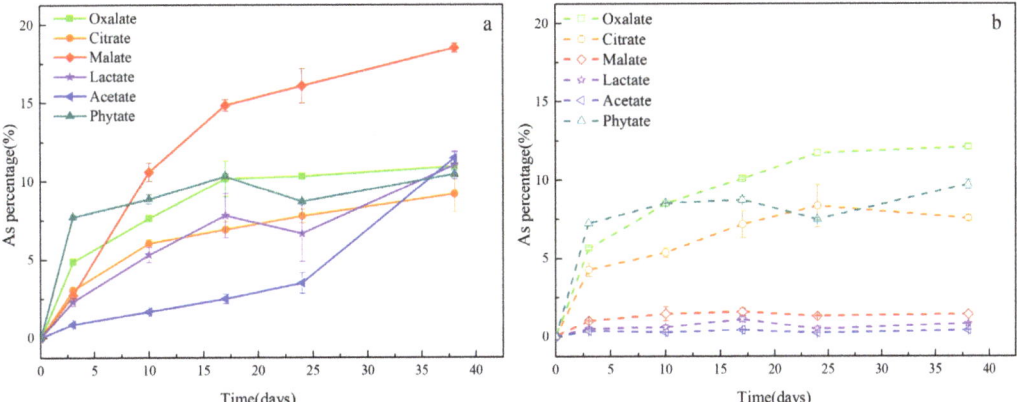

Figure 1. Dissolved As release as a percent of total As in soil under biotic (a) and abiotic (b) conditions with different LMWOAS amendments in unremediated soils.

Biotic and abiotic control groups treated with the same LMWOAS were then further analyzed. In all abiotic controls, minor As release occurred in the malate, lactate, and acetate treatment groups. However, in the oxalate, citrate, and phytate treatments, significant amounts of As were still released under sterilized conditions, and there were small differences from the unsterilized conditions. Compared to the abiotic control assays, the microorganisms inhibited the release of As in oxalate treatments, and the release of As in the other five treatments was promoted (Figure 2). In all biotic cases, As release from the slurries failed to plateau during the experimental period, suggesting the potential for further release of As with longer incubation times.

Low-molecular-weight dissolved organic carbon can enhance the biotic or abiotic reductive dissolution of iron oxides, oxyhydroxides, and hydroxides in the presence or absence of Fe(III)-reducing microbial communities [47]. In addition, HCl-extractable Fe(II) is biologically available to microorganisms, and it could represent predominant biological and chemical sources of Fe(II) species produced by ferric iron reduction and could keep their concentration and valence state stable during the extraction [48]. The HCl-extractable total Fe(II) includes dissolved and solid-phase Fe(II). Since the Fe(II) content of the supernatant constituted a very small proportion of the HCl-extractable total Fe(II), no significant correlation was found between As and Fe(II) in the aqueous solution (Figure S1). Therefore, only the solid-phase Fe(II) concentration was further analyzed. Data for solid-phase Fe(II) for the six samples are shown in Figure 2. In general (except for the oxalate treatment), the solid-phase Fe(II) concentrations of the remaining five biotic control groups were much higher than those of the abiotic control groups. In contrast, there was no clear relationship between the solid-phase Fe(II) or soluble As in different LMWOAS treatments. However, when the data from the six soil samples are plotted together, except for the oxalate treatment, which showed no relationship, a significant relationship between the solid-phase Fe(II) and soluble As for the other individual soil samples was observed ($p < 0.05$). The $R^2$ value for the five soil samples ranged from 0.332 to 0.958 (Figure 3).

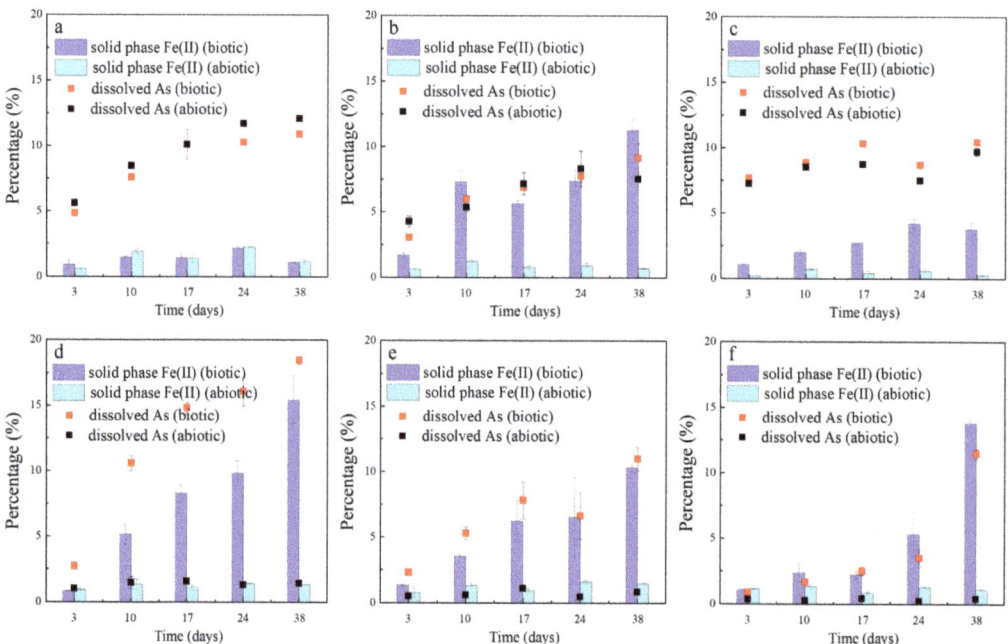

**Figure 2.** Variation in dissolved As in the supernatant and solid-phase Fe(II) as percent extracted of total in soil under biotic and abiotic conditions with different LMWOAS amendments in unremediated soils. (**a**) Oxalate, (**b**) citrate, (**c**) phytate, (**d**) malate, (**e**) lactate, (**f**) acetate.

Song et al. (2010) showed that the primary mechanism by which low-molecular-weight organic acids or their salts promote the release of heavy metals may not reside in the action of acid but rather in the action of organic anions [49]. The solid-phase Fe(II) concentrations of the malate, lactate, and acetate treatments in the abiotic control group always showed very little variation, and the release of As was always low. This result indicates that these three LMWOASs have a low capability of reductive or non-reductive dissolution of iron minerals in soils under sterile conditions. The competitive desorption ability of As was weak under sterile conditions. However, with the application of these three LMWOASs, the participation of microorganisms significantly promoted the reductive dissolution of iron minerals and greatly enhanced the release of As from the soil. The changes in As and solid-phase Fe(II) concentrations in lactate and acetate treatment groups showed a similar tendency to that observed by Wang et al. (2021) [50]. One possible reason that the largest amount of iron reduction and arsenic release was observed in the biotic malate treatments is the mitochondrial Krebs cycle in fungi [51]. Several experiments have demonstrated that malate is closely associated with the migration and transformation of Fe and As in the plant rhizosphere [52,53]. The results show that microorganisms probably affect the behavior of As in the LMWOA–transition zone soil mixtures.

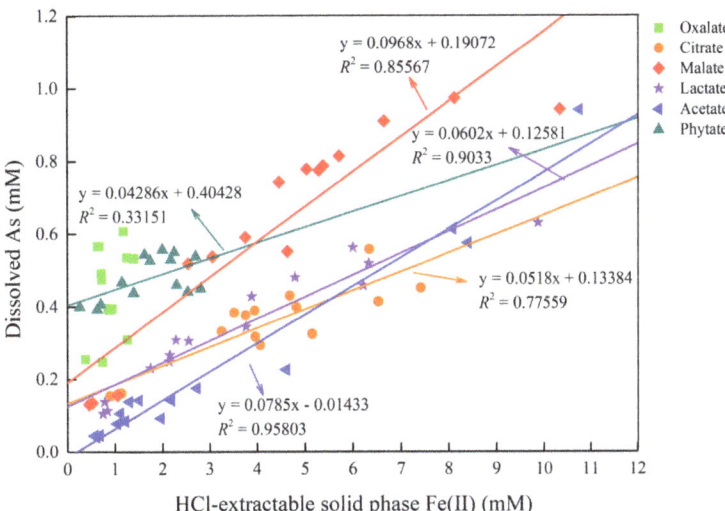

**Figure 3.** Relationships between dissolved As and HCl-extractable solid-phase Fe(II) for unremediated soil with different LMWOAS amendments.

The citrate and phytate treatment groups released large quantities of As under sterile conditions. Additionally, the Fe(II) concentration in the sterile supernatant was also consistently high (Figure S2). Both the solid-phase Fe(II) concentration and As release increased after microbial involvement in the reaction. This indicates that, on the one hand, the hydroxyl group and the carboxyl group of citrate and the orthophosphate moieties of phytate can promote the abiotic release of As and Fe(II) from soils. On the other hand, microbial action can develop a synergy with abiotic action to further promote As release from the LMWOAS–transition zone soil mixtures.

The oxalate treatments showed little difference in the solid-phase Fe(II) concentration in the presence or absence of microorganisms. The differences in Fe(II) concentration suggest that the chemistry of oxalate is mainly responsible for promoting the reaction with iron minerals. The release rate of As gradually decreased in oxalate treatments. The insoluble Fe(III)–organic complexes that formed covered the surface of the binding site and led to a block in the release of ions [54]. Since the release of arsenic is mainly the result of the chemical role of oxalate, the phenomenon becomes more pronounced in oxalate treatments. Additionally, based on the release of As from the solution, microorganisms inhibit the release of As from the soil in the presence of oxalate. This result is consistent with that of Mei et al. (2022), who found that microorganisms may facilitate the release of As from sediments in the presence of citric and malic acids, but they suppress As mobilization in the presence of oxalic acid. After sterilization, the As extraction from sediments by citric and malic acids decreased, whereas the extraction by oxalic acid increased [39].

### 3.2. Arsenic Release under LMWOAS Treatments in Restored Soil

The remediation effect was determined by $NaHCO_3$ extraction after 7 days' remediation by applying excess nZVI. According to the available As concentration extracted before and after the remediation, the repair efficiency was ~70%. Compared with the unrestored soil, there was little change in the ORP range throughout the experiment. However, the pH of the soil slurry increased in restored soil due to the oxidation of nZVI and the release of OH- (Figure 4) [55]. The citrate and phytate treatments showed a greater change (pH and ORP in sterilized incubations are not shown).

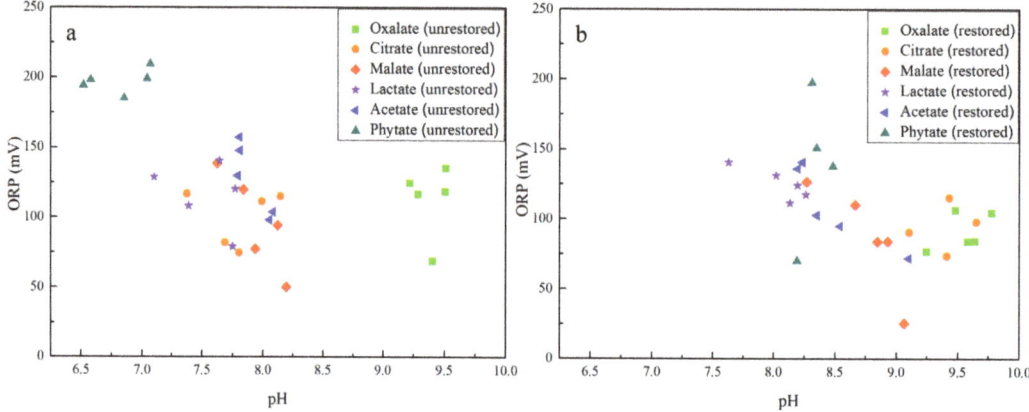

**Figure 4.** Soil pH and ORP distribution under biotic conditions with different LMWOAS amendments in unrestored (**a**) and restored (**b**) soils.

For amendments with different LMWOASs, the concentrations of dissolved As and Fe(II) in the supernatant are shown in Figure 5. Only Fe(II) in the supernatant was analyzed due to the problem of significant interference with HCl-extractable Fe(II) with the application of excessive nZVI.

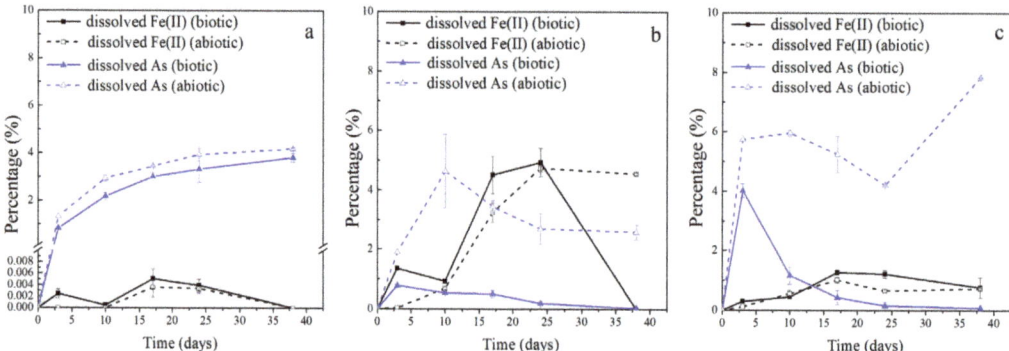

**Figure 5.** Variation in dissolved As and Fe(II) in the supernatant as percent extracted of total in soil under biotic and abiotic conditions with different LMWOAS amendments in remediated soils. (**a**) Oxalate, (**b**) citrate, (**c**) phytate.

In the restored soil, with the application of excess nZVI, the soluble Fe(II) in the malate, lactate, and acetate amendments showed similar trends compared to the unrestored soil (Figure S3). Relatively minor As release occurred, regardless of the presence or absence of microorganisms in the soil. According to these data, we speculate that excess nZVI led to a significant increase in the number of adsorption sites. Hence, the As released by either microbial reductive dissolution or desorption was immediately readsorbed.

Although the release of As initially occurred with oxalate, citrate, and phytate amendments in restored soil, the release was significantly lower than that in unrestored soil. On the one hand, nZVI reduced As mobility. On the other hand, oxalate, citrate, and phytate could maintain As release. The added oxalate extensively formed strong complexes with iron, which can prevent the precipitation of a new iron oxide phase and can inhibit the mechanisms of nZVI repair during As release [56]. Under near-neutral pH conditions, Fe(III)–citrate can accelerate As release in the presence of arsenopyrite [57]. It was reported that phytic acid solubilized 39% of poorly soluble $FeAsO_4$, while citric and oxalic acid solubilized 32 and 10% because of their stronger complex stability with Fe [58]. Moreover, the As(V) re-release amount changed with the aging time of ferrihydrite. The effect of organic ions such as oxalate, citrate, and phytate on the As(V) release tendency is associated with factors such as the As loading rate and ligand concentration in Fe–As complexes and leads to higher As(V) re-release after longer ferrihydrite aging time [59].

At the beginning of the experiment, the magnitude of As release in the abiotic control group occurred in the following order: phytate > oxalate ≈ citrate amendment. In the biotic control groups, the release of As occurred in the following order: phytate > oxalate > citrate amendment. In the oxalate treatment groups, the trend of the release of As and Fe(II) in the restored soil was the same as that in the unrestored soil. The As release increased gradually over time, but microorganisms inhibited its release.

Due to the citrate and phytate and microorganisms acting synergistically, significant dissolution release of Fe(II) occurred in the restored soil with the addition of citrate and phytate. With the participation of microorganisms, the release of As showed an initial increasing trend, followed by a decreasing trend. The reason for this may be because As was initially released by synergistic action. However, because of large amounts of soluble Fe(II) in the solution and the alkaline soil pH, secondary minerals were formed that captured As, leading to its re-immobilization. A similar phenomenon was observed by Wang et al. (2021) and Cai et al. (2020) [50,60]. Compared to the experimental biotic group, the amount of As released was much greater, although the abiotic control group also released significant amounts of Fe(II). In parallel, the sterilized and unsterilized groups had similar pH values (Figure S4). Therefore, we speculate that microorganisms actively participate in the formation of secondary minerals under such conditions.

### 3.3. Effect of Phosphate on as Release in Restored and Unrestored Soils

During previous experiments with LMWOASs, the concentration of P in the supernatant was also measured. Because phytate itself carries a large quantity of P, which interferes with the measurement results, the phytate treatment group was excluded when comparing P concentrations. The remaining LMWOAS groups in the abiotic and biotic groups had similar P concentrations (Figure 6). The maximum P concentration occurred on the 3rd day, and subsequently, on the 10th day, the concentrations decreased and then remained stable. This phenomenon suggests that the release of P is due to the perturbation of the soil environment caused by the addition of the solution. By the third day, there was a significant correlation between the As and P concentrations in the supernatant of the five groups ($p < 0.01$). This likely demonstrates the physicochemical effect of P on the release of As in the initial period of the reaction. Therefore, we investigated whether the role of exogenous P input on As release was facilitated by the participation of microorganisms.

The changes in soluble P and As concentration in the unrestored soil are shown in Figure 7. The results clearly show that high P concentrations increase the initial As release, which was evident in the results for the 1st day. The higher the P concentration applied, the greater the As release, which is probably because phosphate promotes the release of As from the soil via competitive adsorption. However, it should be noted that the release of As still increased when the P concentration leveled off. This may be because the presence of phosphate promotes microbial action, resulting in persistent As release. Arsenate and phosphate create competition for the same transport channel protein. Phosphate stimulated As release due to fast bacterial reproduction because increasing phosphate concentration in the environment appears to decrease the growth inhibition attributed to the presence of arsenate [32].

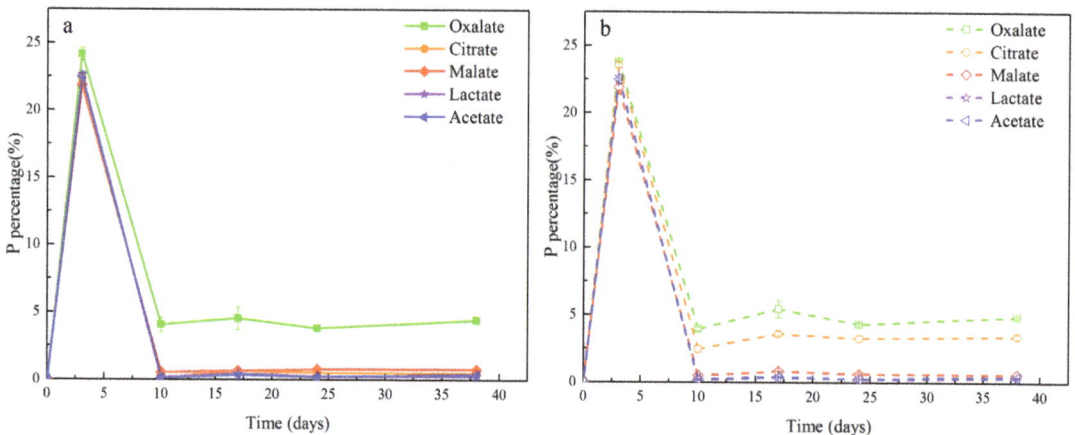

**Figure 6.** Dissolved P release as a percent of total P in soil under (**a**) biotic and (**b**) abiotic conditions with different LMWOAS amendments in unremediated soils.

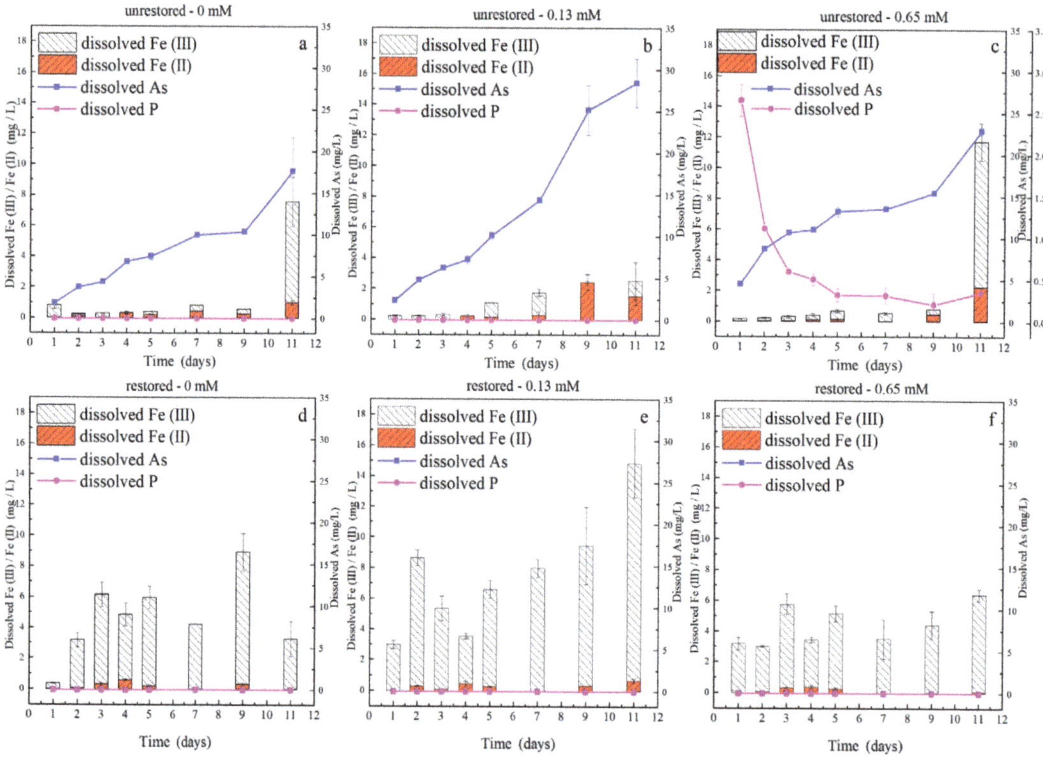

**Figure 7.** Variation in dissolved As, Fe(II), and Fe(III) concentrations in the supernatant under biotic conditions with different phosphate concentrations in unrestored and restored soils. (**a**) Unrestored, 0 mM, (**b**) unrestored, 0.13 mM, (**c**) unrestored, 0.65 mM, (**d**) restored, 0 mM, (**e**) restored, 0.13 mM, (**f**) restored, 0.65 mM.

Although we found a higher initial release of As in the 0.65 mM phosphate groups, the 0.13 mM phosphate groups eventually released the most As at the end of the reaction

(Figure 8). The proportion of soluble Fe(II) in the released dissolved Fe was the highest in the 0.13 mM phosphate groups. One possible explanation for the lower Fe(II) levels in the 0.65 mM phosphate groups is that dissolved Fe(II) reacts with the excess P to precipitate as $Fe_3(PO_4)_2$ (vivianite). A similar phenomenon was observed by Zhang et al., (2017); additionally, the lack of P may lead to lower microbial activity in the 0 mM phosphate groups [14]. A significant positive correlation ($p < 0.01$) was observed between soluble Fe(II) and As in the solutions of the three groups (Figure 9).

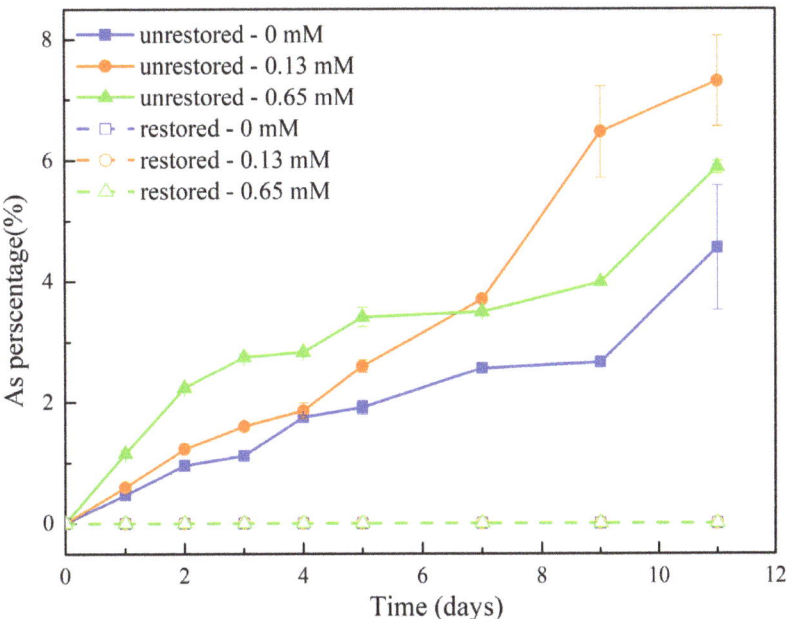

**Figure 8.** Dissolved As release as a percent of total As in soil under biotic conditions in restored and unrestored soils with different phosphate concentrations.

In the unrestored soil, the P concentration in the 0.65 mM phosphate groups gradually decreased and eventually remained stable, but we did not detect the release of P in the supernatant in the 0 and 0.13 mM phosphate groups. It was previously suggested that As is translocated with Fe during redox changes, whereas phosphate is not affected by redox, likely because phosphate is adsorbed onto non-reduced iron oxides [61]. Therefore, it is possible that we could not observe P in the supernatant because the original P in the soil and the low concentration of exogenous P were stabilized in the soil by the soil matrix. While the soil could not completely absorb the high concentration of exogenous P input at the beginning, it was gradually stabilized in the soil as the reaction progressed. In the restored soil, even high concentrations of exogenous P were immobilized by the presence of excess nZVI, so no P release was observed in the supernatant.

No As release was observed in the restored soil even when more iron was released (Figure 7). In particular, in the 0.65 mM phosphate groups, neither As nor P was released into the supernatant. We speculate that the phenomenon is due to the high affinity of the Fe oxide shell of nZVI for $PO_4^{3-}$ through surface complexes including electrostatic attraction and chemical binding (Fe-O-P) [62]. The result showed a similar tendency to that observed by Huang et al. (2019) [63]. The experimental results show that excess nZVI may, to some extent, avoid the competitive desorption caused by phosphate input and the effect of the Fe reduction process promoted by microorganisms.

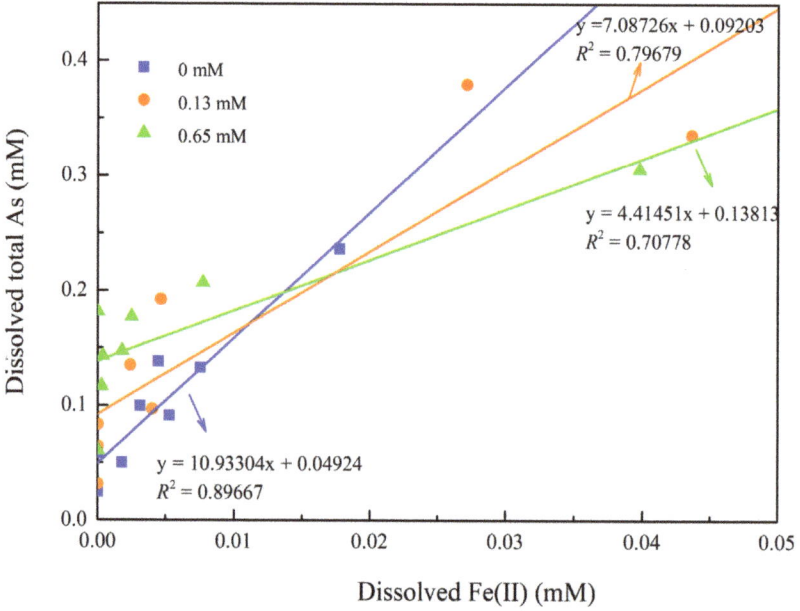

**Figure 9.** Relationship between dissolved As and Fe(II) for unremediated soil with different phosphate concentrations.

## 4. Conclusions

The secondary release of As in rhizospheric soil has been neglected in previous studies and deserves special attention. Microorganisms play the dominant role in the process of As release, and the interactions of As, microorganisms, LMWOASs, and phosphates during the process were investigated in this study. Microorganisms showed complex inhibition or facilitation effects during the release of As. NZVI is widely applied for the remediation of As-contaminated soil, and it suppressed the majority of As release in this study. This study improves our understanding of As mobilization and transformation in the rhizosphere and helps to provide an effective strategy for reducing the secondary release of As from soils due to replanting after restoration.

**Supplementary Materials:** The following supporting information can be downloaded at: https://www.mdpi.com/article/10.3390/ijerph19084512/s1, Figure S1: Relationships between dissolved As and the dissolved Fe(II) for unremediated soil with different LMWOAS amendments; Figure S2. Dissolved Fe(II) release as a percent of total Fe in soil for unremediated soil with different LMWOAS amendments.(a) oxalate, (b) citrate, (c) phytate, (d) malate, (e) lactate, (f) acetate; Figure S3: Variation of dissolved As and Fe(II) in the supernatant as percent extracted of total in soil under biotic and abiotic conditions with different LMWOAS amendments in remediated soils.(a) malate, (b) lactate, (c) acetate; Figure S4: Soil pH under biotic and abiotic conditions with citrate and phytate amendments in restored soils; Table S1: XRD results of phreatic aquifer soil sample; Table S2: Analysis results of phreatic aquifer soil sample; Table S3: Arsenic fraction percentage in the different fractions of experimental soil; Table S4: Reagent Manufacturers.

**Author Contributions:** Conceptualization, methodology, investigation, writing—review and editing, X.L. and X.G.; investigation, methodology, formal analysis, data curation, visualization, writing—original draft preparation, writing—review and editing, Y.Y.; formal analysis, J.Z. and Y.T. and X.S.; investigation, M.L. and X.H. All authors have read and agreed to the published version of the manuscript.

**Funding:** This work was supported by the National Natural Science Foundation of China (42072275).

**Institutional Review Board Statement:** Not applicable.

**Informed Consent Statement:** Not applicable.

**Data Availability Statement:** The data that support the findings of this study are available from the corresponding author, X. Luo (luoxm@cugb.edu.cn), upon reasonable request.

**Acknowledgments:** The authors would like to thank Fei Liu's constructive suggestion. We thank Jan Bloemendal for improving the English language.

**Conflicts of Interest:** The authors declare no conflict of interest.

## References

1. Wan, X.; Lei, M.; Chen, T. Review on remediation technologies for arsenic-contaminated soil. *Front. Environ. Sci. Eng.* **2019**, *14*, 24. [CrossRef]
2. Gong, Y.; Zhao, D.; Wang, Q. An overview of field-scale studies on remediation of soil contaminated with heavy metals and metalloids: Technical progress over the last decade. *Water Res.* **2018**, *147*, 440–460. [CrossRef] [PubMed]
3. Lin, L.-Y.; Yan, X.-L.; Yang, S. Stabilizing effects of Fe-Ce oxide on soil As(V) and P. *Environ. Sci.* **2019**, *40*, 3785–3791.
4. Wu, J.; Li, Z.; Huang, D.; Liu, X.; Tang, C.; Parikh, S.J.; Xu, J. A novel calcium-based magnetic biochar is effective in stabilization of arsenic and cadmium co-contamination in aerobic soils. *J. Hazard. Mater.* **2020**, *387*, 122010. [CrossRef]
5. Zhai, W.; Dai, Y.; Zhao, W.; Yuan, H.; Qiu, D.; Chen, J.; Gustave, W.; Maguffin, S.C.; Chen, Z.; Liu, X.; et al. Simultaneous immobilization of the cadmium, lead and arsenic in paddy soils amended with titanium gypsum. *Environ. Pollut.* **2020**, *258*, 113790. [CrossRef] [PubMed]
6. Ma, L.; Cai, D.; Tu, S. Arsenite simultaneous sorption and oxidation by natural ferruginous manganese ores with various ratios of Mn/Fe. *Chem. Eng. J.* **2020**, *382*, 123040. [CrossRef]
7. Zheng, Q.; Hou, J.; Hartley, W.; Ren, L.; Wang, M.; Tu, S.; Tan, W. As(III) adsorption on Fe-Mn binary oxides: Are Fe and Mn oxides synergistic or antagonistic for arsenic removal? *Chem. Eng. J.* **2020**, *389*, 124470. [CrossRef]
8. Liu, Y.; Wu, T.; White, J.C.; Lin, D. A new strategy using nanoscale zero-valent iron to simultaneously promote remediation and safe crop production in contaminated soil. *Nat. Nanotechnol.* **2021**, *16*, 197–205. [CrossRef]
9. Hou, Q.; Han, D.; Zhang, Y.; Han, M.; Huang, G.; Xiao, L. The bioaccessibility and fractionation of arsenic in anoxic soils as a function of stabilization using low-cost Fe/Al-based materials: A long-term experiment. *Ecotoxicol. Environ. Saf.* **2020**, *191*, 110210. [CrossRef]
10. Li, Z.; Wang, L.; Wu, J.; Xu, Y.; Wang, F.; Tang, X.; Xu, J.; Ok, Y.S.; Meng, J.; Liu, X. Zeolite-supported nanoscale zero-valent iron for immobilization of cadmium, lead, and arsenic in farmland soils: Encapsulation mechanisms and indigenous microbial responses. *Environ. Pollut.* **2020**, *260*, 114098. [CrossRef]
11. Allegretta, I.; Porfido, C.; Martin, M.; Barberis, E.; Terzano, R.; Spagnuolo, M. Characterization of As-polluted soils by laboratory X-ray-based techniques coupled with sequential extractions and electron microscopy: The case of Crocette gold mine in the Monte Rosa mining district (Italy). *Environ. Sci. Pollut. Res.* **2018**, *25*, 25080–25090. [CrossRef] [PubMed]
12. An, J.; Jeong, B.; Nam, K. Evaluation of the effectiveness of in situ stabilization in the field aged arsenic-contaminated soil: Chemical extractability and biological response. *J. Hazard. Mater.* **2019**, *367*, 137–143. [CrossRef] [PubMed]
13. Corsini, A.; Cavalca, L.; Zaccheo, P.; Crippa, L.; Andreoni, V. Influence of microorganisms on arsenic mobilization and speciation in a submerged contaminated soil: Effects of citrate. *Appl. Soil Ecol.* **2011**, *49*, 99–106. [CrossRef]
14. Zhang, Z.; Moon, H.S.; Myneni, S.C.B.; Jaffé, P.R. Phosphate enhanced abiotic and biotic arsenic mobilization in the wetland rhizosphere. *Chemosphere* **2017**, *187*, 130–139. [CrossRef] [PubMed]
15. Fang, J.; Xie, Z.; Wang, J.; Liu, D.; Zhong, Z. Bacterially mediated release and mobilization of As/Fe coupled to nitrate reduction in a sediment environment. *Ecotoxicol. Environ. Saf.* **2021**, *208*, 111478. [CrossRef] [PubMed]
16. Fan, L.; Zhao, F.; Liu, J.; Frost, R.L. The As behavior of natural arsenical-containing colloidal ferric oxyhydroxide reacted with sulfate reducing bacteria. *Chem. Eng. J.* **2018**, *332*, 183–191. [CrossRef]
17. Huang, J.H. Impact of competitive adsorption on microbial arsenate reduction at the water-goethite interface. *Appl. Geochem.* **2018**, *88*, 59–67. [CrossRef]
18. Huang, J.-H. Characterising microbial reduction of arsenate sorbed to ferrihydrite and its concurrence with iron reduction. *Chemosphere* **2018**, *194*, 49–56. [CrossRef]
19. Yamamura, S.; Sudo, T.; Watanabe, M.; Tsuboi, S.; Soda, S.; Ike, M.; Amachi, S. Effect of extracellular electron shuttles on arsenic-mobilizing activities in soil microbial communities. *J. Hazard. Mater.* **2018**, *342*, 571–578. [CrossRef]
20. Zhalnina, K.; Louie, K.B.; Hao, Z.; Mansoori, N.; da Rocha, U.N.; Shi, S.; Cho, H.; Karaoz, U.; Loque, D.; Bowen, B.P.; et al. Dynamic root exudate chemistry and microbial substrate preferences drive patterns in rhizosphere microbial community assembly. *Nat. Microbiol.* **2018**, *3*, 470–480. [CrossRef]
21. Van Hees, P.A.W.; Vinogradoff, S.I.; Edwards, A.C.; Godbold, D.L.; Jones, D.L. Low molecular weight organic acid adsorption in forest soils: Effects on soil solution concentrations and biodegradation rates. *Soil Biol. Biochem.* **2003**, *35*, 1015–1026. [CrossRef]
22. Jones, D.L.; Dennis, P.G.; Owen, A.G.; van Hees, P.A.W. Organic acid behavior in soils-misconceptions and knowledge gaps. *Plant Soil* **2003**, *248*, 31–41. [CrossRef]

23. Tu, S.X.; Ma, L.; Luongo, T. Root exudates and arsenic accumulation in arsenic hyperaccumulating *Pteris vittata* and non-hyperaccumulating *Nephrolepis exaltata*. *Plant Soil* **2004**, *258*, 9–19. [CrossRef]
24. Gerke, J.; Romer, W.; Jungk, A. The excretion of citric and malic acid by proteoid roots of *Lupinus albus* L.; effects on soil solution concentrations of phosphate, iron, and aluminum in the proteoid rhizosphere in samples of an oxisol and a luvisol. *Z. Med. Phys.* **1994**, *157*, 289–294. [CrossRef]
25. Liu, L.; Yang, Y.-P.; Duan, G.-L.; Wang, J.; Tang, X.-J.; Zhu, Y.-G. The chemical-microbial release and transformation of arsenic induced by citric acid in paddy soil. *J. Hazard. Mater.* **2022**, *421*, 126731. [CrossRef]
26. Lee, J.-C.; Kim, E.J.; Baek, K. Synergistic effects of the combination of oxalate and ascorbate on arsenic extraction from contaminated soils. *Chemosphere* **2017**, *168*, 1439–1446. [CrossRef]
27. Mei, K.; Liu, J.; Shi, R.; Guo, X.; Lu, H.; Yan, C. The migrated behavior and bioavailability of arsenic in mangrove sediments affected by pH and organic acids. *Mar. Pollut. Bull.* **2020**, *159*, 111480. [CrossRef]
28. Xu, Y.; Wan, L.; Wang, K.; Liu, C.; Zhang, J. Enhanced mobilization of arsenic from tailing soil by four types of low molecular weight organic acids with different functional groups. *J. Soils Sediments* **2021**, *21*, 3834–3844. [CrossRef]
29. Ji, Y.; Luo, W.; Lu, G.; Fan, C.; Tao, X.; Ye, H.; Xie, Y.; Shi, Z.; Yi, X.; Dang, Z. Effect of phosphate on amorphous iron mineral generation and arsenic behavior in paddy soils. *Sci. Total Environ.* **2019**, *657*, 644–656. [CrossRef]
30. Deng, Y.; Weng, L.; Li, Y.; Chen, Y.; Ma, J. Redox-dependent effects of phosphate on arsenic speciation in paddy soils. *Environ. Pollut.* **2020**, *264*, 114783. [CrossRef]
31. Yan, G.; Chen, X.; Du, S.; Deng, Z.; Wang, L.; Chen, S. Genetic mechanisms of arsenic detoxification and metabolism in bacteria. *Curr. Genet.* **2019**, *65*, 329–338. [CrossRef] [PubMed]
32. Wang, J.; Xie, Z.; Wei, X.; Chen, M.; Luo, Y.; Wang, Y. An indigenous bacterium *Bacillus* XZM for phosphate enhanced transformation and migration of arsenate. *Sci. Total Environ.* **2020**, *719*, 137183. [CrossRef] [PubMed]
33. Onireti, O.O.; Lin, C. Mobilization of soil-borne arsenic by three common organic acids: Dosage and time effects. *Chemosphere* **2016**, *147*, 352–360. [CrossRef] [PubMed]
34. Sun, Y.; Luo, T.; Zhong, S.; Zhou, F.; Zhang, Y.; Ma, Y.; Fu, Q. Long-term effects of low-molecular-weight organic acids on remobilization of Cd, Cr, Pb, and As in alkaline coastal wetland soil. *Env. Pollut. Bioavail.* **2021**, *33*, 266–277. [CrossRef]
35. Onireti, O.O.; Lin, C.; Qin, J. Combined effects of low-molecular-weight organic acids on mobilization of arsenic and lead from multi-contaminated soils. *Chemosphere* **2017**, *170*, 161–168. [CrossRef]
36. Nworie, O.E.; Qin, J.; Lin, C. Differential effects of low-molecular-weight organic acids on the mobilization of soil-borne arsenic and trace metals. *Toxics* **2017**, *5*, 18. [CrossRef]
37. Vitkova, M.; Komarek, M.; Tejnecky, V.; Sillerova, H. Interactions of nano-oxides with low-molecular-weight organic acids in a contaminated soil. *J. Hazard. Mater.* **2015**, *293*, 7–14. [CrossRef]
38. Wang, S.; Mulligan, C.N. Effects of three low-molecular-weight organic acids (LMWOAs) and pH on the mobilization of arsenic and heavy metals (Cu, Pb, and Zn) from mine tailings. *Environ. Geochem. Health* **2013**, *35*, 111–118. [CrossRef]
39. Mei, K.; Wu, G.; Liu, J.; Jiajia, W.; Hong, H.; Lu, H.; Yan, C. Dynamics of low-molecular-weight organic acids for the extraction and sequestration of arsenic species and heavy metals using mangrove sediments. *Chemosphere* **2022**, *286*, 131820. [CrossRef]
40. Chen, Z.; Dong, G.; Gong, L.; Li, Q.; Wang, Y. The role of low-molecular-weight organic carbons in facilitating the mobilization and biotransformation of As(V)/Fe(III) from a realgar tailing mine soil. *Geomicrobiol. J.* **2018**, *35*, 555–563. [CrossRef]
41. Chen, Z.; Li, D.; Luo, X. Research on arsenic form in the gold mine tailings by different leaching processes. *Rock Miner. Anal.* **2014**, *33*, 363–368.
42. Luo, X.M. Case Study on the Migration and Transform of Arsenic in Aquatic Environment Caused by Gold Mining—Dandong, Liaoning. Ph.D. Thesis, China University of Geosciences, Beijing, China, 2007.
43. Wang, Y.; Zeng, X.; Lu, Y.; Bai, L.; Su, S.; Wu, C. Dynamic arsenic aging processes and their mechanisms in nine types of Chinese soils. *Chemosphere* **2017**, *187*, 404–412. [CrossRef] [PubMed]
44. Yamamura, S.; Ike, M.; Fujita, M. Dissimilatory arsenate reduction by a facultative anaerobe, *Bacillus* sp. strain SF-1. *J. Biosci. Bioeng.* **2003**, *96*, 454–460. [CrossRef]
45. Yamamura, S.; Kurasawa, H.; Kashiwabara, Y.; Hori, T.; Aoyagi, T.; Nakajima, N.; Amachi, S. Soil microbial communities involved in reductive dissolution of arsenic from arsenate-laden minerals with different carbon sources. *Environ. Sci. Technol.* **2019**, *53*, 12398–12406. [CrossRef]
46. Fredrickson, J.K.; Gorby, Y.A. Environmental processes mediated by iron-reducing bacteria. *Curr. Opin. Biotechnol.* **1996**, *7*, 287–294. [CrossRef]
47. Liu, Y.; Li, F.-B.; Xia, W.; Xu, J.-M.; Yu, X.-S. Association between ferrous iron accumulation and pentachlorophenol degradation at the paddy soil-water interface in the presence of exogenous low-molecular-weight dissolved organic carbon. *Chemosphere* **2013**, *91*, 1547–1555. [CrossRef]
48. Gao, S.-J.; Cao, W.-D.; Gao, J.-S.; Huang, J.; Bai, J.-S.; Zeng, N.-H.; Chang, D.-N.; Shimizu, K. Effects of long-term application of different green manures on ferric iron reduction in a red paddy soil in Southern China. *J. Integr. Agric.* **2017**, *16*, 959–966. [CrossRef]
49. Song, J.; Yang, J.; Cui, X. Effects of low molecular-weight organic acids/sallts on availability of Lead, Zinc and Arsenic in mixed metal-polluted soil. *J. Soil. Water Conserv.* **2010**, *24*, 108–112,118.

50. Wang, Y.; Zhang, G.; Wang, H.; Cheng, Y.; Liu, H.; Jiang, Z.; Li, P.; Wang, Y. Effects of different dissolved organic matter on microbial communities and arsenic mobilization in aquifers. *J. Hazard. Mater.* **2021**, *411*, 125146. [CrossRef]
51. De Araujo, T.O.; Isaure, M.-P.; Alchoubassi, G.; Bierla, K.; Szpunar, J.; Trcera, N.; Chay, S.; Alcon, C.; da Silva, L.C.; Curie, C.; et al. *Paspalum urvillei* and *Setaria parviflora*, two grasses naturally adapted to extreme iron-rich environments. *Plant Physiol. Biochem.* **2020**, *151*, 144–156. [CrossRef]
52. Wu, F.; Xu, F.; Ma, X.; Luo, W.; Lou, L.; Wong, M.H. Do arsenate reductase activities and oxalate exudation contribute to variations of arsenic accumulation in populations of *Pteris vittata*? *J. Soils Sediments* **2018**, *18*, 3177–3185. [CrossRef]
53. Mei, K.; Liu, J.; Fan, J.; Guo, X.; Wu, J.; Zhou, Y.; Lu, H.; Yan, C. Low-level arsenite boosts rhizospheric exudation of low-molecular-weight organic acids from mangrove seedlings (*Avicennia marina*): Arsenic phytoextraction, removal, and detoxification. *Sci. Total Environ.* **2021**, *775*, 145685. [CrossRef]
54. Chen, C.; Dynes, J.J.; Wang, J.; Sparks, D.L. Properties of Fe-organic matter associations via coprecipitation versus adsorption. *Environ. Sci. Technol.* **2014**, *48*, 13751–13759. [CrossRef] [PubMed]
55. Li, Z.; Wang, L.; Meng, J.; Liu, X.; Xu, J.; Wang, F.; Brookes, P. Zeolite-supported nanoscale zero-valent iron: New findings on simultaneous adsorption of Cd(II), Pb(II), and As(III) in aqueous solution and soil. *J. Hazard. Mater.* **2018**, *344*, 1–11. [CrossRef]
56. Kim, E.J.; Baek, K. Enhanced reductive extraction of arsenic from contaminated soils by a combination of dithionite and oxalate. *J. Hazard. Mater.* **2015**, *284*, 19–26. [CrossRef]
57. Hong, J.; Liu, L.; Tan, W.; Qiu, G. Arsenic release from arsenopyrite oxidative dissolution in the presence of citrate under UV irradiation. *Sci. Total Environ.* **2020**, *726*, 138429. [CrossRef] [PubMed]
58. Liu, X.; Fu, J.W.; Guan, D.X.; Cao, Y.; Luo, J.; Rathinasabapathi, B.; Chen, Y.S.; Ma, L.Q. Arsenic induced phytate exudation, and promoted $FeAsO_4$ dissolution and plant growth in As-hyperaccumulator *Pteris vittata*. *Environ. Sci. Technol.* **2016**, *50*, 9070–9077. [CrossRef] [PubMed]
59. Yang, Z.; Zhang, N.; Sun, B.; Su, S.; Wang, Y.; Zhang, Y.; Wu, C.; Zeng, X. Contradictory tendency of As(V) releasing from Fe–As complexes: Influence of organic and inorganic anions. *Chemosphere* **2022**, *286*, 131469. [CrossRef]
60. Cai, X.; Yin, N.; Wang, P.; Du, H.; Liu, X.; Cui, Y. Arsenate-reducing bacteria-mediated arsenic speciation changes and redistribution during mineral transformations in arsenate-associated goethite. *J. Hazard. Mater.* **2020**, *398*, 122886. [CrossRef]
61. Strawn, D.G. Review of interactions between phosphorus and arsenic in soils from four case studies. *Geochem. Trans.* **2018**, *19*, 10. [CrossRef]
62. Cai, R.; Wang, X.; Ji, X.; Peng, B.; Tan, C.; Huang, X. Phosphate reclaim from simulated and real eutrophic water by magnetic biochar derived from water hyacinth. *J. Environ. Manag.* **2017**, *187*, 212–219. [CrossRef] [PubMed]
63. Huang, R.; Wang, X.; Xing, B. Removal of labile arsenic from flooded paddy soils with a novel extractive column loaded with quartz-supported nanoscale zero-valent iron. *Environ. Pollut.* **2019**, *255*, 113249. [CrossRef] [PubMed]

Article

# Arsenic Accumulation and Physiological Response of Three Leafy Vegetable Varieties to As Stress

Yuan Meng [1,*], Liang Zhang [1,2], Zhi-Long Yao [1], Yi-Bin Ren [1], Lin-Quan Wang [3] and Xiao-Bin Ou [2]

1 College of Agriculture and Forestry, Longdong University, Qingyang 745000, China; txzzl891017@163.com (L.Z.); yzl8844@163.com (Z.-L.Y.); 18093486955@163.com (Y.-B.R.)
2 Gansu Key Laboratory of Protection and Utilization for Biological Resources and Ecological Restoration, Qingyang 745000, China; xbou@zju.edu.cn
3 College of Resources and Environment, Northwest A&F University, Xiangyang 712100, China; linquanw@nwsuaf.edu.cn
* Correspondence: sibyl19910101@163.com

**Citation:** Meng, Y.; Zhang, L.; Yao, Z.-L.; Ren, Y.-B.; Wang, L.-Q.; Ou, X.-B. Arsenic Accumulation and Physiological Response of Three Leafy Vegetable Varieties to As Stress. *Int. J. Environ. Res. Public Health* **2022**, *19*, 2501. https://doi.org/10.3390/ijerph19052501

Academic Editors: Fayuan Wang, Liping Li, Lanfang Han and Aiju Liu

Received: 31 December 2021
Accepted: 18 February 2022
Published: 22 February 2022

**Publisher's Note:** MDPI stays neutral with regard to jurisdictional claims in published maps and institutional affiliations.

**Copyright:** © 2022 by the authors. Licensee MDPI, Basel, Switzerland. This article is an open access article distributed under the terms and conditions of the Creative Commons Attribution (CC BY) license (https://creativecommons.org/licenses/by/4.0/).

**Abstract:** Arsenic (As) in leafy vegetables may harm humans. Herein, we assessed As accumulation in leafy vegetables and the associated physiological resistance mechanisms using soil pot and hydroponic experiments. Garland chrysanthemum (*Chrysanthemum coronarium* L.), spinach (*Spinacia oleracea* L.), and lettuce (*Lactuca sativa* L.) were tested, and the soil As safety threshold values of the tested leafy vegetables were 91.7, 76.2, and 80.7 mg kg$^{-1}$, respectively, i.e., higher than the soil environmental quality standard of China. According to growth indicators and oxidative stress markers (malondialdehyde, the ratio of reduced glutathione to oxidized glutathione, and soluble protein), the order of As tolerance was: GC > SP > LE. The high tolerance of GC was due to the low transport factor of As from the roots to the shoots; the high activity of superoxide dismutase, glutathione peroxidase, and catalase; and the high content of phytochelatin in the roots. Results of this work shed light on the use of As-contaminated soils and plant tolerance of As stress.

**Keywords:** heavy metals; garland chrysanthemum; lettuce; antioxidant defense enzymes; GSH; PCs

## 1. Introduction

Arsenic (As) is a non-metal, but as its toxicity and some of its properties are similar to those of heavy metals, it is generally included in the range of heavy metal pollutants [1]. The density of As is 5.727 g cm$^{-3}$, and it is toxic and biologically non-essential. Many countries, including China, the United States, and Canada, have listed it as an environmental priority pollutant [2]. Direct exposure to As pollution will lead to liver poisoning and kidney damage, further causing hypertension, cardiovascular disease, and even cancer [3]. Environmental pollution such as heavy metal contamination has also been reported to increase the risk of mental disorders in humans [4]. There are even reports that the living environment and diet of captive giant pandas in China contain heavy metals, such as As and cadmium (Cd) [5]. In 2016, the State Council of China issued the "soil pollution prevention and control action plan", which shows that China's soil pollution control has become an increasingly urgent problem.

According to the bulletin of China's national survey, inorganic pollutants are mainly responsible for soil pollution in China. Notably, the exceedance rate of soil As pollution points was 2.7%. It is difficult and impractical to eliminate heavy metal contamination in the soil [6,7]. Assessing how to reasonably and safely use contaminated soil has important theoretical and practical significance. Although the distribution area of heavy metal pollution is wide, the distribution of pollutants in the soil is shallow, and the degree of pollution is light. Thus, the cultivation of food or vegetable crops with a low-enrichment capacity is feasible [8,9]. Heavy metals in the environment enter the human body through the food chain. Vegetables are an indispensable food for human beings, being rich in vitamins, minerals, and crude fiber necessary for the digestive system. Leafy vegetables are believed to be more vulnerable to heavy metal pollution than other crops, such as *Solanaceae, Brassica oleracea*, root vegetables, shallots, and legumes [8,10,11]. They therefore have a higher potential safety risk, and thus their contamination is of great concern worldwide.

The mechanism of Cd enrichment and tolerance in the three leafy vegetables has been explored [12]. The mechanism of plant response to As is similar to that of Cd. Enzymes within the antioxidant system are responsible for scavenging reactive oxygen free radicals in cells caused by heavy metals, and the thiol pool is related to the compartmentalization of heavy metals in cells [13]. However, there are few reports on the response mechanism of garland chrysanthemum (GC), spinach (SP), and lettuce (LE) to As stress. This study aimed to compare the As bioconcentration factors of different leafy vegetables and reveal the associated mechanisms to provide a basis for the rational and effective utilization of heavy metal contaminated soil.

## 2. Materials and Methods

### 2.1. Soil Pot Experiment

The pot experiment was completed under a shelter from 6 March to 25 April 2015. The shelter was ventilated, light-transparent, and rainproof, and its temperature was the same as the temperature outside. The vegetable seeds were purchased from the Shaanxi Hua-xing Green Seed Company (Yangling, Shaanxi 712100, China). On 6 March, spinach (*Spinacia oleracea* L.), lettuce (*Lactuca sativa* L.), and garland chrysanthemum (*Chrysanthemum coronarium* L.) were sown in black pots with a top diameter, bottom diameter, and height of 14, 10, and 12 cm, respectively. There were three small holes in the bottom of each pot, and the pots containing plant seeds were placed in a plastic tray to prevent soil erosion when watering. Water was added to the plastic tray to retain soil moisture during the experiment, and 1.6 g of basal fertilizer (15-15-15) was applied to each pot. Based on previous investigations of heavy metal pollution in soil, cultivated soil (0–20 cm) from vegetable fields with different degrees of pollution in the Baqiao District, Weiyang District, and Lintong District of Xi'an city (China) was collected. A total of 12 treatments (i.e., different As levels; as shown in Figure 1, the As concentration in the soil was 12.63–28.31 mg kg$^{-1}$) were performed with three types of leafy vegetables (three repetitions) and 108 pots.

### 2.2. Hydroponic Test

The hydroponic test was carried out in a light incubator with a light intensity of 20,000 lx. The experiment took place with a photoperiod of 16 h of light (23 °C) and 8 h of darkness (15 °C). The seeds had been soaked in sterile water for 6–8 h prior to the experiment, then germinated in the seedling tray for about 10 days, and finally cultured with 1/10 modified Hogland solution. The hydroponic nutrient solution contained the following [14]: 0.1 mM $NH_4H_2PO_4$, 0.505 mM $KNO_3$, 0.15 mM $Ca(NO_3)_2$, 0.1 mM $MgSO_4$, 1.64 μM $FeSO_4$, 0.91 μM $MnCl_2$, 0.03 μM $CuSO_4$, 0.16 μM $ZnSO_4$, 4.63 μM $H_3BO_3$, 0.06 μM $H_2MoO_4$, and 0.81 μM $Na_2EDTA$. The hydroponic device was a cylindrical container with a diameter and height of 12 and 15 cm, respectively. The outside of the container was wrapped in black plastic to shade the roots. A sponge plate with two holes was placed above the hydroponic container to fix the leafy vegetables. The As concentrations of the hydroponic treatment (four repetitions) were 0 μM (0), 200 μM (As1), and 400 μM (As2). The compound used in the hydroponic test was sodium dihydrogen arsenate ($Na_2HAsO_4$). Vegetable seedlings were treated in the As solution on day 24 and harvested on day 31. Hydroponic As treatment was carried out for 7 d. The leafy vegetable samples were determined immediately after harvesting. The fresh samples were stored in liquid nitrogen for subsequent testing, whereas the dry samples were crushed for dry weight determination.

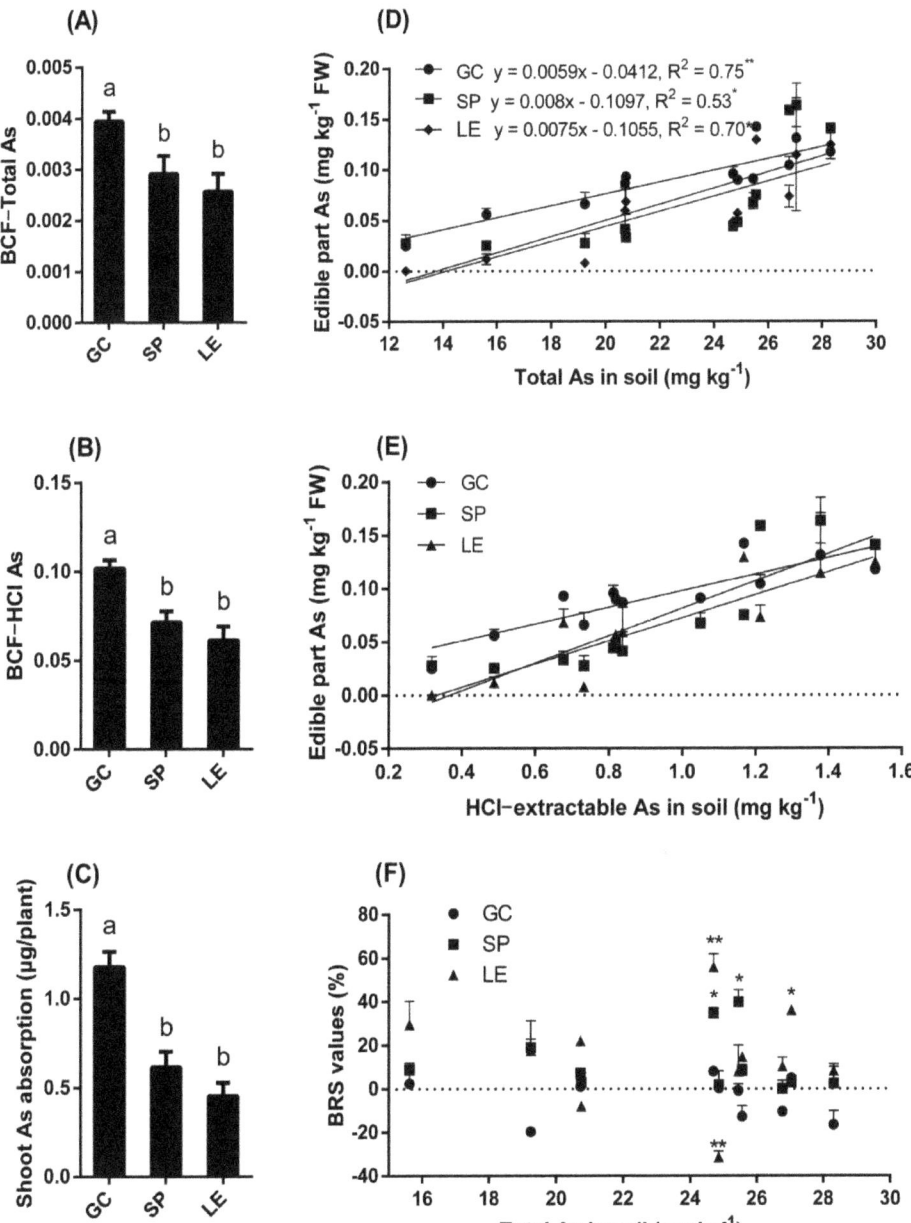

**Figure 1.** Bioconcentration factor (**A,B**), absorption (**C**), vegetable As concentration (**D,E**), and BRS values (**F**) in the soil test. Note: The different lowercase letters in the figure indicate significant differences between different vegetables (5% levels). The symbols * and ** indicate that the difference between values is significant at the respective 5% and 1% levels.

### 2.3. Assay Methods

Plant sampling method: For the soil pot experiment, when harvesting, scissors were used to destroy the pot, and the leafy vegetable roots and soil were taken out completely, soaked in a bucket and rinsed gently with water until all the plants and roots were clean. For the hydroponic experiment, the roots of vegetable plants were cut directly from the stem base and washed with deionized water 4 times. The method of removing metal ions attached to the root surface was to immerse the root in 20 mM $Na_2$-EDTA for 15 min [15].

For the determination of chlorophyll content [16], 80% acetone was used to extract total chlorophyll (a, b) and carotenoids, and the contents were calculated by the formulas: chl a (µg/mL) = 12.7 $OD_{663}$–2.69 $OD_{645}$; chl b (µg/mL) = 22.9 $OD_{645}$–4.68 $OD_{663}$; car = 20.2 $D_{645}$ + 8.02 $D_{663}$.

For the determination of As content [15], the soil and vegetable samples were digested with aqua regia-perchloric acid and 10% nitric acid ($HNO_3$). The soil-available arsenic was extracted with 0.1 mol·$L^{-1}$ hydrochloric acid (HCl). The As content was determined using an inductively coupled plasma mass spectrometer (ICP-MS 7500, Agilent, Santa Clara, CA, USA).

The oxidative stress marker malondialdehyde (MDA) [17] was determined by the 0.5% thiobarbituric acid (TBA) method, which produces a reddish-brown color reaction with MDA in the tissue. Soluble protein was measured with Coomassie Brilliant Blue G-250.

To determine the plant superoxide dismutase (SOD), glutathione peroxidase (POD), and catalase (CAT) enzyme activities [16], the activities of SOD and CAT were quantified on a Shimadzu UV-2450 spectrophotometer (Japan) by measuring the ability to inhibit the photochemical reduction of nitroblue tetrazolium (NBT) and by monitoring the disappearance of $H_2O_2$ at 240 nm. The activity of POD was quantified by catalyzing the oxidation of guaiacol to brown products.

The determination of the thiol pool (cysteine (cys), γ-glutamylcysteine (γ-EC), reduced glutathione (GSH), and phytochelatins (PCs)) was as follows [18]: The determination of cys was measured by an acidic ninhydrin reagent at 560 nm. The γ-EC and PC determination method used mBBr to derivatize thiol, and γ-EC, $PC_2$, and $PC_3$ were used as standard materials for the determination of thiol by high-performance liquid chromatography (HPLC) coupled with parallel mass spectrometry for thiol-containing compounds in the plant samples. For the determination of GSH and GSSG, the absorbance was read at 412 nm by adding phosphate buffer (100 mM, pH 7.0) and 5′5-dithiobis-2-nitrobenzoic acid (DTNB) to the supernatant.

*2.4. Standards and Calculation Formulas*

The criteria for judging soil pollution were as follows: the "Agricultural Land Soil Pollution Risk Screening Value" (pH > 7.5, As: 25 mg $kg^{-5}$; 6.5 < pH ≤ 7.5, As: 30 mg $kg^{-1}$; 5.5 < pH ≤ 6.5, As: 40 mg $kg^{-1}$; pH ≤ 5.5, As: 40 mg $kg^{-1}$) according to China's Soil Environmental Quality Standard (GB 15618-2018).

The calculation of the critical value of heavy metal safety was done as follows: First, the concentration of As in the edible vegetables (y, mg $kg^{-1}$) and the concentration of As in the soil (x, mg $kg^{-1}$) were used to fit the linear equation, and then the limit value of heavy metal in the leafy vegetables [the "Limits in Foods" standard (GB2762-2017) stipulates that the limit value of As is 0.5 mg $kg^{-1}$ FW] was substituted into the equation. Finally, the obtained x value was the safety threshold of soil As.

$$BCF1 = \frac{a}{b}$$

where $BCF_1$ is the bioconcentration factor of the soil total As; a is the concentration of As in the vegetable shoot (mg $kg^{-1}$ FW); b is the total concentration of the soil As (mg $kg^{-1}$).

$$BCF2 = \frac{a}{c}$$

where BCF$_2$ is the bioconcentration factor of the soil HCl-extracted As; a is the concentration of As in the vegetable shoot (mg kg$^{-1}$ FW); c is the soil HCl-extracted As concentration (mg kg$^{-1}$).

$$\mathrm{TI}(\%) = \frac{\mathrm{DWt}}{\mathrm{DWck}} * 100$$

where TI is aboveground/root/total tolerance index; DWt is the average dry weight of shoots/roots/total plant in the As treatment group; DWck is the shoot/root/total average dry weight in the control group.

$$\mathrm{BRS}(\%) = \frac{\mathrm{BIOt} - \mathrm{BIOck}}{\mathrm{BIOck}} * 100$$

where BRS is biomass response to stress; BIOt is the biomass in the As treatment group; BIOck is the biomass of the control vegetable.

$$\mathrm{TF} = \frac{x}{y}$$

where TF is the transfer factor (i.e., the concentration ratio); x is the As concentration in the shoots; y is the As concentration in the roots.

$$\mathrm{Absorption} = \frac{d \times e}{1000}$$

where Absorption is the As absorption of each leafy vegetable (μg); d is the vegetable As content (mg kg$^{-1}$); e is dry weight of the vegetable (mg).

$$\mathrm{SR\ ratio} = \frac{g}{h}$$

where SR ratio (i.e., the accumulation ratio) is the accumulation ratio of shoot As to root As; g is As absorption (shoots); h is As absorption (roots).

*2.5. Data Processing Methods*

As shown in Figure 1D, the univariate linear regression was performed using SAS V8 software (North Carolina State University, Raleigh, NC, USA). ANOVA (analysis of variance) was performed, and Duncan's test was used for multiple comparisons in the SAS program. As shown in Figure 1F, the 'TTEST' process of SAS software was run, which was used for the significance testing of the difference in unpaired data means. SPSS 20.0 (IBM Inc., Armonk, NY, USA) was used for the two-way ANOVA. GraphPad Prism6 (La Jolla, CA, USA) was used to construct the graphs.

## 3. Results

*3.1. The As Stress Test of Leafy Vegetables in the Soil*

The bioconcentration factor of GC was significantly higher than that of SP and LE (Figure 1A,B). The amount of As absorbed in the edible parts of GC was significantly higher than that of SP and LE (Figure 1C). Figure 1D,E show the relationship between the As content in the edible parts of the vegetables and the total As content in the soil or HCl-extracted As content, and there was a linear positive correlation between the two. The equation of the line is shown in Figure 1D. The heavy metal limit (0.5 mg kg$^{-1}$ FW) was substituted into the equation. The calculated soil safety critical values of GC, SP, and LE were 91.7, 76.2, and 80.7 mg kg$^{-1}$, respectively. Figure 1F shows the BRS values of the vegetables in the soil. When the As content in the soil was 24.86 mg kg$^{-1}$, the BRS value of LE was −31.41%. When the As content in the soil was 24.69, 25.44, and 27.04 mg kg$^{-1}$ (the respective soil organic matter content was 14.0, 13.1, and 11.0 g kg$^{-1}$), the BRS values of SP and LE were positive, which may be due to the higher organic matter content in the soil.

The biomass of the vegetables showed a significant downward trend (Figure 2A,C,E) with the increase in As concentration in the hydroponic solution. However, under the As1 treatment, there was no significant change in the total biomass and the shoot and root biomass of GC compared to the CK treatment. The As TI of the whole plant, shoots, and roots of GC was significantly higher than that of SP and LE (Figure 2B,D,F). The TI of SP was slightly higher than that of LE, but not significantly so.

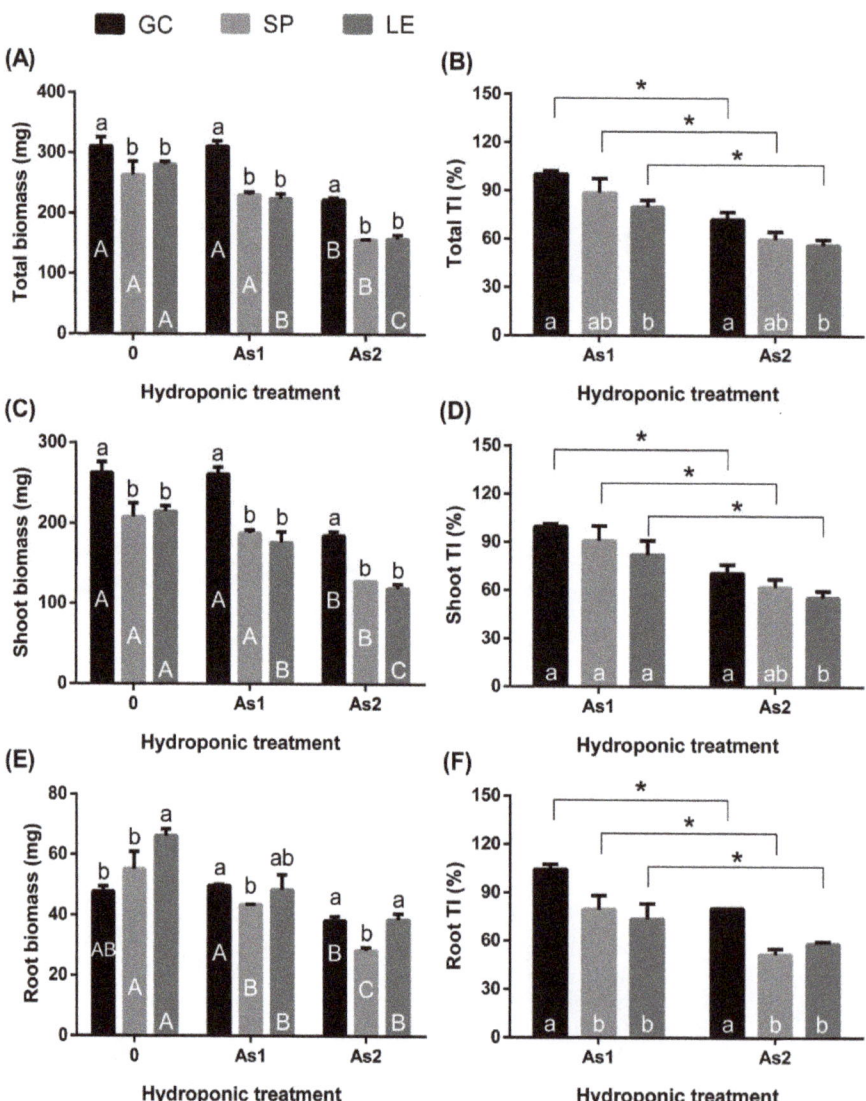

**Figure 2.** Biomass (**A,C,E**) of GC, SP, and LE and the TI of the plants to As in the hydroponic test (**B,D,F**). Note: Different lowercase letters in the figure indicate comparisons between different vegetables under the same treatment; different uppercase letters indicate the comparison of the same vegetable under different treatments (5% levels). The symbol * indicates that the difference in values between As1 and As2 was significant at the respective 5% levels. The error bars in the graph indicate the SEM (standard error of the mean).

As shown in Figure 3, the chla, chlb, car, and chla+b of the vegetables decreased with the increase in As content in the hydroponic treatment (Figure 3A–C,E). The decrease in the chla, chlb, car chla/b, and chla+b content in LE was the smallest among all of the vegetables, and the results were not significant. It can be seen from Figure 3D that under the As1 treatment the chla/b value of the vegetables was significantly higher than that of CK and As2.

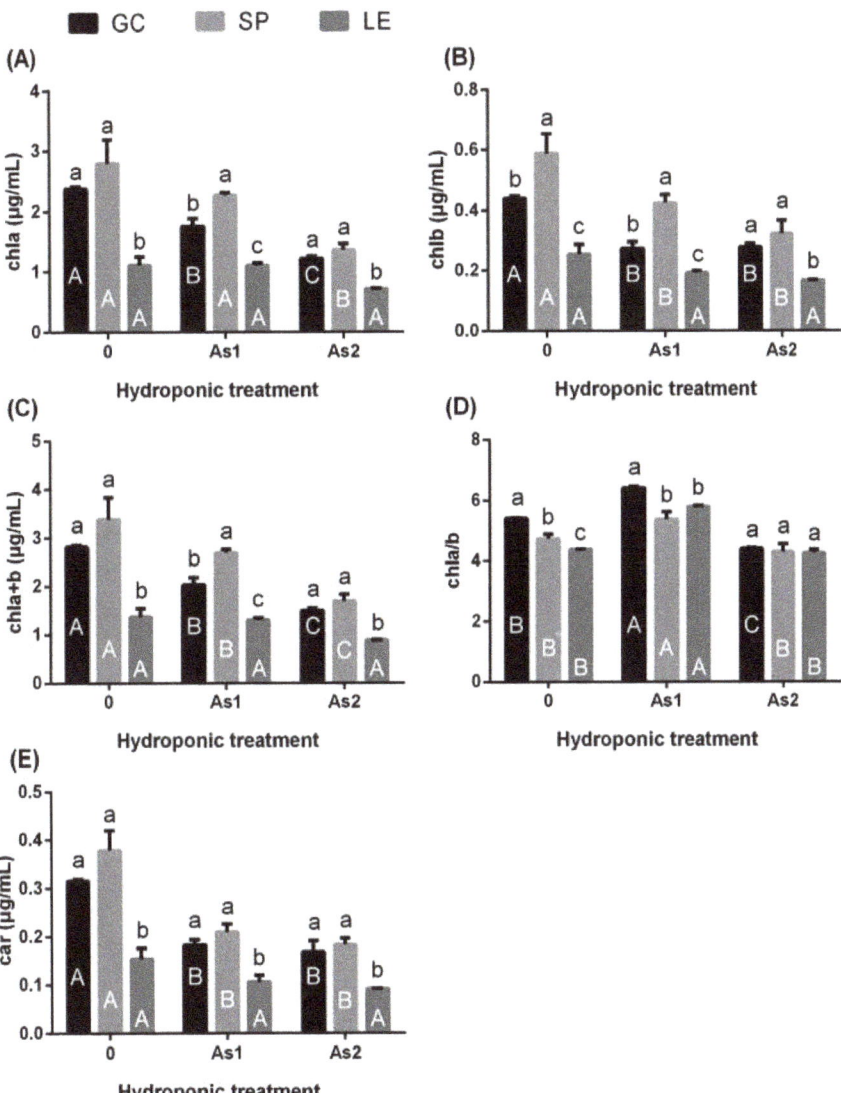

**Figure 3.** The content of chlorophyll (chla, chlb) (**A**,**B**) and carotenoids (car) (**E**) and the values of chla+b (**C**) and chla/b (**D**) of GC, SP, and LE under hydroponic As treatment. Note: Different lowercase letters in the figure indicate comparisons between different vegetables under the same treatment; different uppercase letters indicate the comparison of the same vegetable under different treatments (5% levels). The error bars in the graph indicate the SEM (standard error of the mean).

## 3.2. As Enrichment and Distribution in Plants

As the concentration of As in the hydroponic solution increased, the As concentration in the shoots and roots of the vegetables increased (Figure 4A,C). The concentration of As in the shoots and roots of SP was high. By contrast, the concentration of As was low in the shoots of GC but high in the roots of GC. The As concentrations in the shoots and roots of LE were both low. The absorption of As by the shoots and roots of LE was very low, while that of GC was high (Figure 4B,D,E). Figure 4F shows the TF of As from the roots to the shoots of the vegetables. The TF of GC was the lowest, and there was no significant difference in TF between SP and LE.

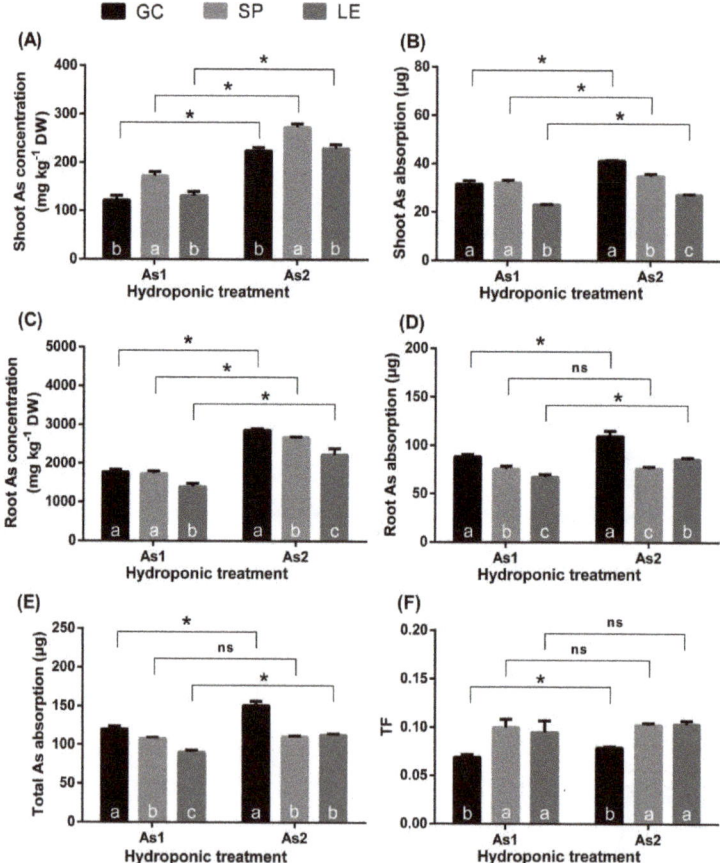

**Figure 4.** The As concentration (**A,C**), As absorption amount (**B,D,E**), and the root-to-ground TF of As (**F**) under hydroponic culture. Note: Different lowercase letters in the figure indicate comparisons between different vegetables under the same treatment (5% levels). The symbol * indicates that the difference between values is significant at the respective 5% levels. The error bars in the graph indicate the SEM (standard error of the mean).

## 3.3. Oxidative Stress Biomarkers and Antioxidative Defense Enzymes

The SOD activity in the shoots of GC and LE decreased compared with CK, while there was an increasing trend in SP (Figure 5A). Under As treatment, the activity of SOD in the roots of GC increased, whereas SP and LE remained unchanged (Figure 5B). The As treatment increased the POD enzyme activity in the shoots of GC and SP, while that of LE remained unchanged (Figure 5C). The POD activity in the roots of GC increased, while that in SP and LE remained unchanged (Figure 5D). The activity of CAT in the shoots of GC and LE remained unchanged, while that in SP decreased significantly (Figure 5E). Compared with the CK treatment, the CAT activities of the roots of the three vegetables were significantly decreased (Figure 5F). Under As treatment, the activities of SOD (Figure 5A) and POD (Figure 5C) in the shoots of SP were significantly higher than those of GC and LE. The SOD activity of the roots of SP was higher than that of LE but lower than that of GC (Figure 5B). The activities of SOD, POD, and CAT in the roots of GC were the highest among the three vegetables (Figure 5B,D,F). Under As treatment, there was no significant difference in CAT enzyme activity in the aboveground parts of the three vegetables (Figure 5E,F).

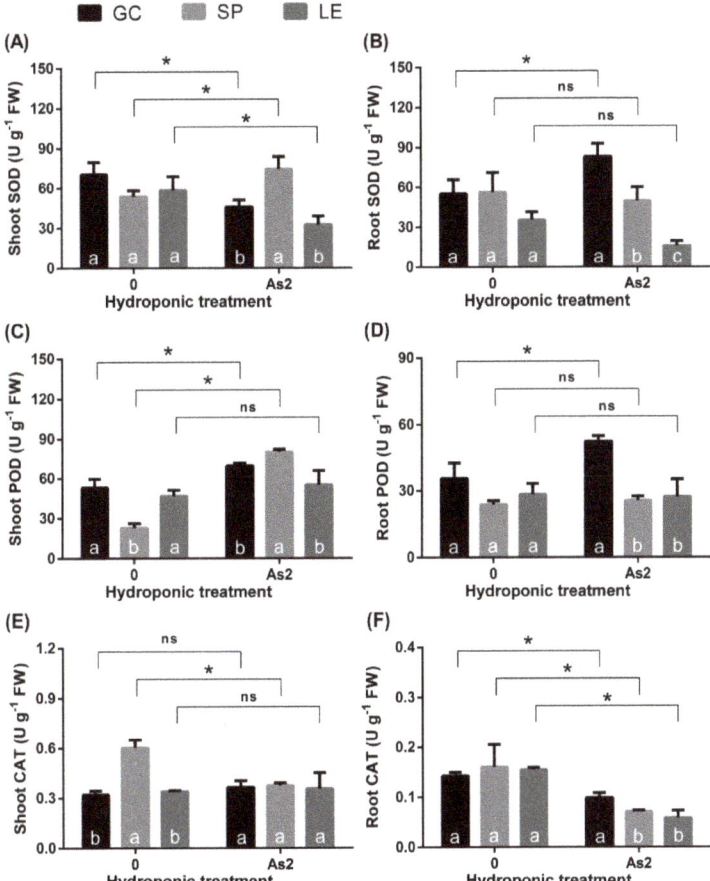

**Figure 5.** SOD (**A**,**B**), POD (**C**,**D**), and CAT (**E**,**F**) activities in the shoots and roots of GC, SP, and LE in the hydroponic As treatment. Note: Different lowercase letters in the figure indicate comparisons between different vegetables under the same treatment (5% levels). The symbol * indicates that the difference between values is significant at the respective 5% levels. The error bars in the graph indicate the SEM (standard error of the mean).

Figure 6 indicates that compared with CK the As treatment increased the MDA content in the shoots and roots of the vegetables (Figure 6A,B). The MDA content in the upper part of LE was significantly higher than that in GC, and the MDA content in the roots of SP was the lowest. Furthermore, under As treatment, the proportion of GSH/GSSG in the upper part of SP was significantly higher than that in the CK treatment (Figure 6C) and was significantly higher than that of GC and LE. The As treatment increased the ratio of GSH/GSSG in the roots of GC and SP and decreased the ratio of GSH/GSSG in the roots of LE (Figure 6D). The proportion of GSH/GSSG in the roots of LE was significantly lower than that in the other vegetables. The As treatment significantly reduced the content of soluble protein in the vegetables (Figure 6E), and the soluble protein content in LE was significantly lower than that in GC and SP.

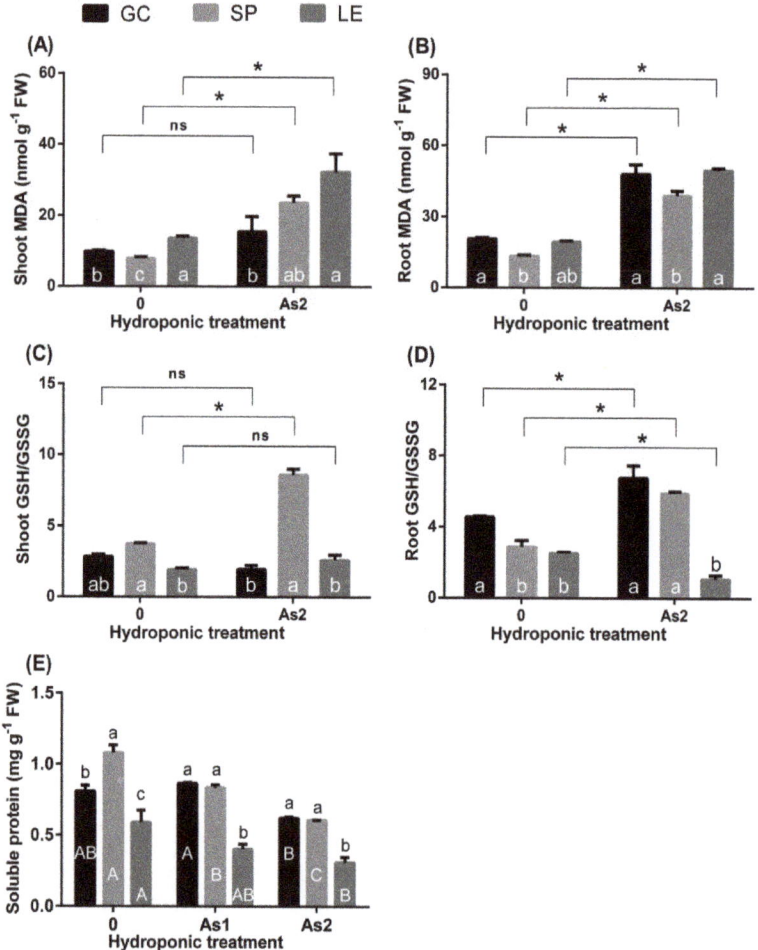

**Figure 6.** MDA (**A,B**), GSH/GSSG ratio (**C,D**), and soluble protein (**E**) content in the vegetables under hydroponic As treatment. Note: Different lowercase letters in the figure indicate comparisons between different vegetables under the same treatment; different uppercase letters indicate the comparison of the same vegetable under different treatments (5% levels). The symbol * indicates that the difference between values is significant at the respective 5% levels. The error bars in the graph indicate the SEM (standard error of the mean).

*3.4. Thiol Pool Concentration*

Figure 7 shows the contents of the vegetable thiol pool (cys, γ-EC, GSH, and PCs). Under As treatment, the cys content in the shoots and roots of GC and SP remained unchanged compared to the control, while that in the roots of LE decreased (Figure 7A,B). The roots of LE had the lowest amount of cys. In the As treatment, the shoot γ-EC content of

GC and SP remained stable, while that of LE declined markedly compared with the control (Figure 7C). The shoot γ-EC content of LE was lower than that of the other vegetables. There was a decreasing trend in the root γ-EC content of SP and LE under As exposure (Figure 7D). Compared with the treatment without heavy metals, the shoot GSH content of LE decreased under As stress (Figure 7E). By contrast, in the As treatment, there were no obvious changes (Figure 7F). The order of the PC content in aboveground parts was GC < SP < LE (Figure 7G). Differences were observed in the roots under As treatment, i.e., GC had a higher PC content (Figure 7H).

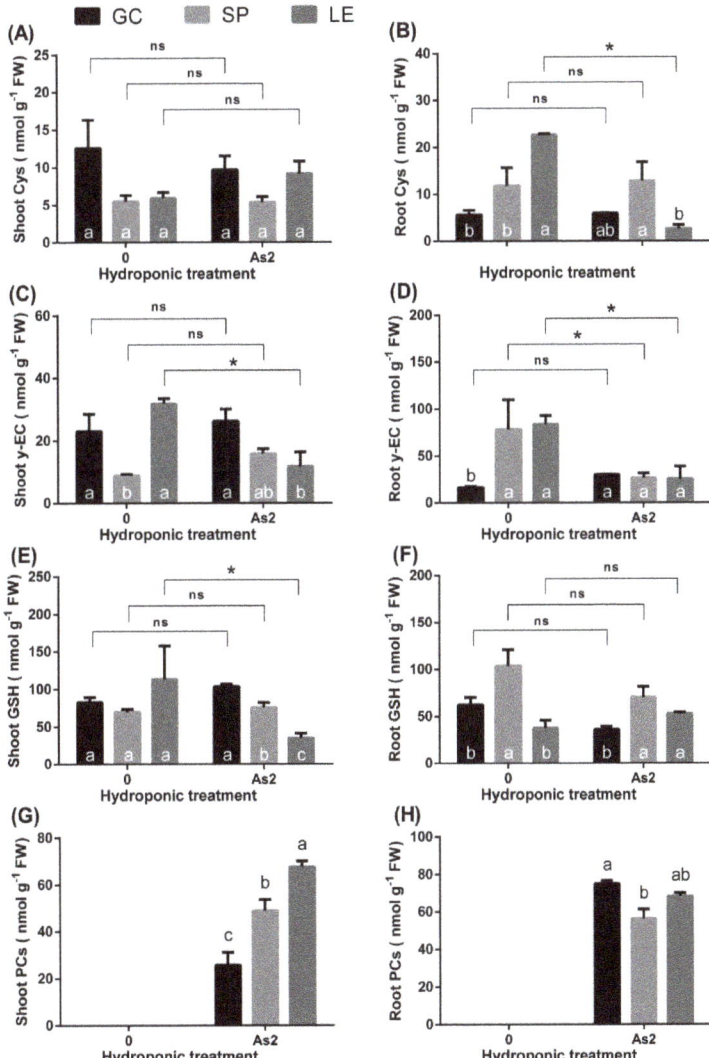

**Figure 7.** Content of thiol [cys (**A**,**B**), γ-EC (**C**,**D**), GSH (**E**,**F**), PCs (**G**,**H**)] in the shoots and roots of vegetables in the hydroponic experiments. Note: Different lowercase letters in the figure indicate comparisons between different vegetables under the same treatment (5% levels). The symbol * indicates that the difference between values is significant at the respective 5%. The error bars in the graph indicate the SEM (standard error of the mean).

## 4. Discussion
### 4.1. Leafy Vegetables Suitable for Cultivation on As-Contaminated Soil

Figure 1 shows that the soil safety thresholds of As in GC, SP, and LE were 91.7, 76.2, and 80.7 mg kg$^{-1}$, respectively, which are much higher than the soil environmental quality standard for agricultural land (25 mg kg$^{-1}$). The safety critical value of heavy metals refers to when the concentration of heavy metals in the soil is higher than this value, indicating that the concentration of heavy metals in the edible part of leafy vegetables exceeds the limit value of the pollutants. Based on this, the three leafy vegetables planted on soil slightly polluted with As can be safely consumed. In the soil tests, after 50 days of cultivation of GC, SP, and LE on soil slightly polluted with As, the amount of As enriched in the plant was 1.18, 0.62, and 0.45, μg plant$^{-1}$, respectively (Figure 1). The World Health Organization (WHO) stipulates that the allowable intake of inorganic As for adults is about 128 μg person$^{-1}$ D$^{-1}$. For children aged 0–6 years, the short-term and long-term As exposure should be less than 15 and 5 μg person$^{-1}$ D$^{-1}$ [19]. According to this regulation, as long as the daily number of GC, SP, and LE that is consumed does not exceed 108, 206, and 284, the health of adults will not be harmed by the consumption of these leafy vegetables. For children, the leafy vegetable intake number limits of short-term As exposure of GC, SP, and LE are 13, 24, and 33, respectively, while the limits for long-term As exposure are 4, 8, and 11, respectively. This assumes that cooking does not affect the As content in the vegetables; that is, the As content in vegetables is equal to the amount of As ingested by the human body through the vegetables.

### 4.2. Tolerance Metabolism
#### 4.2.1. As Tolerance of the Leafy Vegetables

GC can endure As stress better than SP and LE. This was confirmed by our hydroponic experiment results. Growth [20] and oxidative stress [21] are two tolerance criteria widely accepted by researchers. Furthermore, oxidative stress markers (MDA, GSH/GSSG, and soluble protein) can measure the degree of oxidative stress [22,23]. MDA reflects the degree of cell membrane lipid peroxidation, while GSH/GSSG is an indicator of cellular redox potential [24]. Accordingly, the higher the proportion of GSH/GSSG, the higher the ability of plants to scavenge reactive oxygen species in the cells [25]. LE had the highest MDA content and the lowest GSH: GSSG in both the shoots and roots. The content of its soluble protein was much lower than those of the other leafy vegetables. These findings indicate that the degree of oxidative stress caused by As in LE was the highest among the three vegetables, and thus LE is an As-sensitive vegetable. As for GC, its root and shoot TI was the highest among the three vegetables (Figure 2), indicating that it is an As-tolerant variety.

#### 4.2.2. Absorption and Transport of As in Plants

In the hydroponic experiment, the absorption of As in the whole, aboveground, and root parts of GC was the highest among the three vegetables (Figure 2). In addition, the soil pot experiment also confirmed that the As bioconcentration factor and the amount of As in the edible parts of GC were higher than those of SP and LE (Figure 1C,E). More As entering GC cells means more damage [21]. However, GC had the highest As TI. This means that the As uptake and accumulation ability of plants does not completely determine the tolerance of plants [20,24]. This situation is also clearly reflected in LE, wherein the concentration and absorption of As in the shoots and roots of LE were at the lowest level, but its TI was the lowest among the three vegetables.

The canopy is the direct foundation of plant growth because the leaves are the main organs of photosynthesis and provide the energy necessary for the growth of the entire plant. Additionally, roots are an important line of defense for heavy metals in the soil. The tolerance of plants to heavy metals is highly related to the transport of As to the shoots (TF). The As TF from the roots to shoots of GC was significantly lower than that of SP and LE (Figure 4). Similar conclusions were found in a study on *Oenothera odorata* [21].

4.2.3. Enzyme Activity of the Antioxidant System

Heavy metal ions usually cause harm by interfering with or replacing the function of essential nutrient elements in plants [26] or causing the accumulation of reactive oxygen radicals in cells [25]. Enzymes of the antioxidant system can effectively alleviate the accumulation of reactive oxygen species; therefore, a higher enzyme activity of the antioxidant system leads to a stronger tolerance of As in leafy vegetables. SOD is the first line of defense for converting $O_2^-$ to $H_2O_2$, while POD and CAT are mainly responsible for the clearance of $H_2O_2$ in cells [25,26]. Under Cd stress, the tolerance of different leafy vegetables is related to the change of antioxidant enzyme activity, while the As tolerance is more related to the value of antioxidant enzyme activity. This shows that the antioxidant system of leafy vegetables has different responses to Cd and As. The specific reasons and underlying mechanisms need to be further studied. Under Cd stress, the activity of superoxide dismutase, peroxidase, and catalase in the roots of LE decreased significantly. Nevertheless, these parameters in the roots of SP and LE remained steady or were even enhanced [12]. In this experiment, under As stress, the roots of GC had the highest SOD, POD, and CAT activities (Figure 5), which improved the As tolerance of this vegetable. By contrast, the LE plants had the lowest enzyme activity in both the roots and shoots. The activities of SOD and POD in the shoots of SP were higher, whereas those in the roots were lower. This indicates that its As tolerance is higher than that of LE.

4.2.4. Thiol Pool

Studies have shown that plants can alleviate heavy metal stress by immobilizing heavy metals in cell vacuoles through partitioning [26]. Plants usually produce a variety of chelating agents (PCs) to mask and reduce toxicity [13,27]. It is believed that As in plant cells is sequestered by GSH and PCs and fixed in the vacuoles for detoxification purposes [1,13,28]. In addition, the content of PCs (heavy metal chelators) can explain the difference in the As tolerance of vegetables to some extent. It appears that the high As tolerance of GC was due to its high content of PCs and their ability to chelate and fix As in the roots; thus, the transport of As from the roots to the shoots was reduced. Studies have found that heavy metal sensitive plants sequester heavy metals in the leaves rather than in the roots [29]. Our study found that the content of PCs was high in the shoots of LE but very low in the roots. However, the superiority of shoot PCs in LE did not alleviate the growth inhibition caused by As. A similar conclusion was drawn in the study of white lupin [30]. In this study, the PC content of the roots of SP were lower than that of L, which does not explain its higher tolerance. However, there is research on O. odorata suggesting that PC content is significantly positively correlated with As concentration in the roots [21]. These contradictory results indicate the diversity of plant resistance to heavy metal stress, which appears to partly depend on the plant variety.

In this study, the concentrations of GSH and PCs in the leafy vegetables showed the opposite trend, i.e., the higher the content of GSH, the lower the content of PC (Figure 7E–H). The decrease in the GSH concentration may be due to the synthesis of more PCs. Heavy metal induces the expression of intracellular phytochelatin synthase and other related genes [31] and forms a thiol pool according to the following sequence cys→γ-EC→GSH→PCs [1,13]. Research has found that the high content of the thiol pool induced by high SP treatment in the roots increased As accumulation [15]. Under As treatment, the content of cys in the roots of LE exhibited a downward trend, and there were no significant changes in the content of cys in the roots and shoots of the other leafy vegetables (Figure 7A,B). This finding shows that the content of cys in the leafy vegetables was sufficient, and As does not cause a big change. Under As treatment, the γ-EC content in the shoots and roots of GC did not fluctuate significantly, while the γ-EC content in the roots of LE and SP decreased (Figure 7C,D). This may be because GC has more abundant γ-EC than SP and LE for the synthesis of GSH.

## 5. Conclusions

Screening for leafy vegetable species and varieties suitable for soils slightly polluted with heavy metals is beneficial for soil utilization. The present results showed that GC, SP, and LE can be planted in soil slightly polluted by As and can be consumed safely. The soil safety thresholds of As in GC, SP, and LE were 91.7, 76.2, and 80.7 mg kg$^{-1}$, which are much higher than the soil environmental quality standard in agricultural land. GC is As tolerant, and LE is As sensitive. In addition, both the antioxidant enzyme system and thiol pool were found to play an important role in the As stress of leafy vegetables.

**Author Contributions:** Conceptualization, Y.M. and L.Z.; methodology, Y.M. and L.Z.; software, L.Z. and Y.-B.R.; validation; Y.-B.R.; formal analysis, X.-B.O. and Z.-L.Y.; original draft preparation, Y.M. and L.Z.; writing—reviewing and editing, Y.M.; funding acquisition, Y.M. and Z.-L.Y.; supervision, L.-Q.W. All authors have read and agreed to the published version of the manuscript.

**Funding:** Funding was received from the National Natural Science Foundation of China (Grant No. 32001200), Gansu Province Science & Technology Foundation (20JR10RA136) and the Doctoral Fund Project of Longdong University. The Special Program of Central Government Guiding Local Science & Technology Development (QNKB2-2) and Gansu Province Key Research & Development Program (20YF3NA019) also provided financial support for this work.

**Institutional Review Board Statement:** Not applicable.

**Informed Consent Statement:** Not applicable.

**Data Availability Statement:** Not applicable.

**Acknowledgments:** We thank LetPub for its linguistic assistance during the preparation of this manuscript.

**Conflicts of Interest:** The authors declare no conflict of interest.

## References

1. Mendoza-Cózatl, D.G.; Jobe, T.O.; Hauser, F.; Schroeder, J.I. Long-distance transport, vacuolar sequestration, tolerance, and transcriptional responses induced by cadmium and arsenic. *Curr. Opin. Plant Biol.* **2011**, *14*, 554–562. [CrossRef] [PubMed]
2. Jennings, A.A. Analysis of worldwide regulatory guidance values for the most commonly regulated elemental surface soil contamination. *J. Environ. Manag.* **2013**, *118*, 72–95. [CrossRef] [PubMed]
3. Valko, M.; Klaudia, J.; Rhodes, C.J.; Kuča, K.; Musílek, K. Redox- and non-redox-metal-induced formation of free radicals and their role in human disease. *Arch. Toxicol.* **2016**, *90*, 1–37. [CrossRef]
4. Luigi, A.; Francesco, B.; Raffaele, G.; Compton, M.T. Environmental pollution and risk of psychotic disorders: A review of the science to date. *Schizophr. Res.* **2017**, *181*, 55–59.
5. Chen, Y.P.; Lorraine, M.; Liu, Q.; Song, Y.; Zheng, Y.J.; Ellison, A.M.; Ma, Q.Y.; Wu, X.M. Captive pandas are at risk from environmental toxins. *Front. Ecol. Environ.* **2016**, *14*, 363–367. [CrossRef]
6. Yang, Q.L.; Wu, Z.Z.; Chen, J.L.; Liu, X.G.; Wang, W.H.; Liu, Y.W. Research Status of Phytoremediation of Heavy Metals Contaminated Soil and Prospects of Water and Fertilizer Regulating Technology. *Ecol. Environ.* **2015**, *24*, 1075–1084.
7. Chen, W.P.; Xie, T.; Li, X.N.; Wang, R.D. Thinking of Construction of Soil Pollution Prevention and Control Technology System in China. *Acta Pedol. Sin.* **2018**, *55*, 557–568.
8. Yang, Y.; Zhang, F.S.; Li, H.F.; Jiang, R.F. Accumulation of cadmium in the edible parts of six vegetable species grown in Cd-contaminated soils. *J. Environ. Manag.* **2008**, *90*, 1117–1122. [CrossRef]
9. Prasanna, K.; Saman, S. Arsenic accumulation in rice (*Oryza sativa*, L.) is influenced by environment and genetic factors. *Sci. Total Environ.* **2018**, *642*, 485–496.
10. Alexander, P.D.; Alloway, B.J.; Dourado, A.M. Genotypic variations in the accumulation of Cd, Cu, Pb and Zn exhibited by six commonly grown vegetables. *Environ. Pollut.* **2016**, *144*, 736–745. [CrossRef]
11. Zhuang, P.; McBride, M.B.; Xia, H.P.; Li, N.Y.; Li, Z. Health risk from heavy metals via consumption of food crops in the vicinity of Dabaoshan mine, South China. *Sci. Total Environ.* **2008**, *407*, 1551–1561. [CrossRef] [PubMed]
12. Meng, Y.; Zhang, L.; Wang, L.; Zhou, C.; Shangguan, Y.; Yang, Y. Antioxidative enzymes activity and thiol metabolism in three leafy vegetables under Cd stress. *Ecotoxicol. Environ. Saf.* **2019**, *173*, 214–224. [CrossRef] [PubMed]
13. Verbruggen, N.; Hermans, C.; Schat, H. Mechanisms to cope with arsenic or cadmium excess in plants. *Curr. Opin. Plant Biol.* **2009**, *12*, 364–372. [CrossRef] [PubMed]
14. Smeets, K.; Ruytinx, J.; Van Belleghem, F.; Semane, B.; Lin, D.; Vangronsveld, J.; Cuypers, A. Critical evaluation and statistical validation of a hydroponic culture system for Arabidopsis thaliana. *Plant Physiol. Biochem.* **2008**, *46*, 212–218. [CrossRef] [PubMed]

15. Garima, D.; Amit, P.S.; Amit, K.; Pradyumna, K.S.; Smita, K.; Sanjay, D.; Prabodh, K.T.; Vivek, P.; Gareth, J.N.; Om, P.D.; et al. Sulfur mediated reduction of arsenic toxicity involves efficient thiol metabolism and the antioxidant defense system in rice. *J. Hazard. Mater.* **2015**, *298*, 241–251.
16. Li, X.; Zhang, L.; Li, Y.; Ma, L.; Bu, N.; Ma, C. Changes in photosynthesis, antioxidant enzymes and lipid peroxidation in soybean seedlings exposed to UV-B radiation and/or Cd. *Plant Soil* **2012**, *352*, 377–387. [CrossRef]
17. Fernández, R.; Bertrand, A.; Reis, R.; Mourato, M.P.P.; Martins, L.L.L.; González, A.; Reis, R.; González, A.; Mourato, M.P.P.; Bertrand, A.; et al. Growth and physiological responses to cadmium stress of two populations of *Dittrichia viscosa* (L.). *J. Hazard. Mater.* **2012**, *244–245*, 555–562. [CrossRef]
18. Jobe, T.O.; Sung, D.Y.; Akmakjian, G.; Pham, A.; Komives, E.A.; Mendoza-Cózatl, D.G.; Schroeder, J.I. Feedback inhibition by thiols outranks glutathione depletion: A luciferase-based screen reveals glutathione-deficient γ-ECS and glutathione synthetase mutants impaired in cadmium-induced sulfate assimilation. *Plant J.* **2012**, *70*, 783–795. [CrossRef]
19. Joyce, S.T.; Robert, B.; Rosalind, A.S. Health effect levels for risk assessment of childhood exposure to arsenic. *Regul. Toxicol. Pharmacol.* **2004**, *39*, 99–110.
20. Dubey, A.K.; Kumar, N.; Sahu, N.; Verma, P.K.; Chakrabarty, D.; Behera, S.K.; Mallick, S. Response of two rice cultivars differing in their sensitivity towards arsenic, differs in their expression of glutaredoxin and glutathione S transferase genes and antioxidant usage. *Ecotoxicol. Environ. Saf.* **2016**, *124*, 393–405. [CrossRef]
21. Kim, D.Y.; Park, H.; Lee, S.H.; Koo, N.; Kim, J.G. Arsenate tolerance mechanism of Oenothera odorata from a mine population involves the induction of phytochelatins in roots. *Chemosphere* **2009**, *75*, 505–512. [CrossRef] [PubMed]
22. Dong, W.Q.; Sun, H.J.; Zhang, Y.; Lin, H.J.; Chen, J.R.; Hong, H.C. Impact on growth, oxidative stress, and apoptosis-related gene transcription of zebrafish after exposure to low concentration of arsenite. *Chemosphere* **2018**, *211*, 648–652. [CrossRef] [PubMed]
23. Sarkar, S.; Mukherjee, S.; Chattopadhyay, A.; Bhattacharya, S. Differential modulation of cellular antioxidant status in zebrafish liver and kidney exposed to low dose arsenic trioxide. *Ecotoxicol. Environ. Saf.* **2017**, *135*, 173–182. [CrossRef] [PubMed]
24. Panda, A.; Rangani, J.; Kumari, A.; Parida, A.K. Efficient regulation of arsenic translocation to shoot tissue and modulation of phytochelatin levels and antioxidative defense system confers salinity and arsenic tolerance in the Halophyte Suaeda maritima. *Environ. Exp. Bot.* **2017**, *143*, 149–171. [CrossRef]
25. Mittler, R. Oxidative stress, antioxidants and stress tolerance. *Trends Plant Sci.* **2002**, *7*, 405–410. [CrossRef]
26. Sytar, O.; Kumar, A.; Latowski, D.; Kuczynska, P.; Strzałka, K.; Prasad, M.N. Heavy metal-induced oxidative damage, defense reactions, and detoxification mechanisms in plants. *Acta Physiol. Plant.* **2013**, *35*, 985–999. [CrossRef]
27. Clemens, S. Toxic metal accumulation, responses to exposure and mechanisms of tolerance in plants. *Biochimie* **2006**, *88*, 1707–1719. [CrossRef]
28. Khare, R.; Kumar, S.; Shukla, T.; Ranjan, A.; Trivedi, P.K. Differential sulphur assimilation mechanism regulates response of Arabidopsis thaliana natural variation towards arsenic stress under limiting sulphur condition. *J. Hazard. Mater.* **2017**, *337*, 198–207. [CrossRef]
29. Čabala, R.; Slováková, L.; Zohri, M.E.; Frank, H. Accumulation and translocation of Cd metal and the Cd-induced production of glutathione and phytochelatins in *Vicia faba* L. *Acta Physiol. Plant.* **2011**, *33*, 1239–1248. [CrossRef]
30. Va'zquez, S.; Goldsbrough, P.; Carpena, R.O. Comparative analysis of the contribution of phytochelatins to cadmium and arsenic tolerance in soybean and white lupin. *Plant Physiol. Biochem.* **2009**, *47*, 63–67. [CrossRef]
31. Kumari, A.; Pandey, N.; Pandey-Rai, S. Protection of Artemisia annua roots and leaves against oxidative stress induced by arsenic. *Biol. Plant.* **2017**, *61*, 367–377. [CrossRef]

Article

# Effect of Soil Solution Properties and Cu²⁺ Co-Existence on the Adsorption of Sulfadiazine onto Paddy Soil

Ziwen Xu [1], Shiquan Lv [1], Shuxiang Hu [1], Liang Chao [2], Fangxu Rong [1], Xin Wang [1], Mengyang Dong [1], Kai Liu [2], Mingyue Li [2] and Aiju Liu [2,*]

[1] School of Agricultural Engineering and Food Science, Shandong University of Technology, Zibo 225049, China; 17864388719@163.com (Z.X.); lvshiquan24@163.com (S.L.); HSX1419791336@163.com (S.H.); rong3557622763@163.com (F.R.); wangxinsdlg@163.com (X.W.); Dongmengyer@163.com (M.D.)

[2] School of Resources and Environmental Engineering, Shandong University of Technology, Zibo 255049, China; chaoliang718@163.com (L.C.); kliu@sdut.edu.cn (K.L.); myli@sdut.edu.cn (M.L.)

* Correspondence: aijvliu@sdut.edu.cn

**Abstract:** Paddy soils are globally distributed and saturated with water long term, which is different from most terrestrial ecosystems. To better understand the environmental risks of antibiotics in paddy soils, this study chose sulfadiazine (SDZ) as a typical antibiotic. We investigated its adsorption behavior and the influence of soil solution properties, such as pH conditions, dissolved organic carbon (DOC), ionic concentrations (IC), and the co-existence of $Cu^{2+}$. The results indicated that (1) changes in soil solution pH and IC lower the adsorption of SDZ in paddy soils. (2) Increase of DOC facilitated the adsorption of SDZ in paddy soils. (3) $Cu^{2+}$ co-existence increased the adsorption of SDZ on organic components, but decreased the adsorption capacity of clay soil for SDZ. (4) Further FTIR and SEM analyses indicated that complexation may not be the only form of $Cu^{2+}$ and SDZ co-adsorption in paddy soils. Based on the above results, it can be concluded that soil solution properties and co-existent cations determine the sorption behavior of SDZ in paddy soils.

**Keywords:** sulfadiazine; $Cu^{2+}$ co-existence; paddy soils; adsorption; soil properties

**Citation:** Xu, Z.; Lv, S.; Hu, S.; Chao, L.; Rong, F.; Wang, X.; Dong, M.; Liu, K.; Li, M.; Liu, A. Effect of Soil Solution Properties and Cu²⁺ Co-Existence on the Adsorption of Sulfadiazine onto Paddy Soil. *Int. J. Environ. Res. Public Health* **2021**, *18*, 13383. https://doi.org/10.3390/ijerph182413383

Academic Editor: Paul B. Tchounwou

Received: 10 November 2021
Accepted: 17 December 2021
Published: 19 December 2021

**Publisher's Note:** MDPI stays neutral with regard to jurisdictional claims in published maps and institutional affiliations.

**Copyright:** © 2021 by the authors. Licensee MDPI, Basel, Switzerland. This article is an open access article distributed under the terms and conditions of the Creative Commons Attribution (CC BY) license (https://creativecommons.org/licenses/by/4.0/).

## 1. Introduction

Overuse and the uncontrolled disposal of antibiotics have caused severe environmental problems, especially for the soil environment [1–3]. As one of the main crops in the world, rice is widely planted and feeds the majority of the world's population; while over 92% of rice production, as a primary staple food, is in Asia [4]. In China, the planting area of rice is near 30 million ha, which occupies over 20% of the total farmlands [5]. Therefore, paddy soils have a high chance of exposure to antibiotic pollutants, such as sulfonamides [1]. However, most studies focused on the adsorption and transport behaviors of antibiotics in upland farmland, while few investigations were about the behavior of antibiotics in paddy soils.

The co-existence of heavy metals and antibiotic pollution in the soil environment is receiving more and more attention. With their wide antibacterial spectrum, excellent curative effect, and low cost, sulfonamide antibiotics (SAs) are widely used in disease treatment and prevention for humans and animals [6]. Sulfadiazine (SDZ), one of the most commonly used SA chemicals, has been widely detected in various environmental mediums, especially in soils [1,7,8]. Previous studies proved that Cu ions could coordinate with SDZ [9–11], altering their molecular speciation and environmental behavior [11,12]. For example, it was reported that $Cu^{2+}$ co-addition could improve the adsorption of antibiotics to organic matters [13] and soils [14]. On the other hand, there was competitive adsorption between co-existent metal ions and the ionizable antibiotics, which inhibited the adsorption ability of antibiotics on soil components [15,16] and increase their transportation risk in the environment [17]. However, limited investigations have been conducted on the effect of $Cu^{2+}$ on SA sorption in paddy soils.

To obtain more knowledge about the environmental risk of SAs in paddy soils, we chose Sulfadiazine (SDZ) as a representative antibiotic. The batch sorption experiments were conducted in soil suspensions with different pH, ion concentrations (IC), and dissolved organic carbon (DOC), as well as co-existent $Cu^{2+}$. The aim of this study was: (1) to explore the sorption kinetics and isotherm of SDZ in paddy soils under different soil solution conditions; (2) to characterize the sorption behavior and mechanisms of SDZ on various soil components; and (3) to investigate the effects of co-existent $Cu^{2+}$ on the sorption of SAs in paddy soils.

## 2. Materials and Method

### 2.1. Paddy Soil Sampling and Characterization

Five soil samples were collected from the rice production area located in Huang gang, Hubei, of China. About 20 kg of topsoil (0–10 cm) was collected from sample sites occupying an area of approximately 500 m$^2$ and transported to the lab in a cabinet with an ice pack. After thorough mixing, soils were air-dried, ground, and sieved through a 2-mm sieve for the following soil property analysis and sorption batch experiments. The soil texture was silty loam with 2.55% clay, 89.67% silt, and 7.79% sand. Soil pH, organic matter (OM), cation exchange capacity (CEC), and Zeta potential are listed in Table 1.

**Table 1.** Physical and chemical properties of paddy soils in this study.

| Soil Components | pH | OM [1] (mg g$^{-1}$) | CEC [2] (cmol kg$^{-1}$) | Zeta-Potential ($\zeta$, mV) |
|---|---|---|---|---|
| Crude soil | 6.8 | 23.68 ± 2.46 | 5.84 ± 0.11 | −33.79 ± 2.06 |
| Organic particle | 7.3 | 50.55 ± 3.12 | 8.15 ± 0.19 | −59.99 ± 5.25 |
| Soil Clay | 5.3 | – | 3.84 ± 0.04 | −64.55 ± 3.35 |

[1], OM: organic matter; [2], CEC: cation exchange capacity.

### 2.2. Chemicals and Reagents

In this study, sulfadiazine (4-amino-N-5-methylisoxazol-3-yl)-benzene sulfonamide, (CAS: 723-46-6, purity >98%, MW 250 g mol$^{-1}$) was purchased from Sigma-ALDRICH (Shanghai, China), whose lgK$_{ow}$ and pKa values were −0.09, 2.00 (pK$_{a1}$) and 6.50 (pK$_{a2}$), respectively [18]. SDZ was first dissolved in methanol (grade: HPLC, Sigma, Shanghai, China) and diluted to 100 mg kg$^{-1}$, with ultra-pure water used for a stock solution. $Cu^{2+}$ stock solution (10 g·L$^{-1}$) was prepared from $CuCl_2·2H_2O$ (min. purity 99.99%, Sinopharm, Shanghai, China) in ultra-pure water. All the other chemicals, including Hydrochloric acid (HCl) (Sinopharm, Shanghai, China), sodium hydroxide (NaOH) (Sinopharm, Shanghai, China), calcium chloride ($CaCl_2$) (Sinopharm, Shanghai, China), and ammonium dihydrogen phosphate (ADP) (Sinopharm, Shanghai, China), etc., used in this study were analytical grade, apart from acetonitrile (ACN) (Sigma, Shanghai, China), which was HPLC grade and used for the HPLC analysis of SDZ.

### 2.3. Preparation of Soil Organic Particles and Soil Clay

In this study, the soil organic particle fraction was obtained with the wet sieving method suggested by Elliott [19]. Briefly, put 500 g of air-dried soil on a sieve of 2 mm, and separate aggregates by shaking the sieve up and down for 2 min with 50 repetitions, after being submerged in water for 5 min. Aggregates passed through a 0.25-mm sieve were collected and designated as a soil organic particle fraction. The physical and chemical properties of the prepared organic particles are listed in Table 1.

The soil clay fraction was extracted with 0.1 M $Na_4P_2O_7$ (pH = 7) and 0.1 M $H_2O_2$ according to [20]. After being treated with $Na_4P_2O_7$ (Sigma, Shanghai, China) and 0.1 M $H_2O_2$ (Sinopharm, Shanghai, China), the sample mixture was centrifuged and separated into three distinct layers. After removing the top layer using suction, the residual material was separated into two fractions by repeat centrifuging. After removing the top liquid layer, recover the clay mineral fraction of 1 M NaCl (Sinopharm, Shanghai, China), after repeat centrifuging. The physical and chemical properties of the prepared clay are listed in Table 1.

### 2.4. Batch Sorption Experiment

Sorption experiments were conducted in 100-mL plastic centrifuge tubes with Teflon lids, according to the OECD 106 method and Jiang [21]. Briefly, weigh 1.000 g dry paddy soil into 25 mL of 0.01 M $CaCl_2$ (Sinopharm, Shanghai, China) containing a specific concentration of SDZ. All sorption experiments were carried out with an oscillator (Changzhou Guohua Electric Appliance Co., Ltd., Changzhou, China) at 150 rpm in darkness. The sample tubes containing an initial concentration of 8 $\mu mol\ L^{-1}$ SDZ were shaken from 0 h to 72 h for the kinetic sorption experiment, and those containing a series of initial concentrations of 4 $\mu mol\ L^{-1}$, 8 $\mu mol\ L^{-1}$, 12 $\mu mol\ L^{-1}$, and 16 $\mu mol\ L^{-1}$ were shaken for 72 h for the isotherm experiment. At the end of the sorption experiments, 2 mL supernatant was collected from each tube and centrifuged at 10000 r for 1 min. Then, the supernatant was filtered through a 0.45 $\mu m$ Whatman filter (Whatman$^{TM}$, Germany) for the subsequent quantification with HLPC. Each experiment was conducted in triplicates, and a blank treatment of SDZ solution without soil was used as a control, which was used to evaluate the loss of SDZ caused by sorption onto the tube walls and its degradation.

*Environmental factors:* The pH of soil solution was adjusted to 5.0, 7.0, and 9.0, respectively, with 1 M HCl (Sinopharm, Shanghai, China) and NaOH (Sinopharm, Shanghai, China) to analyze the influence of the medium pH. Fulvic acid (FA, $\geq$90%, CAS: 1415-93-6, Aladdin, Shanghai, China) was added at rates of 1 $g\ L^{-1}$, 3 $g\ L^{-1}$, and 5 $g\ L^{-1}$ to the soil solution to analyze the effect of soil DOC on the adsorption of SDZ on soils. At the same time, the effect of solution ionic concentration was evaluated with $Ca^{2+}$ ions at the concentrations of 0.05 M and 0.1 M. The sorption characteristics of SDZ on various soil fractions was tested with the same procedure as the above soils, and the solid phase was replaced with the soil organic particles and clays prepared in 1.3.

*Effect of co-existent $Cu^{2+}$:* For the sorption experiments of SDZ with $Cu^{2+}$ co-existence, the added concentrations of $Cu^{2+}$ were 200 and 500 $mg\ L^{-1}$, respectively. The others were the same as the soil sorption experiment.

### 2.5. Analysis Method

The pH, CEC, and OM of the soil samples were analyzed according to the description of Lu (1999) [22]. The zeta potentials of the crude soil, organic particle fraction, and soil clay fraction were measured with a zeta potential analyzer (JS94H, Shanghai Zhongchen Digital Technic Apparatus Co., Ltd., Shanghai, China) after being dispersed in a solution of 0.01 M NaCl (Sinopharm, Shanghai, China).

*HLPC analysis of SDZ:* SDZ concentration in the filtrate was quantified at 270 nm by a HPLC/UV (Agilent/Bruker HP1100/Esqure2000, Agilent/Bruker Co., Ltd., Walter cloth, Germany) using a C18 column. The mobile phase was ADP (0.01 M, pH = 2.8) (Sinopharm, Shanghai, China): CAN (Sinopharm, Shanghai, China ) (80:20, $v$:$v$) at a flow rate of 0.5 mL $min^{-1}$. The injection volume was 10 $\mu L$.

The concentration of SDZ was increased to 40 $\mu mol\ L^{-1}$ in the soil solutions and they were then centrifuged after the adsorption reached equilibrium. The centrifuged samples were washed three times with ultra-water, dried, and powdered with a mortar and pestle for the subsequent scanning with FTIR spectrum and SEM visualization. The FTIR scanning was conducted with a FTIR Microscope-Spectrometer (Nicolet5700, Thermo, Shanghai, China) in the range 4000–400 $cm^{-1}$. The microstructure observation of soil samples was conducted with a SEM (Quanta250, FEI, Shanghai, China) at magnifications of ×20,000.

## 2.6. Data Calculation

The amount of adsorbed SDZ was calculated using the following equation:

$$Y = \frac{V(C_0 - C_e)}{M} \tag{1}$$

where $Y$ is the absorbed amount of SDZ (μmol kg$^{-1}$); $V$ (L) is the volume of soil solution; $C_0$ and $C_e$ (μmol L$^{-1}$) represent the concentration of SDZ at the beginning and end of the equilibrium sorption experiments in solution, respectively; and $M$ (kg) is the dry soil weight added to the background solution.

The sorption kinetics of SDZ in paddy soils in various soil solutions were fitted with the following kinetic model equations:

$$q_t = q_e\left(1 - e^{-k_1 t}\right) \tag{2}$$

$$q_t = \frac{k_2 q_e^2 t}{1 + k_2 q_e t} \tag{3}$$

The sorption isotherm of SDZ in paddy soils in various soil solutions were fitted with the Linear model equation (4) and Freundlich model equation (5):

$$q_e = K_d C_{eq} \tag{4}$$

$$q_e = K_F C_E^{1/n} \tag{5}$$

For Equations (2)–(5), $q_t$ (μmol kg$^{-1}$) is the amount of SDZ sorption to paddy soil at time t (h); $q_e$ (μmol kg$^{-1}$) is the amount of SDZ sorption to paddy soil when sorption reaches the equilibrium; and $k_1/k_2$ is the constant of the kinetics sorption velocity. $C_e$ (μmol L$^{-1}$) is the concentration of SDZ in the soil supernatant when sorption reaches equilibrium; $K_d$ (L kg$^{-1}$) is the coefficient of SDZ distribution between the liquid and solid phases in the equilibrium system; $K_F$ is the Freundlich sorption coefficient; and $n$ is the nonlinearity factor.

## 3. Results

### 3.1. Effect of Soil Solution Properties on the Sorption of SDZ

Batch sorption experiments were conducted to analyze the sorption characteristics of SDZ on paddy soils under different experimental systems. The results are shown in Figure 1.

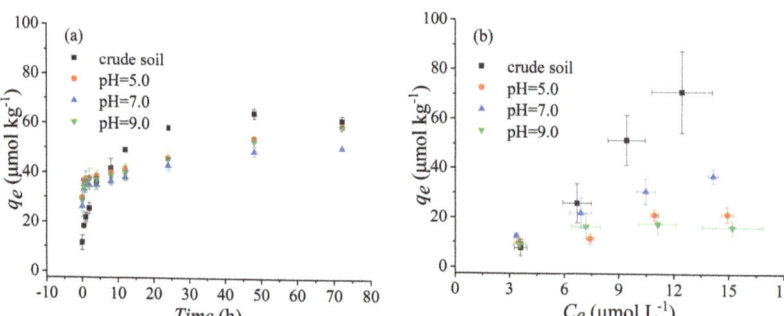

**Figure 1.** Cont.

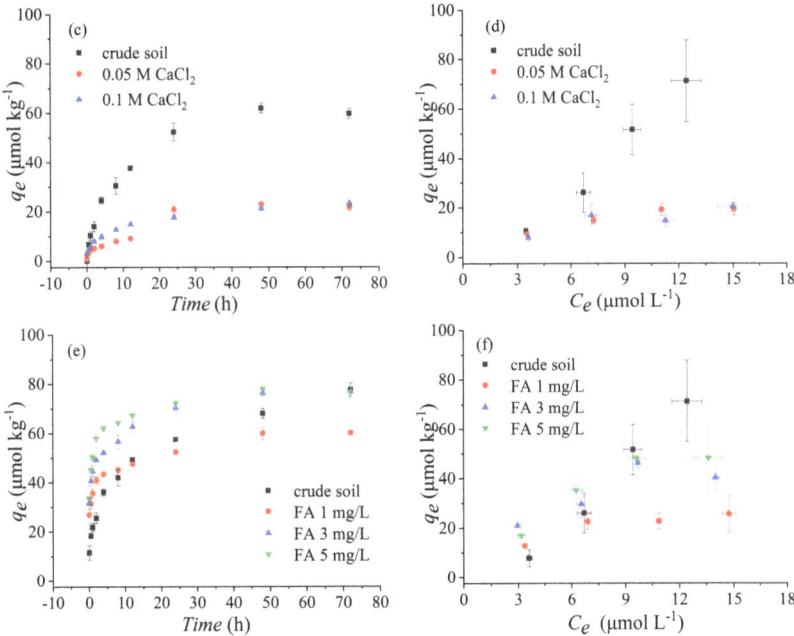

**Figure 1.** The sorption kinetics (**a,c,e**), and isotherm (**b,d,f**) of SDZ in soil solution with different pH, IC, and DOM conditions.

As plotted in Figure 1b, the adsorption capacity of SDZ in crude soils was decreased, with the pH changing to 5.0, 7.0, and 9.0, and $q_e$ was over 60 µmol kg$^{-1}$ in crude soil but lower than 50 µmol kg$^{-1}$ with the pH change (Table 2). As for the ion strength, the adsorption capacity of SDZ on paddy soils was significantly reduced, to about 20 µmol kg$^{-1}$, when the CaCl$_2$ concentration increased to 0.05 M and 0.1 M CaCl$_2$ (Figure 1d). The variation in the fitting parameters of the adsorption isotherms (Table 3) indicated that the soil solution pH and ion strength could change the sorption affinity of SDZ in paddy soil. However, there was a significant increase in SDZ adsorption capacity with increasing soil solution DOM (Figure 1f and Table 2). In addition, the values of K$_1$ and K$_2$ were enhanced with the increasing of pH and DOM (Table 2), which indicated that the increase of pH and DOM accelerated the reaction process and increased the velocity of SDZ adsorption on paddy soil (Figure 1a,c,e). Whatever the changes in soil solution properties in the present study, the adsorptions of SDZ had a good fit with the pseudo-first, pseudo-second-order, and the Freundlich equation, as the R parameter of each selected fitting model was always over 0.99 at a level of $p < 0.05$ (list in Tables 2 and 3). This might suggest that there was more than one dominant mechanism responsible for the adsorption of SDZ.

**Table 2.** Fitting results of kinetics data to pseudo-first and pseudo-second-order equations for SDZ adsorption in various soil solutions (mean values ± standard error).

| Treatments | The Pseudo-First-Order | | | The Pseudo-Second-Order | | |
|---|---|---|---|---|---|---|
| | $q_e$ (µmol kg$^{-1}$) | $k_1$ | R | $q_e$ (µmol kg$^{-1}$) | $k_2$ | R |
| Crude soil | 65.84 ± 0.36 | 1.40 ± 0.59 | 0.9980 | 69.48 ± 0.76 | 0.02 ± 0.00 | 0.9901 |
| pH 5.0 | 60.16 ± 0.37 | 0.19 ± 0.02 | 0.9990 | 65.72 ± 0.32 | 0.02 ± 0.00 | 0.9996 |
| pH 7.0 | 41.56 ± 0.30 | 1.69 ± 0.69 | 0.9986 | 43.96 ± 0.33 | 0.20 ± 0.07 | 0.9989 |
| pH 9.0 | 45.72 ± 0.40 | 1.04 ± 0.37 | 0.9977 | 48.40 ± 0.45 | 0.13 ± 0.05 | 0.9983 |

**Table 2.** Cont.

| Treatments | The Pseudo-First-Order | | | The Pseudo-Second-Order | | |
|---|---|---|---|---|---|---|
| | $q_e$ (μmol kg$^{-1}$) | $k_1$ | R | $q_e$ (μmol kg$^{-1}$) | $k_2$ | R |
| FA 1 mg L$^{-1}$ | 52.08 ± 0.33 | 0.33 ± 0.15 | 0.9986 | 74.24 ± 0.39 | 0.04 ± 0.01 | 0.9990 |
| FA 3 mg L$^{-1}$ | 69.24 ± 0.44 | 0.45 ± 0.08 | 0.9979 | 74.24 ± 0.39 | 0.04 ± 0.01 | 0.9990 |
| FA 5 mg L$^{-1}$ | 71.12 ± 0.29 | 0.29 ± 0.13 | 0.9989 | 74.52 ± 0.25 | 0.07 ± 0.01 | 0.9995 |
| 0.05 M CaCl$_2$ | 22.88 ± 0.27 | 0.06 ± 0.01 | 0.9997 | 28.28 ± 0.49 | 0.49 ± 0.01 | 0.9998 |
| 0.1 M CaCl$_2$ | 19.68 ± 0.20 | 0.16 ± 0.02 | 0.9999 | 22.00 ± 0.19 | 0.19 ± 0.01 | 0.9999 |

**Table 3.** Fitting results of the Linear and Freundlich models for adsorption curves of SDZ in various soil solutions (mean values ± standard error).

| Treatments | Linear | | Freundlich | | |
|---|---|---|---|---|---|
| | $k_d$ (L kg$^{-1}$) | R | $K_F$ (μmol$^{1-N}$ L$^N$ kg$^{-1}$) | $n$ | R |
| Crude soil | 2.17 ± 0.47 | 0.9537 | 18.56 ± 1.95 | 1.48 ± 0.09 | 0.9990 |
| pH 5.0 | 1.65 ± 0.11 | 0.9743 | 12.57 ± 0.75 | 1.22 ± 0.26 | 0.9755 |
| pH 7.0 | 1.82 ± 0.15 | 0.9841 | 17.47 ± 0.86 | 1.32 ± 0.46 | 0.9952 |
| pH 9.0 | 1.42 ± 0.17 | 0.9251 | 11.85 ± 0.66 | 1.32 ± 0.42 | 0.9578 |
| FA 1 mg L$^{-1}$ | 2.07 ± 0.21 | 0.9436 | 24.46 ± 0.68 | 1.89 ± 0.20 | 0.9644 |
| FA 3 mg L$^{-1}$ | 3.70 ± 0.33 | 0.9540 | 25.56 ± 0.73 | 1.89 ± 0.86 | 0.9925 |
| FA 5 mg L$^{-1}$ | 4.22 ± 0.36 | 0.9589 | 28.45 ± 0.96 | 1.96 ± 0.30 | 0.9950 |
| 0.05M CaCl$_2$ | 1.48 ± 0.13 | 0.9538 | 10.86 ± 0.35 | 1.90 ± 0.40 | 0.9772 |
| 0.1 M CaCl$_2$ | 4.22 ± 0.36 | 0.9589 | 10.90 ± 0.54 | 1.27 ± 0.40 | 0.9454 |

### 3.2. Effect of $Cu^{2+}$ Co-Existing on the Sorption of SDZ

The adsorption kinetics and isotherms of SDZ were compared for the crude soil, organic particles, and clay with the presence or absence of $Cu^{2+}$ (Figure 2). Whether $Cu^{2+}$ was present or not, the adsorption of SDZ on the three soil fractions could reach equilibrium within about 48 h (Figure 2a,c,e). However, with the co-existence of $Cu^{2+}$, the adsorption capacity of SDZ on the crude soil and the organic fraction was significantly increased, which did not happen with the clay (Figure 2f). The values of $K_F$ and $n$ were also enhanced with $Cu^{2+}$ co-addition for the crude soils and organic particles. Morover, with the co-existence of $Cu^{2+}$, the adsorption characteristics of SDZ on each soil fraction still fit well with the selected kinetic equations and isotherm equations, as most of the coefficients of R were >0.99 (Tables 4 and 5), although the adsorption of SDZ was decreased for clay soils (Table 5). Moreover, the subsequent linear model fitting also found that the adsorption affinity ($k_d$, Table 5) of clay to SDZ was much lower than the crude soil and organic particles, and the presence of Cu further decreased this adsorption affinity. Nevertheless, this was increased for the crude soil and organic particles.

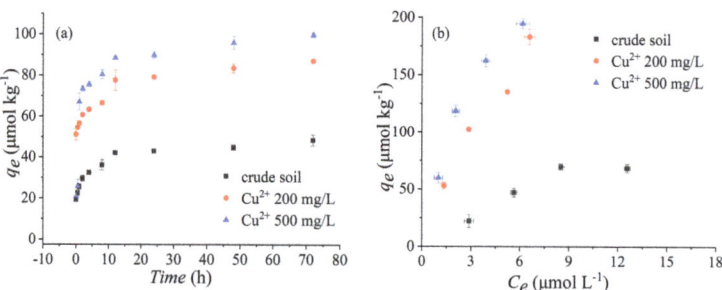

**Figure 2.** Cont.

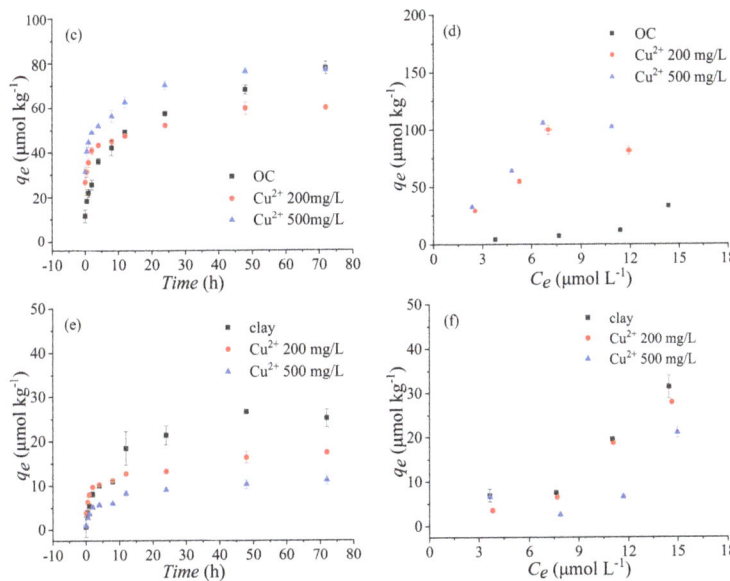

**Figure 2.** The sorption kinetics (**a,c,e**) and isotherms(**b,d,f**) of SDZ in the crude soil, organic particles, and clay with $Cu^{2+}$ co-existing in the soil solution.

**Table 4.** Fitting results of kinetics data to pseudo-first and pseudo-second-order equations for SDZ adsorption on soil composition.

| Treatments | The Pseudo-First-Order | | | The Pseudo-Second-Order | | |
|---|---|---|---|---|---|---|
| | $qe$ (μmol $L^{-1}$) | $k_1$ | $R$ | $qe$ (μmol $L^{-1}$) | $k_2$ | $R$ |
| Crude soil | 42.76 ± 0.26 | 0.54 ± 0.09 | 0.9992 | 45.64 ± 0.23 | 0.07 ± 0.01 | 0.9996 |
| $Cu^{2+}$ 200 mg $L^{-1}$ | 77.92 ± 0.42 | 0.76 ± 0.18 | 0.9977 | 82.40 ± 0.40 | 0.06 ± 0.01 | 0.9988 |
| $Cu^{2+}$ 500 mg $L^{-1}$ | 91.16 ± 0.37 | 0.58 ± 0.10 | 0.9984 | 93.92 ± 0.71 | 0.07 ± 0.01 | 0.9178 |
| Organic particles | 70.64 ± 0.75 | 0.09 ± 0.01 | 0.9974 | 81.64 ± 0.97 | 0.01 ± 0.00 | 0.9982 |
| $Cu^{2+}$ 200 mg $L^{-1}$ | 79.52 ± 0.51 | 0.34 ± 0.06 | 0.9975 | 104.56 ± 0.49 | 0.03 ± 0.01 | 0.9987 |
| $Cu^{2+}$ 500 mg $L^{-1}$ | 81.60 ± 0.40 | 0.39 ± 0.07 | 0.9984 | 106.08 ± 0.36 | 0.04 ± 0.01 | 0.9992 |
| Clay | 19.40 ± 0.11 | 0.25 ± 0.04 | 0.9999 | 20.36 ± 0.11 | 0.14 ± 0.03 | 1.0000 |
| $Cu^{2+}$ 200 mg $L^{-1}$ | 15.24 ± 0.26 | 0.09 ± 0.01 | 0.9997 | 18.96 ± 0.36 | 0.02 ± 0.00 | 0.9997 |
| $Cu^{2+}$ 500 mg $L^{-1}$ | 13.76 ± 0.14 | 0.66 ± 0.14 | 0.9998 | 15.08 ± 0.13 | 0.22 ± 0.05 | 0.9999 |

**Table 5.** Fitting results of the Linear and Freundlich models for adsorption curves of SDZ on soil composition (mean values ± standard error).

| Treatments | Linear | | Freundlich | | |
|---|---|---|---|---|---|
| | $k_d$ (L $kg^{-1}$) | $R$ | $K_F$ (μmol$^1$ $L^N$ $kg^{-1}$) | $n$ | $R$ |
| Crude soil | 6.57 ± 0.38 | 0.9801 | 18.65 ± 1.56 | 1.45 ± 0.29 | 0.9992 |
| $Cu^{2+}$ 200 mg $L^{-1}$ | 28.62 ± 1.09 | 0.9914 | 35.76 ± 2.34 | 1.20 ± 0.12 | 1.0000 |
| $Cu^{2+}$ 500 mg $L^{-1}$ | 36.62 ± 2.62 | 0.9706 | 40.68 ± 2.66 | 1.87 ± 0.25 | 1.0000 |
| Organic particles | 1.69 ± 0.20 | 0.9266 | 15.43 ± 1.35 | 1.46 ± 0.06 | 0.9938 |
| $Cu^{2+}$ 200 mg $L^{-1}$ | 9.05 ± 0.94 | 0.9408 | 30.21 ± 1.98 | 1.81 ± 0.32 | 0.9987 |
| $Cu^{2+}$ 500 mg $L^{-1}$ | 11.55 ± 0.85 | 0.9692 | 32.53 ± 2.01 | 1.13 ± 0.30 | 0.9996 |

Table 5. Cont.

| Treatments | Linear | | Freundlich | | |
|---|---|---|---|---|---|
| | $k_d$ (L kg$^{-1}$) | R | $K_F$ (μmol$^1$ L$^N$ kg$^{-1}$) | n | R |
| Clay | 1.87 ± 0.13 | 0.9725 | 13.69 ± 0.96 | 1.63 ± 0.08 | 0.9940 |
| Cu$^{2+}$ 200 mg L$^{-1}$ | 1.66 ± 0.12 | 0.9715 | 12.56 ± 0.58 | 1.54 ± 0.04 | 0.9979 |
| Cu$^{2+}$ 500 mg L$^{-1}$ | 1.00 ± 0.15 | 0.8920 | 8.57 ± 0.35 | 1.50 ± 0.14 | 0.9445 |

### 3.3. Co-Adsorption Mechanism of Cu and SDZ on Different Soil Components

Figure 3 shows the FTIR spectra of soil OC and clay particles with/without SDZ/Cu adsorption. Compared to the treatment without SDZ, attenuation was observed at wavenumbers from 1050 to 950 cm$^{-1}$ (C-O-C) on the OC particles after SDZ adsorption. Compared to the single adsorption of SDZ, a stretching vibration in the range 1700–1600 cm$^{-1}$ (C=C or -NH) in OC components became stronger with Cu co-addition. These results suggest that the -NH, C-O-C, or C=C groups play an important role in the SDZ adsorption to soil particles [23]. For the clays, a weak peak, ranging from 3000 cm$^{-1}$ to 2900 cm$^{-1}$ (-CH), occurred after SDZ adsorption, and this signal became stronger in clay particles with the co-presence of Cu. Moreover, an attenuation of the wavenumbers from 1700 to 1600 cm$^{-1}$ and an obvious -OH stretching vibration (3678.0 cm$^{-1}$) were observed for the Cu-SDZ co-adsorption, compared to the single adsorption of SDZ. This might suggest that hydrogen bonding plays an essential role in the co-adsorption of Cu and SDZ on soil clays.

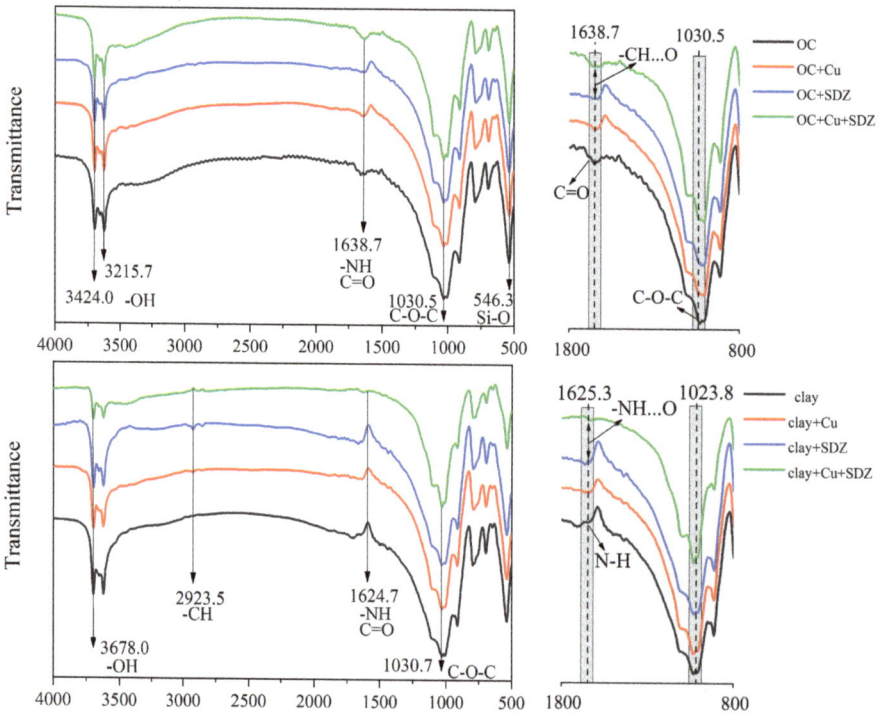

**Figure 3.** The FTIR spectra of OC and clay before and after the reaction with SDZ and Cu$^{2+}$.

The surface morphologies of the soil OC and clay particles were scanned using SEM (Figure 4). The SEM images show that soil OC particles surface was rougher than the clay soils, with an irregular and bulky shape. However, there was no obvious change in the surface morphology of the OC and clay particles with the sorption of Cu and SDZ. According to the mapping results, the Cu was very hard to detect in OC and clay without $Cu^{2+}$ treatment, but it covered the surface of the samples treated with $Cu^{2+}$, with the signal becoming stronger for the combined organic soil treated with SDZ. This indicated that $Cu^{2+}$ adsorbs easily on organic soil and clays, and SDZ co-addition increases the sorption of Cu on organic soil. From the above results of the FTIR analysis and the batch adsorption experiments, it was indicated that co-adsorption by the complexation of $Cu^{2+}$-SDZ is one method of $Cu^{2+}$ and SDZ adsorption to soil particles. While the co-adsorption of $Cu^{2+}$ and SDZ might mainly be conducted by their independent adsorption to different sites on the soil particles.

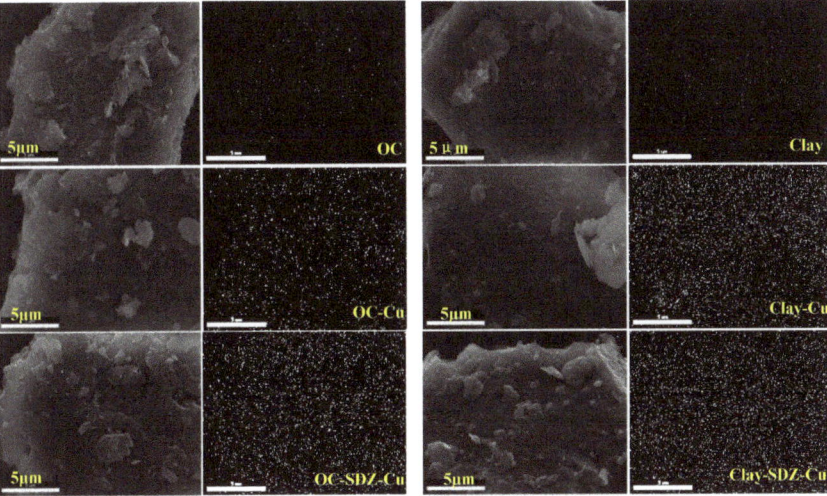

**Figure 4.** SEM images of OC and clay before and after reaction in $Cu^{2+}$ and SDZ, elemental mappings showing the distributions of Cu.

## 4. Discussion

*4.1. Effects of Soil Solution Properties on the Sorption of SDZ in Paddy Soils*

In addition to the basic properties of antibiotics, soil physical and chemical properties also play an essential role in the sorption of antibiotics to soil constituents [24]. In this study, the adsorption kinetics and isotherms processes of SDZ were all well fitted (R > 0.99) by the first/second-order equations and the Freundlich equation, whatever the variation of the soil properties. This indicated that the variation of soil solution properties did not change the process and mechanism of SDZ adsorption on paddy soils, which might be ascribed to the hydrophobic distribution that contributed to the sorption of SDZ on paddy soil [21,25], in view of the chemical character of SDZ, which was amphoteric and weak acid polar, with little water-solubility. It was also proven that the adsorption process of SDZ in soils was driven by weak hydrophobic forces, as in neutral or anionic specie of natural soil. The present study also indicated further that the SDZ chemicals were easily adsorbed to the soil organic matter, as the SDZ adsorption increased with soil organic matter addition (Figure 1f), and there was more obvious signal noise for soil OC constituents than the clay after being treated with SDZ (Figure 3). These results indirectly suggests that hydrophobic distribution has a beneficial effect on the adsorption of SDZ in paddy soils.

The pH determined whether SDZ existed as cations, zwitterions, or anions in the soils [26]. Hence, SDZ exhibited pH-dependent adsorption on soil constituents [27,28]. The variation of soil solution pH significantly decreased the adsorption capacity of SDZ on paddy soils, based on the adsorption coefficients of SDZ (Table 3); which is similar to the previous results, that the increase of pH from 4 to 8 significantly decreased adsorption coefficients of SAs from 30 to 1 [29]. This might be explained by the fact that the disturbance of the acid–base balance of crude soils would increase the electrostatic repulsion between anionic SDZ- and the negatively charged soil surface [25,30], and by the lower lipophilic interactions between the uncharged chemical molecules and soil particles [31]. As for the ion strength factor, a lower adsorption potential of SDZ was found with increasing $Ca^{2+}$ concentration (Figure 1), which was dissimilar to the previous reports that multivalence ions can improve antibiotic adsorption by covalent bonding action [32]. This inconsistent result might indicate that a high ion strength would weaken the function of the ion-bridge for SDZ adsorption on paddy soils, as it has a higher OM content (Table 1).

### 4.2. Effect of $Cu^{2+}$ Co-Existing on the Sorption of SDZ

With increased detection of heavy metals and antibiotics in soils, more attention has been paid to the process of antibiotics related to the co-existent metal ions, aside from the various soil properties [33–35]. It is widely thought that heavy metal cations may influence the mobility of antibiotics in soils, through ion exchange or complexation. In the present study, the adsorption isotherms of SDZ were well described by the Freundlich model (R = 0.9445–1.0000) in a binary system of Cu–SDZ. Moreover, both the adsorption capacity and affinity of SDZ were augmented, as $Cu^{2+}$ was increased, based on the changing trend of $k_F$ and $n$ values (Table 5). These results indicated that the presence of co-existent $Cu^{2+}$ did not change the mechanism of SDZ adsorption to soil particles but could alter the adsorption capacity of SDZ to soil particles with different compositions.

The further FTIR and SEM analyses also detected the co-adsorption of Cu and SDZ on soil constituents (Figures 3 and 4). It was suspected that a $Cu^{2+}$ bridge between soil particles and SDZ [11,36] or the formation of SDZ–Cu complexes [37,38] occurred for the co-adsorption of Cu and SDZ on paddy soils. However, this might be mainly carried out by their independent adsorption to different sites of the soil particles, as the weak signal changes of the FTIR spectrum were found primarily in soil OC constituents after co-treating with Cu and SDZ (Figure 3). It was reported that the presence of heavy metal ions could result in changes in soil properties [39–41], which might shape the sorption of antibiotics on soils. Therefore, the decrease of SDZ adsorption on clay in the present study might be attributed to the precipitation of soil mineral colloid induced by the presence of $Cu^{2+}$ [39], which decreased the size or blocked the pores of the soil clay particles [42].

### 5. Conclusions

The change of soil solution pH and ion concentration lowered the SDZ adsorption on paddy soil, but it was increased with the addition of organic matter. Whatever the changes of soil properties, the sorption kinetics of SDZ on paddy soils could be well described by the pseudo-first and second-order equations, and the Freundlich equation could fit the sorption isotherms well. $Cu^{2+}$ coexistence in the soil increased the adsorption of SDZ on crude soil and its organic components, but decreased the adsorption capacity of clay soil for SDZ. Based on a further analysis of FTIR and SEM, it could be concluded that the co-adsorption of $Cu^{2+}$ and SDZ on soil constituents might be conducted by the complexation of $Cu^{2+}$ and SDZ, in addition to their independent adsorption to different active sites of the soil particles.

**Author Contributions:** Conceptualization, A.L.; Data curation, Z.X.; Formal analysis, Z.X.; Investigation, Z.X., S.H., L.C., F.R., X.W. and M.L.; Methodology, Z.X. and A.L.; Project administration, A.L.; Resources, A.L.; Software, M.D. and K.L.; Supervision, M.L. and A.L.; Validation, Z.X., M.D., K.L. and M.L.; Visualization, S.L., L.C., F.R. and X.W.; Writing—original draft, Z.X.; Writing—review & editing, Z.X., K.L., M.L. and A.L. All authors have read and agreed to the published version of the manuscript.

**Funding:** This work was funded by Natural Science Foundation of China (No. 41877122 and No.41671322), the Natural Science Foundation of Shandong (No.ZR2020ZD19) & the Key R&D program of Shandong (No.2019GSF109058).

**Institutional Review Board Statement:** Not applicable.

**Informed Consent Statement:** Not applicable.

**Conflicts of Interest:** The authors declare no conflict of interest.

## References

1. Kim, J.H.; Kuppusamy, S.; Kim, S.Y.; Kim, S.C.; Kim, H.T.; Lee, Y.B. Occurrence of sulfonamide class of antibiotics resistance in Korean paddy soils under long-term fertilization practices. *J. Soils Sediments* **2017**, *17*, 1618–1625. [CrossRef]
2. Pan, M.; Chu, L.M. Occurrence of antibiotics and antibiotic resistance genes in soils from wastewater irrigation areas in the Pearl River Delta region, southern China. *Sci. Total Environ.* **2017**, *624*, 145–152. [CrossRef] [PubMed]
3. Zhi, D.; Yang, D.; Zheng, Y.; Yang, Y.; He, Y.; Luo, L.; Zhou, Y. Current progress in the adsorption, transport and biodegradation of antibiotics in soil. *J. Environ. Manag.* **2019**, *251*, 109598. [CrossRef] [PubMed]
4. Arunakumara, K.K.I.U.; Walpola, B.C.; Yoon, M.-H. Current status of heavy metal contamination in Asia's rice lands. *Rev. Environ. Sci. Bio/Technol.* **2013**, *12*, 355–377. [CrossRef]
5. Hu, L.; Qiu, J.; Wang, L.; Tang, H.; Li, C.; Ranst, E.V. Modelling impacts of alternative farming management practices on greenhouse gas emissions from a winter wheat-maize rotation system in China. *Agric. Ecosyst. Environ.* **2010**, *135*, 24–33.
6. Sarmah, A.K.; Meyer, M.; Boxall, A.B. A global perspective on the use, sales, exposure pathways, occurrence, fate and effects of veterinary antibiotics (VAs) in the environment. *Chemosphere* **2006**, *65*, 725–759. [CrossRef] [PubMed]
7. Conde-Cid, M.; Álvarez-Esmorís, C.; Paradelo-Núñez, R.; Nóvoa-Muñoz, J.C.; Arias-Estévez, M.; Álvarez-Rodríguez, E.; Fernández-Sanjurjo, M.J.; Núñez-Delgado, A. Occurrence of tetracyclines and sulfonamides in manures, agricultural soils and crops from different areas in Galicia (NW Spain). *J. Clean. Prod.* **2018**, *197*, 491–500. [CrossRef]
8. Wei, R.; Ge, F.; Zhang, L.; Hou, X.; Cao, Y.; Gong, L.; Chen, M.; Wang, R.; Bao, E. Occurrence of 13 veterinary drugs in animal manure-amended soils in Eastern China. *Chemosphere* **2016**, *144*, 2377–2383. [CrossRef]
9. Kremer, E.; Facchin, G.; Estévez, E.; Alborés, P.; Baran, E.; Ellena, J.; Torre, M. Copper complexes with heterocyclic sulfonamides: Synthesis, spectroscopic characterization, microbiological and SOD-like activities: Crystal structure of [Cu(sulfisoxazole)2(H2O)4]·2H2O. *J. Inorg. Biochem.* **2006**, *100*, 1167–1175. [CrossRef]
10. Ftouni, H.; Sayen, S.; Boudesocque, S.; Dechamps-Olivier, I.; Guillon, E. Structural study of the copper(II)–enrofloxacin metallo-antibiotic. *Inorg. Chim. Acta* **2012**, *382*, 186–190. [CrossRef]
11. Morel, M.-C.; Spadini, L.; Brimo, K.; Martins, J.M. Speciation study in the sulfamethoxazole–copper–pH–soil system: Implications for retention prediction. *Sci. Total Environ.* **2014**, *481*, 266–273. [CrossRef]
12. Xu, Y.; Yu, W.; Ma, Q.; Zhou, H. Interactive effects of sulfadiazine and Cu(II) on their sorption and desorption on two soils with different characteristics. *Chemosphere* **2015**, *138*, 701–707. [CrossRef]
13. Pan, B.; Qiu, M.; Wu, M.; Zhang, D.; Peng, H.; Wu, D.; Xing, B. The opposite impacts of Cu and Mg cations on dissolved organic matter-ofloxacin interaction. *Environ. Pollut.* **2012**, *161*, 76–82. [CrossRef]
14. Graouer-Bacart, M.; Sayen, S.; Guillon, E. Macroscopic and molecular approaches of enrofloxacin retention in soils in presence of Cu(II). *J. Colloid Interface Sci.* **2013**, *408*, 191–199. [CrossRef] [PubMed]
15. Jia, D.-A.; Zhou, D.-M.; Wang, Y.-J.; Zhu, H.-W. Adsorption and co-sorption of Cu(II) and tetracycline on two soils with different characteristics. *Geoderma* **2008**, *146*, 224–230. [CrossRef]
16. Bao, Y.; Wan, Y.; Zhou, Q.; Li, W.; Liu, Y. Competitive adsorption and desorption of oxytetracycline and cadmium with different input loadings on cinnamon soil. *J. Soils Sediments* **2013**, *13*, 364–374. [CrossRef]
17. de la Torre, A.; Iglesias, I.; Carballo, M.; Ramírez, P.; Muñoz, M.J. An approach for mapping the vulnerability of European Union soils to antibiotic contamination. *Sci. Total Environ.* **2012**, *414*, 672–679. [CrossRef]
18. Leal, R.M.P.; Alleoni, L.; Tornisielo, V.L.; Regitano, J.B. Sorption of fluoroquinolones and sulfonamides in 13 Brazilian soils. *Chemosphere* **2013**, *92*, 979–985. [CrossRef] [PubMed]
19. Elliott, E.T. Aggregate Structure and Carbon, Nitrogen, and Phosphorus in Native and Cultivated Soils. *Soil Sci. Soc. Am. J.* **1986**, *50*, 627–633. [CrossRef]
20. Dumat, C.; Cheshire, M.V.; Fraser, A.R.; Shand, C.A.; Staunton, S. The effect of removal of soil organic matter and iron on the adsorption of radiocaesium. *Eur. J. Soil Sci.* **2010**, *48*, 675–683. [CrossRef]
21. Jiang, Y.; Zhang, Q.; Deng, X.; Nan, Z.; Liang, X.; Wen, H.; Huang, K.; Wu, Y. Single and competitive sorption of sulfadiazine and chlortetracycline on loess soil from Northwest China. *Environ. Pollut.* **2020**, *263*, 114650. [CrossRef] [PubMed]
22. Lu, R.K. *Analysis Method of Soil Agricultural Chemistry*; China Agricultural Science and Technology Press: Beijing, China, 2000. (In Chinese)
23. Al-Degs, Y.; El-Barghouthi, M.I.; Issa, A.A.; Khraisheh, M.A.; Walker, G.M. Sorption of Zn(II), Pb(II), and Co(II) using natural sorbents: Equilibrium and kinetic studies. *Water Res.* **2006**, *40*, 2645–2658. [CrossRef] [PubMed]

24. Wang, S.; Wang, H. Adsorption behavior of antibiotic in soil environment: A critical review. *Front. Environ. Sci. Eng.* **2015**, *9*, 565–574. [CrossRef]
25. Srinivasan, P.; Sarmah, A.K.; Manley-Harris, M. Co-contaminants and factors affecting the sorption behaviour of two sulfonamides in pasture soils. *Environ. Pollut.* **2013**, *180*, 165–172. [CrossRef] [PubMed]
26. Chen, K.-L.; Liu, L.-C.; Chen, W.-R. Adsorption of sulfamethoxazole and sulfapyridine antibiotics in high organic content soils. *Environ. Pollut.* **2017**, *231*, 1163–1171. [CrossRef]
27. Ötker, H.M.; Akmehmet-Balcıoğlu, I. Adsorption and degradation of enrofloxacin, a veterinary antibiotic on natural zeolite. *J. Hazard. Mater.* **2005**, *122*, 251–258. [CrossRef]
28. Wegst-Uhrich, S.R.; Navarro, D.A.; Zimmerman, L.; Aga, D.S. Assessing antibiotic sorption in soil: A literature review and new case studies on sulfonamides and macrolides. *Chem. Cent. J.* **2014**, *8*, 5. [CrossRef] [PubMed]
29. Lertpaitoonpan, W.; Ong, S.K.; Moorman, T. Effect of organic carbon and pH on soil sorption of sulfamethazine. *Chemosphere* **2009**, *76*, 558–564. [CrossRef]
30. Hu, S.; Zhang, Y.; Shen, G.; Zhang, H.; Yuan, Z.; Zhang, W. Adsorption/desorption behavior and mechanisms of sulfadiazine and sulfamethoxazole in agricultural soil systems. *Soil Tillage Res.* **2019**, *186*, 233–241. [CrossRef]
31. Rath, S.; Fostier, A.H.; Pereira, L.A.; Dioniso, A.C.; Ferreira, F.D.O.; Doretto, K.M.; Peruchi, L.M.; Viera, A.; Neto, O.F.D.O.; Bosco, S.M.D.; et al. Sorption behaviors of antimicrobial and antiparasitic veterinary drugs on subtropical soils. *Chemosphere* **2019**, *214*, 111–122. [CrossRef]
32. Zhang, J.; Dong, Y. Influence of strength and species of cation on adsorption of norfloxacin in typical soils of China. *Huan Jing Ke Xue* **2007**, *28*, 2383–2388. [PubMed]
33. Maszkowska, J.; Białk-Bielińska, A.; Mioduszewska, K.; Wagil, M.; Kumirska, J.; Stepnowski, P. Sorption of sulfisoxazole onto soil—An insight into different influencing factors. *Environ. Sci. Pollut. Res.* **2015**, *22*, 12182–12189. [CrossRef]
34. Gao, J.; Pedersen, J.A. Adsorption of Sulfonamide Antimicrobial Agents to Clay Minerals. *Environ. Sci. Technol.* **2005**, *39*, 9509–9516. [CrossRef] [PubMed]
35. Park, J.Y.; Huwe, B. Effect of pH and soil structure on transport of sulfonamide antibiotics in agricultural soils. *Environ. Pollut.* **2016**, *213*, 561–570. [CrossRef]
36. Laak, T.; Gebbink, W.A.; Tolls, J. The effect of pH and ionic strength on the sorption of sulfachloropyridazine, tylosin, and oxytetracline to soil. *Environ. Toxicol. Chem.* **2010**, *25*, 904–911. [CrossRef] [PubMed]
37. Wei, C.; Song, X.; Wang, Q.; Liu, Y.; Lin, N. Influence of coexisting Cr(VI) and sulfate anions and Cu(II) on the sorption of F-53B to soils. *Chemosphere* **2019**, *216*, 507–515. [CrossRef] [PubMed]
38. Lu, X.; Deng, S.; Wang, B.; Huang, J.; Wang, Y.; Yu, G. Adsorption behavior and mechanism of perfluorooctane sulfonate on nanosized inorganic oxides. *J. Colloid Interface Sci.* **2016**, *474*, 199–205. [CrossRef] [PubMed]
39. Pei, Z.; Shuang, Y.; Li, L.; Li, C.; Guo, B. Effects of copper and aluminum on the adsorption of sulfathiazole and tylosin on peat and soil. *Environ. Pollut.* **2014**, *184*, 579–585. [CrossRef] [PubMed]
40. Wang, X.; Yang, K.; Tao, S.; Xing, B. Sorption of Aromatic Organic Contaminants by Biopolymers: Effects of pH, Copper (II) Complexation, and Cellulose Coating. *Environ. Sci. Technol.* **2007**, *41*, 185–191. [CrossRef] [PubMed]
41. Pei, Z.-G.; Shan, X.-Q.; Zhang, S.-Z.; Kong, J.-J.; Wen, B.; Zhang, J.; Zheng, L.-R.; Xie, Y.-N.; Janssens, K. Insight to ternary complexes of co-adsorption of norfloxacin and Cu(II) onto montmorillonite at different pH using EXAFS. *J. Hazard. Mater.* **2011**, *186*, 842–848. [CrossRef] [PubMed]
42. Lu, Y.; Pignatello, J.J. Sorption of Apolar Aromatic Compounds to Soil Humic Acid Particles Affected by Aluminum(III) Ion Cross-Linking. *J. Environ. Qual.* **2004**, *33*, 1314–1321. [CrossRef] [PubMed]

Article

# Stabilization/Solidification of Heavy Metals and PHe Contaminated Soil with β-Cyclodextrin Modified Biochar (β-CD-BC) and Portland Cement

Geng Li, Haibo Li *, Yinghua Li *, Xi Chen, Xinjing Li, Lixin Wang, Wenxin Zhang and Ying Zhou

School of Resources and Civil Engineering, Northeastern University, 3-11 Wenhua Road, Heping District, Shenyang 110819, China; 1970915@stu.neu.edu.cn (G.L.); chenxineu@mail.neu.edu.cn (X.C.); 2001049@stu.neu.edu.cn (X.L.); 1970919@stu.neu.edu.cn (L.W.); 1901077@stu.neu.edu.cn (W.Z.); 20192118@stu.neu.edu.cn (Y.Z.)
* Correspondence: lihaibo@mail.neu.edu.cn (H.L.); liyinghua@mail.neu.edu.cn (Y.L.)

**Abstract:** Conventional stabilization/solidification materials have defects in the simultaneous treatment of heavy metals (HMs) and phenanthrene (PHe). In order to solve this problem, a new functional material β-cyclodextrin modified biochar (β-CD-BC) was prepared by integrating the properties of biochar (BC) and the hydrophilic and hydrophobic properties of the β-CD surface and combined with Portland cement (PC) to cure and stabilize HMs and PHe. The effect of key parameters on the treatment effect was discussed by response surface method. The results showed that the minimum leaching concentration if HMs was 16.81 mg·L$^{-1}$, and the leaching concentration of PHe can be as low as 0.059 µg/kg under the conditions of β-CD-BC and Portland cement ratio of 9.75% and 11.4%, curing for 22.85 d. The weak acid soluble state reduced from 9~13% to 0.5~6%, the residual state was increased from 37~61% to 77~87%. The unconfined compressive strength of sample is more than 50 kPa. The results of this study can provide a new technical scheme for long-term curing and stabilization of HMs and PHe.

**Keywords:** β-CD modified BC; stabilization/solidification; response surface methodology; synchronous adsorption investigations; PHe

## 1. Introduction

Heavy metals (HMs) and phenanthrene (PHe) widely exist in the natural environment [1], which can not only accumulate in the human body through skin and respiratory tract and endanger human health, but also migrate in the soil and causing human water risk [2]. The research and development of soil remediation technology have not attracted enough attention. Therefore, the study is of great significance for the solidification/stabilization (S/S) of HMs and PHe in soil.

At present, the commonly used treatment technologies of HMs and PHe contaminated soil include S/S [3], leaching [4], combined Phyto microbial remediation [5,6], and thermal desorption [7]. In the United States, 80% of the S/S projects of Superfund use cement as the curing agent. S/S technology has attracted much attention due to its advantages of simple operation, low cost, and good treatment effect. The selection of repair materials for this technology is the key. At present, common S/S materials at home and abroad include cement, lime, organic matter, biochar (BC), montmorillonite, fly ash, etc. [8]. These materials reduce the mobility and bioavailability of HMs by physical and chemical reactions such as adsorption, precipitation, redox, and ion exchange with HMs [9]. The latest research mainly focuses on composite materials made from the above materials. The study found that using fly ash, quicklime, and blast furnace slag as composite materials can significantly stabilize HMs such as Zn, Pb, Cu, and Cr in contaminated soil [10]. Based on X-ray diffraction, TGA, and SEM tests, Sora et al. [11] proved that organic pollutants can affect the type and

quantity of cement hydration products and make the solidified body present more internal pores. Wang et al. [12] found that fly ash can improve the leaching toxicity of cement solidified organic pollutants. Still, it has a weak fixation effect on HMs, which proves that the fixation mechanism of organic matter and HMs is different. Kogbara et al. found that increasing the cement content can enhance the fixation of HMs, but has no apparent effect on TPH [13].

BC is an environmentally friendly material, cheap, and has many oxygen functional groups, large surface area, and well-developed pore structure [14]. It has the ability and potential to combine with different types of pollutants. BC can stabilize HMs, but it is feeble in stabilizing PHe. β-cyclodextrin (β-CD) could simultaneously achieve efficient cleanup of PHe and Cr, Cd, Cu, Pb, and Zn through avoiding the competitive behaviors between them, which were due to the different adsorption for Cr, Cd, Cu, Pb, and Zn (i.e., electrostatic attraction and complexation) and PHe (i.e., host–guest supramolecular and π-π interactions) [15–17]. B-CD-BC can simultaneously solidify and stabilize HMs and Phe in soil.

To solve the above problems, BC, PC, and β-CD-BC use as S/S agents. Response surface optimization experiment was used to carry out an S/S experiment on heavy metal organic matter, to explore the influence of various factors on the leaching concentration of heavy metals and PHe, and to determine the optimal process parameters. The unconfined compressive strength of the solidified body was selected, and the mechanism of S/S was explored through the analysis of the morphological characteristics of HMs, to provide the basis for the risk control and remediation of the soil polluted by the heavy metal organic compounds.

## 2. Materials and Methods

### 2.1. Materials and Reagents

Chemical reagents including β-CD(CAS: 7585-39-9), epichlorohydrin (EPI)(CAS:106-89-8), HCl(CAS:7647-01-0), NaOH(CAS:1310-73-3) and Pb(NO$_3$)$_2$(CAS:1099-74-8), CuSO$_4$(CAS:7758-98-7), ZnCl$_2$(CAS:7646-85-7), Cr(NO$_3$)$_3$(CAS:13548-38-4), Cd(NO$_3$)$_2$ (CAS:10325-94-7) were all purchased from Xin Ke. (Shenyang, China). All reagents were of analytical grade. Rice husks were acquired in Shenyang, Liaoning province of China.

The soil sample was collected from the surface layer (0–30 cm) of an unpolluted calcareous soil at the College of Agriculture Shenyang. Table 1 presents physical and chemical properties of soil.

Table 1. Physical and chemical properties of soil.

| pH | Organic Matter/g·kg$^{-1}$ | Particle Composition/% | | | TN/g·kg$^{-1}$ | TP/g·kg$^{-1}$ | CEC/cmol·kg$^{-1}$ |
| --- | --- | --- | --- | --- | --- | --- | --- |
| | | Sand | Power | Clay | | | |
| 7.6 | 26.31 | 26.86 | 49.41 | 25.12 | 1.05 | 6.53 | 12.6 |

### 2.2. Preparation and Modification of Biochar

Pyrolysis produces of rice husk was carried out in atmosphere furnace(OFT-1200X-S, Hefei, China) under nitrogen (N$_2$) flow of 0.2 L/min with a heating rate of 10 °C/min at 400 °C for 3 h. After the reactor was set to cool to room temperature, BC was obtained. Before dry preservation, BC was ground through a 100 mesh sieve.

Modification BC: As shown in Figure 1, 10.00 g of β-CD and 4.80 g of EPI in a molar ratio of 4:1 were added to 200 mL of 5% NaOH solution. The mixture was mixed for 4 h (320 rpm) at a temperature (25 °C) to afford the modification solution. Then, 5.00 g BC was weighed out and mixed with 100 mL of modification solution. The mixture was stirred at room temperature for 1.5 h (320 rpm), washed with deionized water, and filtered under suction several times, until the pH of BC was neutral. Then BC was left to stand overnight. The product was dried at 75 °C in an oven until its weight was stable, yielding β-CD functionalized biochar (β-CD-BC) [18].

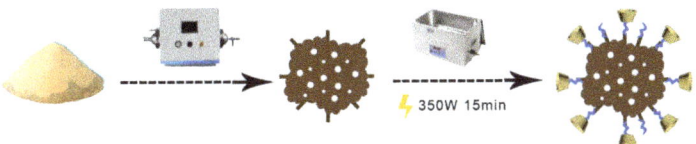

**Figure 1.** β-CD-BC preparation process.

## 2.3. Experimental Method

### 2.3.1. Solidification/Stabilization Experiment

A single factor level experiment was designed to determine the factors and level of orthogonal experiment. To explore the influence of the addition amount of BC and β-CD-BC, 10% PC and 30% deionized water were added, cured in a 24 °C dry environment for 14 days, with the mass fraction of 2.5, 5, 7.5, 10, 12.5% BC and β-CD-BC added. To explore the effect of curing time, 10% PC, 30% deionized water, and 10% BC were added and fixed in a 24 °C incubator for 5, 10, 20, 30, and 40 days, respectively. The evaporated water was supplemented by a weighing method every three days.

Using the Box–Behnken design model of design-expert software, the selection of factors follows the principle of influence on as leaching (reagent ratio and curing time). In this study, response surface factor level experiments (Table 2) were used for the orthogonal design (Table 3). Taking BC + PC compound as the control, the key parameters such as optimal β-CD-BC + PC compound ratio and curing time were explored.

**Table 2.** Box–Behnken factors and levels of experimentation.

| Specimen | Curing Time (d) | Biochar Ratio (%) | Cement Ratio (%) |
|---|---|---|---|
| 1 | 10 | 5 | 6 |
| 2 | 20 | 7.5 | 12 |
| 3 | 30 | 10 | 18 |

**Table 3.** Box–Behnken experimental design table.

| Sample | A Curing Time (d) | B Biochar Ratio (%) | C Cement Ratio (%) |
|---|---|---|---|
| 1 | 10 | 5.0 | 12 |
| 2 | 10 | 7.5 | 6 |
| 3 | 10 | 7.5 | 18 |
| 4 | 10 | 10.0 | 12 |
| 5 | 20 | 5.0 | 6 |
| 6 | 20 | 5.0 | 18 |
| 7 | 20 | 7.5 | 12 |
| 8 | 20 | 10.0 | 6 |
| 9 | 20 | 10.0 | 18 |
| 10 | 30 | 5.0 | 12 |
| 11 | 30 | 7.5 | 6 |
| 12 | 30 | 7.5 | 18 |
| 13 | 30 | 10.0 | 12 |

The leaching concentration of HMs was determined according to the standard toxic leaching procedure TCLP (Toxic Characteristic Leaching Procedure) recommended by the

U.S. Environmental Protection Agency. The leaching rate of HMs was calculated according to the following formula:

$$\eta = \frac{c_0 - c_t}{c_0} \times 100\% \tag{1}$$

where: $\eta$ is the leaching rate; $c_0$ is the leaching concentration of as in the original soil; $c_t$ is the leaching concentration after solidification and stabilization.

2.3.2. Unconfined Compressive Strength (UCS) Test

The procedure of UCS tests was according to ASTM D2166-91. Then, the specimen was placed on the bottom plate of test setup. The wheel was then turned to lower down the upper plate. When the upper plate was in contact with the top surface of the specimen, the indication of dial gauge was adjusted to zero. Then, the wheel was turned automatically at a speed of 0.06 mm/min (corresponding to 0.5% strain of dial gauge) to compress the specimen to failure.

2.3.3. Analytical Method

The surface morphology of BC and β-CD-BC was observed by SEM (Sigma 300, Zeiss, Jena, Germany). The mineral composition of β-CD-BC was analyzed by X-ray diffraction (Ultima IV, Rigaku, Tokyo, Japan). The total arsenic was determined by the aqua regia perchloric acid method, and the metal speciation procedure applied for HMs was recommended by Liang et al. [19]. The data were processed by design expert.8.0.6.1, Excel2003, and Origin 8.0 software.

## 3. Results and Discussion

### 3.1. Characterization of BC and β-CD-BC

The Figure 2a–c shows the scanning electron microscope images of BC and β-CD-BC. The modified BC still retains the rich pore structure of BC, and the morphology of BC itself has not changed. White matter appears on the surface of β-CD-BC. The white point is more uniformly dispersed through images.

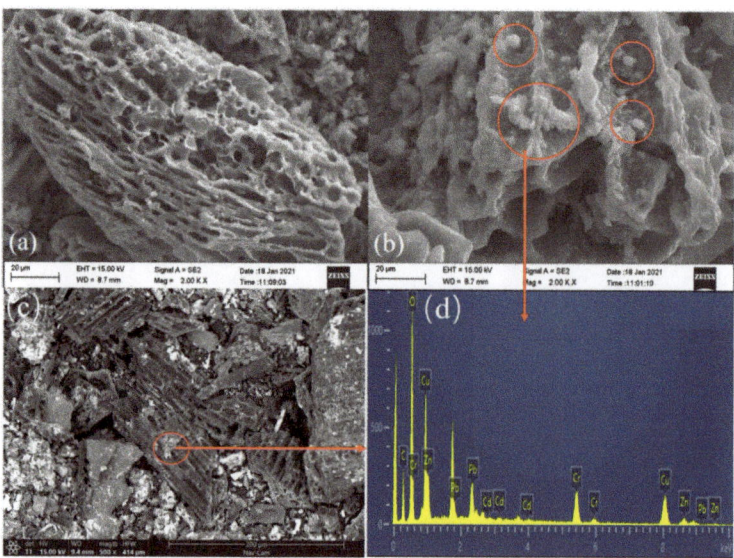

**Figure 2.** SEM images, (**a**) BC, (**b**) β-CD-BC, (**c**) β-CD-BC.EDS images, (**d**) β-CD-BC.

Shown in Figure 2b,c are β-CD-BC adsorbing HMs. It can be seen that there is no obvious change in the surface structure of BC adsorbed with HMs at different scales. As shown in the Figure 2d EDS analysis, the effect of β-CD-BC adsorbing HMs is obvious, and there are a large number of heavy metals on the surface of β-CD-BC. It can be considered that the β-CD did not agglomerate in a large amount, and the modification was relatively successful. Meanwhile, after β-CD functionalization BC lost its well-aligned structure, and a wealth of white substances appeared and covered the surface and pores of BC [20].

Figure 3 shows FTIR spectra that were created for evaluating the functional groups of β-CD, BC, and β-CD-BC. The influential band for −OH stretching in alkyl or aryl at 3434 cm$^{-1}$, the bands at around 1631, 1030 cm$^{-1}$ of BC can be assigned to the C-O stretching or O-H banding, and the band is identified for −CH at 608 cm$^{-1}$. Additionally, β-CD-BC has similar analogous functional groups. The new bands at 1630 and 1460 cm$^{-1}$ represent the COO- and CeN, respectively. Moreover, in comparison with pristine β-CD, the new band at around 608 cm$^{-1}$ of BC could be indorsed to COO- stretching vibration. It could also be observed in β-CD-BC [21], confirming the introduction of the carboxyl groups onto the modified adsorbent. All the results proved that the BC had been successfully functionalized by β-CD [22].

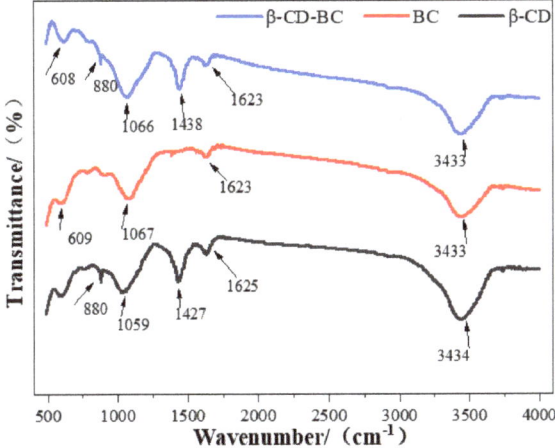

**Figure 3.** FTIR spectra of BC and β-CD-BC.

It can be seen from Table 4 that the specific surface area of the BC modified by β-CD increased a lot, indicating that there is no accumulation during the modification process, and the β-CD is relatively uniformly dispersed on the surface of the BC. Additionally, the pore volume of β-CD-BC was higher than BC, while the average pore size of β-CD-BC was lower than BC. When subjected to β-CD, changes in surface properties are accompanied by changes in structure and composition, such as cation exchange capacity (CEC). The CEC of β-CD-BC is higher than that of BC (Table 3). The difference between the two functional BC is mainly related to the surface functional groups.

**Table 4.** Surface characteristics of biochars.

| Specimen | Specific Surface Area | Pore Volume (cm$^3$/g) | Average Pore Size (nm) |
|---|---|---|---|
| β-CD-BC | 115.3880 | 0.141730 | 4.91317 |
| BC | 51.3219 | 0.077551 | 4.91317 |

## 3.2. Solidification/Stabilization Experiment

### 3.2.1. Single Factor Experiment

Leachability of PHe

As shown in Figure 4a, BC and β-CD-BC were added to the soil for maintenance at the mass ratios of 2.5%, 5%, 7.5%, 10%, and 12.5%. The leaching rate of PHE in the cured BC/PC samples was higher than that in the blank group, which may be due to the hydrophilic environment of PC, which increased the leaching concentration of PHH in the soil. With the increase of β-CD-BC, the leaching concentration of PHe in the soil gradually decreases until it becomes stable, because the cyclodextrin cavity on the surface of BC is hydrophobic and has a good adsorption effect on PHe. Through the above experiments, β-CD-BC can effectively reduce the leaching concentration of PHe in the soil and has a good stabilizing ability. As shown in Figure 4b, the PHe leaching concentration of the sample cured by β-CD-BC + PC reached a stable level after 20 d.

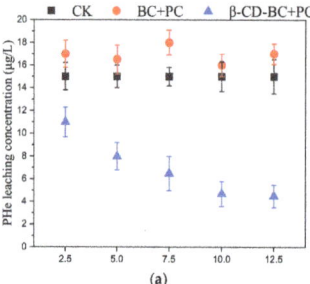 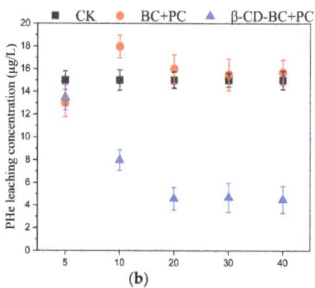

**Figure 4.** Effects of different BC and β-CD-BC (**a**) addition amounts and (**b**) curing times on PHe leaching concentration.

Leachability of Heavy Metals

As shown in Figure 5, BC and β-CD-BC were added to the soil for curing according to the mass ratios of 2.5%, 5%, 7.5%, 10%, and 12.5%. The leaching concentration of HMs in BC + PC samples was higher than that in the blank group. The oxygen-containing functional groups on the surface of BC can chelate and cooperate with HMs in soil, to reduce the leaching concentration of HMs in soil. BC has developed a pore structure and can adsorb HMs on the surface. The leaching concentration of HMs in the β-CD-BC + PC cured sample is lower than that of the blank group and BC + PC. This is because the hydroxyl and carboxyl functional groups on the surface of BC are modified by β-CD, which greatly improves the performance of the material. The chelation between HMs and hydrophilic functional groups enhances their stability. Meanwhile, PC can change the pH value of the soil, increase the content of -OH in the soil, and cause the co-precipitation of HMs and -OH [23]. Therefore, β-CD-BC significantly enhances the adsorption capacity of HMs.

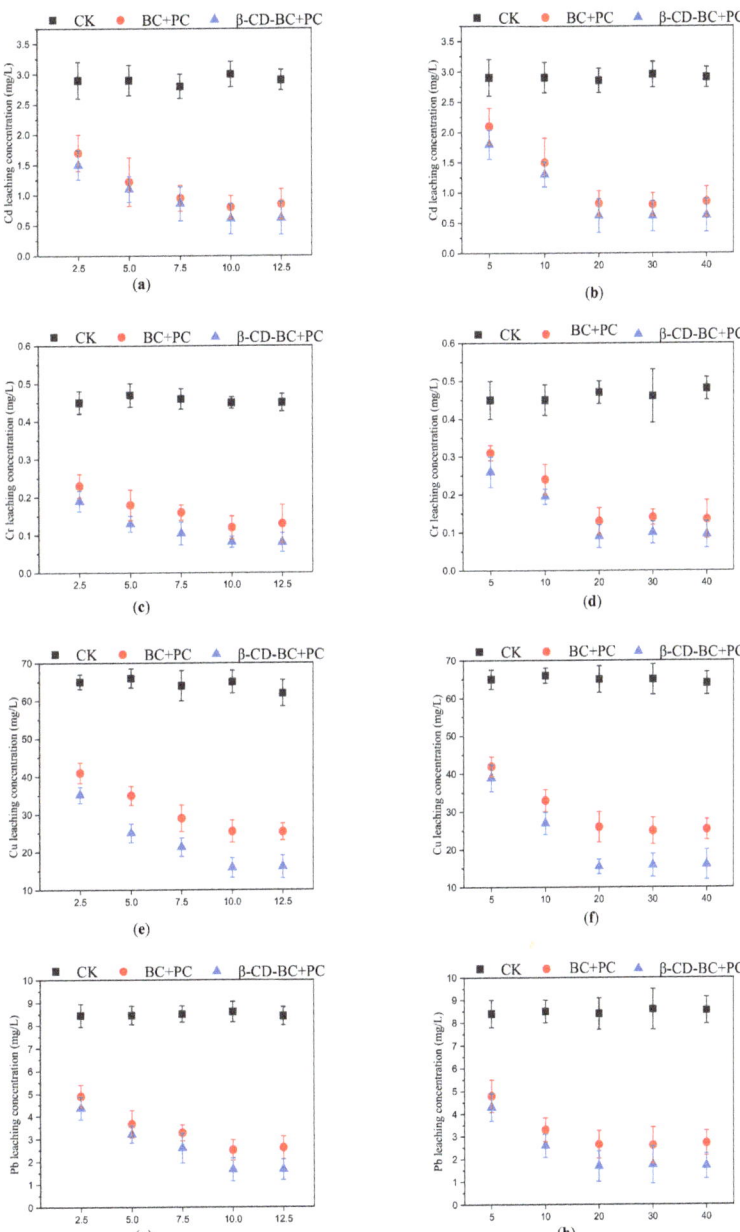

**Figure 5.** *Cont.*

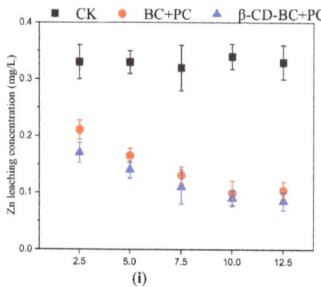
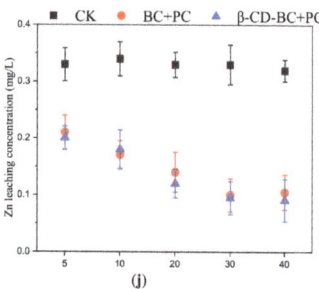

**Figure 5.** Effects of different BC and β-CD-BC addition amounts and curing times on HMs leaching concentration. (**a**) Addition amount and (**b**) Curing time for Cd leaching concentration; (**c**) Addition amount and (**d**) Curing time for Cr leaching concentration; (**e**) Addition amount and (**f**) Curing time for Cu leaching concentration; (**g**) Addition amount and (**h**) Curing time for Pb leaching concentration; (**i**) Addition amount and (**j**) Curing time for Zn leaching concentration.

3.2.2. The Response Surface Experiments

The Results of Box–Behnken Design and the Variance Analysis of Model

The fitting equation of PHe is $Y = 0.12 - 0.028A - 0.041B + 0.023C - 2.5 \times 10^{-3}AB - 0.016AC + 5 \times 10^{-3}BC + 0.011A^2 + 0.022B^2 - 0.012C^2$

Curing time (10–30d), Portland cement (6–18%), and BC (5–10%) are three reaction parameters that affect the leaching concentration of PHe. The results of ANOVA are shown in Table 5. The results of variance analysis show that the $p$ values of β-CD-BC, PC, and curing time are less than 0.05, which indicates that the fitting effect of the model is good, and the response surface approximation model can be used to optimize the leaching concentration of heavy metal and PHe. According to the results of the analysis of variance in Table 5, the $p$ values of A, B, C, and AC are all less than 0.05, indicating that curing time, BC ratio, and cement blending ratio are important factors for curing stability. It plays an important role in the leaching concentration of PHe. The $p$ value of BC is the largest, which indicates that the effect of adding β-CD-BC and PC on the leaching concentration is relatively weak and has little impact on the final result. It also shows that cement enhances the leaching of some PHe and can be fixed by cyclodextrin on the surface of BC. The $p$ values of the three influencing factors are 0.0012, 0.0002, and 0.0064, respectively. The order of influence on PHe leaching concentration is B (Biochar ratio) > A (Curing time) > C (Cement ratio).

**Table 5.** Analysis of variance for response surface quadratic mode of PHe.

| Source | df | Sum of Squares | Mean Squares | F Value | Prob > F |
|---|---|---|---|---|---|
| Model | 9 | $2.6 \times 10^{-2}$ | $2.883 \times 10^{-3}$ | 12.47 | 0.00016 |
| A | 1 | $6.328 \times 10^{-3}$ | $6.328 \times 10^{-3}$ | 27.36 | 0.0012 |
| B | 1 | $1.200 \times 10^{-2}$ | $1.200 \times 10^{-2}$ | 51.95 | 0.0002 |
| C | 1 | $3.403 \times 10^{-3}$ | $3.403 \times 10^{-3}$ | 14.72 | 0.0064 |
| AB | 1 | $2.500 \times 10^{-5}$ | $2.500 \times 10^{-5}$ | 0.11 | 0.7519 |
| AC | 1 | $1.056 \times 10^{-3}$ | $1.056 \times 10^{-3}$ | 4.57 | 0.0699 |
| BC | 1 | 0.000 | 0.000 | 0.000 | 1.0005 |
| A2 | 1 | $2.780 \times 10^{-4}$ | $2.780 \times 10^{-4}$ | 1.20 | 0.3092 |
| B2 | 1 | $2.502 \times 10^{-3}$ | $2.502 \times 10^{-3}$ | 10.82 | 0.0133 |
| C2 | 1 | $3.701 \times 10^{-4}$ | $3.701 \times 10^{-4}$ | 1.60 | 0.2463 |
| Residual | 7 | $1.619 \times 10^{-3}$ | $5.396 \times 10^{-4}$ | | |

The fitting equation of HMs is Y = 25.11 − 2.42A − 8.75B − 5.96C + 3.47AB − 0.62AC + 2.44BC + 4.06A$^2$ − 1.03B$^2$ + 2.11C$^2$.

According to the results of variance analysis in Table 6, the $p$ values of A, B, and C are all less than 0.05, indicating that the addition of β-CD-BC, PC, and curing time are essential factors in the solidification and stabilization process of HMs, which play a decisive role in the leaching concentration of heavy metals. The influence order of the three factors on HMs leaching is B (Biochar ratio) > C (Cement ratio) > A (curing time). The leaching concentration of heavy metals decreases with the increase of curing time under the condition of using the same proportion of reagents in comparison samples 1, 10, 2, 11, 3, 12, and 4, 13. Compared with samples 5–9, under the condition of curing for 20 days, the leaching concentration of HMs decreased with the increase of PC and BC, respectively.

**Table 6.** Analysis of variance for response surface quadratic mode of HMs.

| Source | df | Sum of Squares | Mean Squares | F Value | Prob > F |
|---|---|---|---|---|---|
| Model | 9 | 111.10 | 123.46 | 21.50 | 0.0003 |
| A | 1 | 46.80 | 46.80 | 8.15 | 0.0245 |
| B | 1 | 612.15 | 612.15 | 106.60 | <0.0001 |
| C | 1 | 284.53 | 284.53 | 49.55 | 0.0002 |
| AB | 1 | 48.02 | 48.02 | 8.36 | 0.0233 |
| AC | 1 | 1.55 | 1.55 | 0.27 | 0.6194 |
| BC | 1 | 23.81 | 23.81 | 4.15 | 0.0811 |
| A2 | 1 | 69.36 | 69.36 | 12.08 | 0.0103 |
| B2 | 1 | 4.46 | 4.46 | 0.78 | 0.4076 |
| C2 | 1 | 18.81 | 18.81 | 3.28 | 0.1132 |
| Residual | 7 | 40.20 | 5.74 | | |

As shown in Figure 5a,b, after curing for 7 days, the leaching amount of samples 1, 2, 3, and 4 increased with the increase of cement and BC amount. When the leaching agent is mixed with the sample, materials with lower cement content may form agglomerates, thereby reducing the contact area between the phases. In contrast, samples with higher cement content do not form clumps, thereby maintaining a higher interface contact area and forming a more hydrophilic environment, so the contact between the leachate and the stabilized/cured sample is better.

As is shown in Figure 6a, because β-CD-BC has hydrophobic interaction sites it can adsorb PHe well. When the curing time is about 14 days, the concentration of sample 6 is lower because less PC is added, and β-CD-BC can adsorb with PHe and reduce the concentration in the leachate. The higher concentration of sample 7 compared with sample 11 is due to the increased PHe leaching caused by the environment provided by cement, β-CB BC may not have enough time and PHe suction. Sample 13 contains more PC, which increases the concentration of PHe in the leachate, and the content of cyclodextrin in β-CD-BC is less, which provides fewer hydrophobic sites and cannot adsorb PHe. As is shown in Figure 6b, because BC does not have hydrophobic interaction sites it cannot adsorb PHe well. With the increase of PC content, the hydrophilic environment is caused, and the hydrophobicity of the sample decreases with the increase of the amount of hydrophilic cement, which increases the leaching concentration of PHe. The adsorption capacity of PHe is low only by π-π interaction.

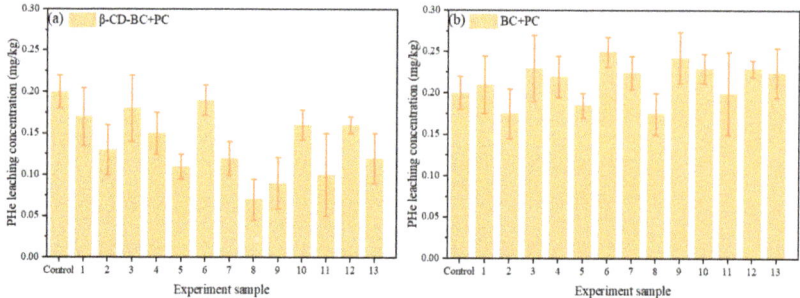

**Figure 6.** Leaching concentration of PHe in Box–Behnken design for (**a**) β-CD-BC + PC and (**b**) BC + PC.

It can be seen from Figure 7 that under the condition of the same reagent addition ratio, the HMs leaching concentration of comparison samples 1 and 10, samples 2 and 11, samples 3 and 12, samples 4 and 13 decreases with the increase of curing time. When the control samples were cured for 20 days, the HMs leaching concentration decreased with the increase of PC and BC. With comparing (a) and (f), (b) and (g), (c) and (H), (d) and (I), respectively, (e) and (J), it can be found that the leaching concentration of HMs treated with β-CD-BC is lower than BC.

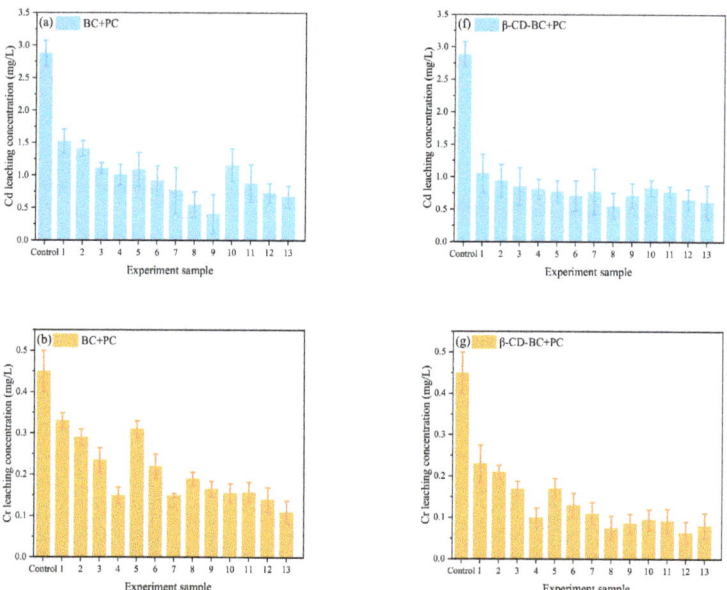

**Figure 7.** *Cont.*

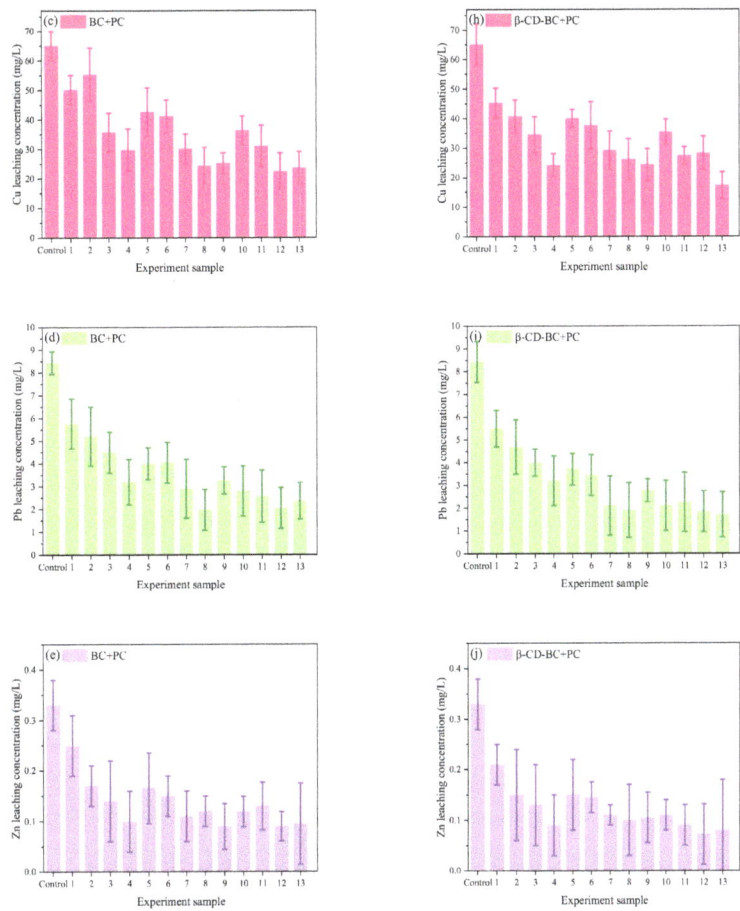

**Figure 7.** Leaching concentration of HMs in Box–Behnken design. Cd leaching concentration of (**a**) BC + PC and (**f**) β-CD-BC + PC; Cr leaching concentration of (**b**) BC + PC and (**g**) β-CD-BC + PC; Cu leaching concentration of (**c**) BC + PC and (**h**) β-CD-BC + PC; Pb leaching concentration of (**d**) BC + PC and (**i**) β-CD-BC + PC; Zn leaching concentration of (**e**) BC + PC and (**j**) β-CD-BC + PC.

Response Surface Analysis and Experimental Verification

Response surface analysis of PHe:

Figure 8a shows that the contour lines of the A and B factors present an oval shape, indicating an interaction between the two factors. Reasonable use of the relationship between A and B can effectively reduce the leaching concentration of PHe. Figure 8a shows that from the relationship between β-CD-BC and curing time, the response surface seen in the B direction is steeper and the contour lines are denser, which indicates that the effect of β-CD-BC and BC on the leaching concentration of PHe is more significant than that of curing time. Figure 8b shows the relationship between PC and curing time. It is seen that the response surface in A and C directions tends to be flat, indicating that curing time and PC dosage have the same effect on the leaching concentration of PHe. In Figure 8c the relationship is shown between the content of β-CD-BC and PC; it is seen that the response surface of B direction tends to be flat, but the contour lines of the two are oval, which indicates that the interaction between the two is vital. This is because PC provides a relatively hydrophilic environment for the soil, so that PHe in the soil precipitates, which is adsorbed on the β-CD-BC surface by cyclodextrin through hydrophobicity. Through the analysis of $p$ value and response surface, the relationship between β-CD-BC and PC addition should be given priority.

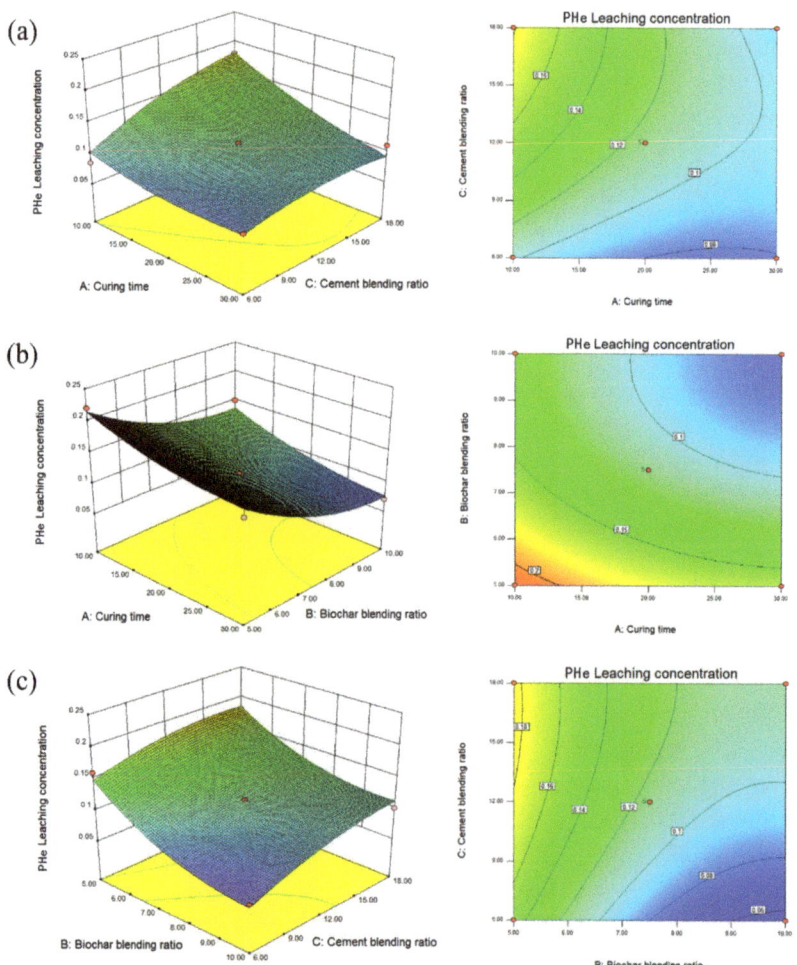

**Figure 8.** Leaching concentration of PHe. (a) Curing time and biochar blending ratio. (b) Biochar blending ratio and cement blending ratio. (c) Curing time and cement blending ratio.

Response surface analysis of heavy metals: As can be seen from Figure 9a,b, the contour is elliptical, indicating a specific interaction between maintenance time and the amount of BC added. With the extension of curing time, HMs were adsorbed in soil. The contour line in Figure 9b does not show an apparent ellipse, indicating that the interaction between BC and curing time is fragile. The relationship between β-CD-BC and curing time can be obtained from Figure 9a. It can be seen that the response surface in direction B is steeper and the contour line is denser. It shows that the effect of β-CD-BC addition on the leaching concentration of HMs is more significant than that of curing time. Increasing the amount of BC in the soil can better solidify and stabilize HMs. From Figure 9b, the relationship between PC and curing time can be obtained. It can be seen that the response surface in C is steep, while the response surface in direction a tends to be flat, which indicates that PC has a more significant effect on the leaching concentration of HMs than PC, indicating that PC adsorbs and precipitates HMs. Through the analysis of $p$ value and response surface, when optimizing the factors, priority should be given to β-CD-BC and PC.

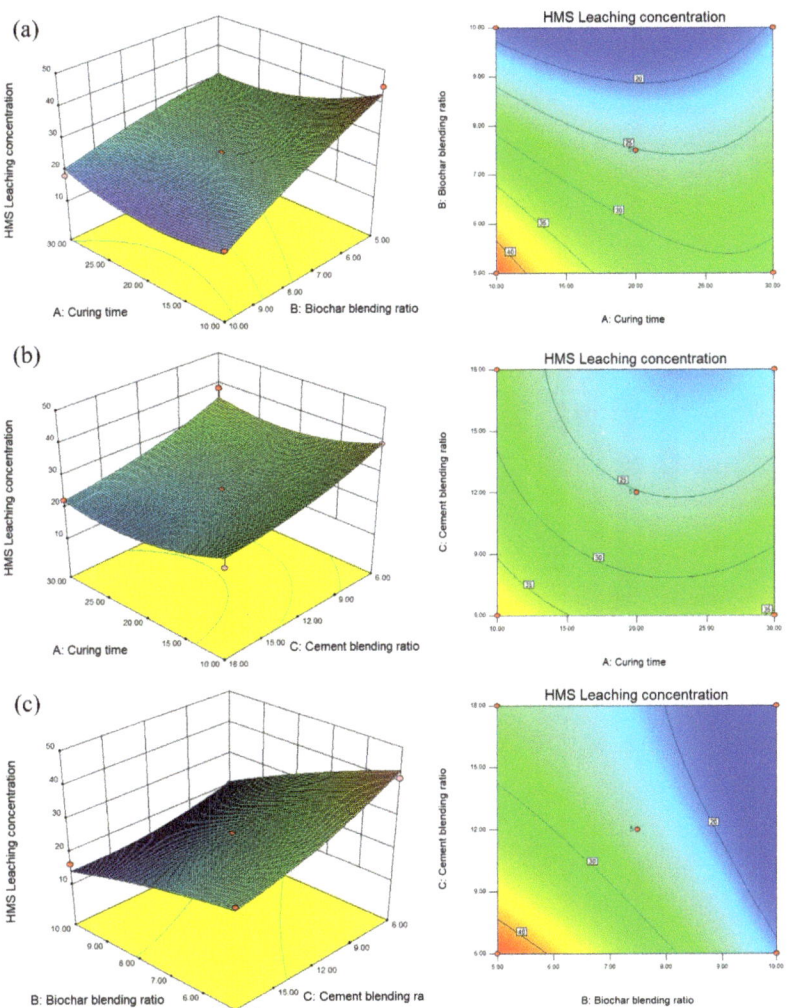

**Figure 9.** Leaching concentration of HMs. (**a**) Curing time and biochar blending ratio. (**b**) Biochar blending ratio and cement blending ratio. (**c**) Curing time and cement blending ratio.

Verification test: The results show that the curing time is 22.85 d, β-CD-BC content of 9.75% and PC content of 11.4% are the best parameters for curing stability. To verify the reliability of the model, according to the optimal parameter combination, the theoretical leaching concentration of HMs in soil was as low as 16.81 mg/kg and that of PHe was as low as 0.054 μg/kg, which is close to the predicted value. The results show that the optimization method and model are reliable.

Figure 10 shows speciation characteristics of HMs. The changes of Cr, Cd, Cu, Pb, and Zn in the soil after adding BC and β-CD-BC + PC are shown in Figure 10. The proportion of

Cd residual in the untreated soil was 37%, while the proportion of Cd treated with BC + PC and β-CD-BC + PC was 65% and 73%, respectively. Meanwhile, the weak acid soluble state and reducible state of soil Cd treated with BC + PC and β-CD-BC + PC decreased from 13% and 9% to 3–6% and 4–7%, respectively. The re sidual state of Cr, Cu, Pb, and Zn in the untreated soil was reduced compared with the soil treated with BC + PC and β-CD-BC + PC, respectively. At the same time, the weak acid extraction and reducible states of Cr, Cu, Pb, and Zn in the treated soil were reduced. This is because the added BC and β-CD-BC are adsorbed with HMs, and the surface of the material contains surface functional groups (-OH, -COOH, etc.) that can undergo complex chelation. The PC added to the soil can not only immobilize HMs in the soil, but also change the pH of the soil, causing HMs to precipitate, thereby reducing mobility. It can be found from Figure 10 that the ratio of Cr, Cu, Pb, and Zn in the residue state of the soil treated with β-CD-BC is lower than that of the residue state in BC/CP. The weak acid soluble state of metals in the soil treated with β-CD-BC is the lowest, because a large amount of cyclodextrin is attached to the surface of the modified BC, and the surface of the cyclodextrin cavity has a hydrophilic effect and contains a large number of hydroxyl groups. It can complex with HMs in the soil, change the form of HMs, and produce better stabilization effects on these.

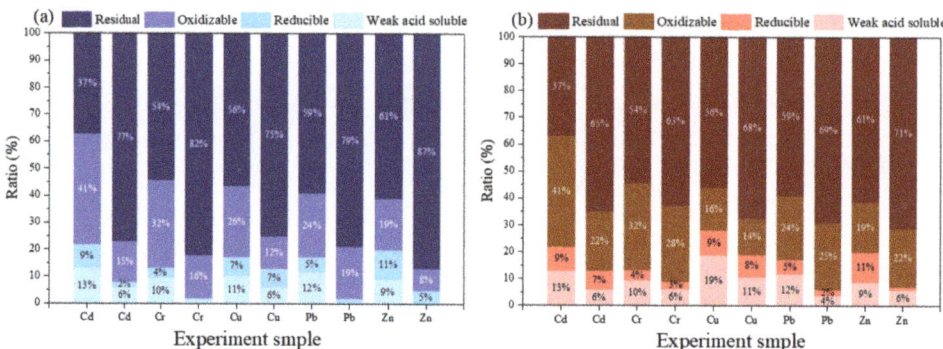

**Figure 10.** Speciation characteristics of HMs, (**a**) β-CD-BC + PC, (**b**) BC + PC.

*3.3. Synchronous Adsorption Investigations of Pb, Cu, Cr, Cd, Zn, and PHe System*

As shown in Figure 11, there is synchronous adsorption of Pb, Cu, Cr, Cd, Zn, and PHe. To realize the simultaneous solidification and stabilization of HMs and PHe in soil, BC was modified to have the hydrophobic properties of β-CD. Meanwhile, the oxygen-containing functional groups on the surface of BC increased and complexed with HMs, which enhanced the curing and stability of the material. After modification, the leaching concentration of Pb, Cu, Cr, Cd, and Zn is reduced, and the curing and stabilizing effect on PHe is also better. The main reason is that Pb, Cu, Cr, CD, Zn, and PHe have different absorption mechanisms in the soil. The cyclodextrin of β-CD-BC can wrap PHe in the cavity of the adsorbent through hydrophobic interaction, and the aromatic structure on the surface of the BC can provide potential π-π interaction [24,25] and PHe accumulation, which enhances the adsorption of PHe [26]. The adsorption of Pb, Cu, Cr, Cd, and Zn is related to the complexation and electrostatic interaction between the oxygen-containing groups of β-CD-BC [27]. The developed pore structure on BC surface can cause the surface adsorption of HMs. Meanwhile, PC can change the pH of the soil, causing the precipitation of HMs and -OH in the soil [23].

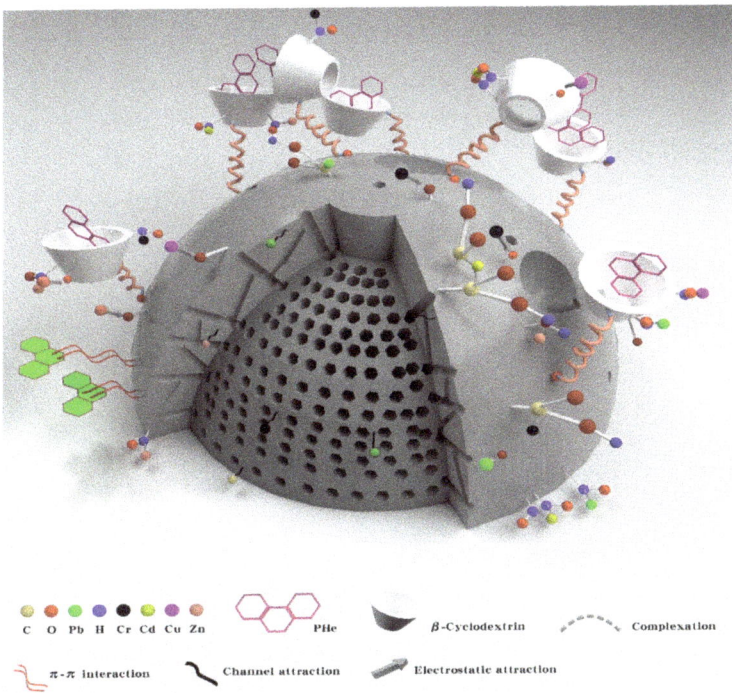

**Figure 11.** Synchronous adsorption of Pb, Cu, Cr, Cd, Zn, and PHe.

In short, as shown in Figure 11, β-CD-BC achieves the adsorption of PHe through hydrophobic effect and π-π interaction, and enhances the adsorption of Pb, Cu, Cr, Cd, and Zn through complexation and electrostatic attraction [28]. This avoids the competition between the two pollutants for adsorption sites and improves the adsorption performance and adsorption capacity of BC.

*3.4. Unconfined Compressive Strength (UCS) Test*

The UCS of the samples are given in Figure 12. It can be seen from Figure 12 that the compressive strength value of the cured body with the same addition of 6% of samples 1, 6, 11, and 16 shows a trend of increasing first and then being stable with the increase of curing time. When the curing time is 7 days, the compressive strength of the solidified sample1 is 33 kPa. The compressive strength value of the solidified sample is lower than the detection limit, which cannot meet the solidification requirements. This may be due to the relatively high water content in the solidified sludge, which results in the dispersion of cement particles and reduces the hydration reaction [29]. The cementing ability of the generated hydrated calcium silicate makes the compressive strength of the solidified sample low. As the curing time increases to 14 d, 21 d, 28 d, the compressive strength of the cured body increases to 52–57 kPa, adding 12% of No. 2, 5, 12, and 15 samples, adding 18% of No. 3, 8, 9, 14 samples, and adding 24% of No. 4, 7, 10, and 13 samples. With the increase of curing time, the compressive strength of solidified body first increases and then stabilizes. The compressive strength of the sludge solidified body reaches the maximum when the solidification time is 21 d [30].

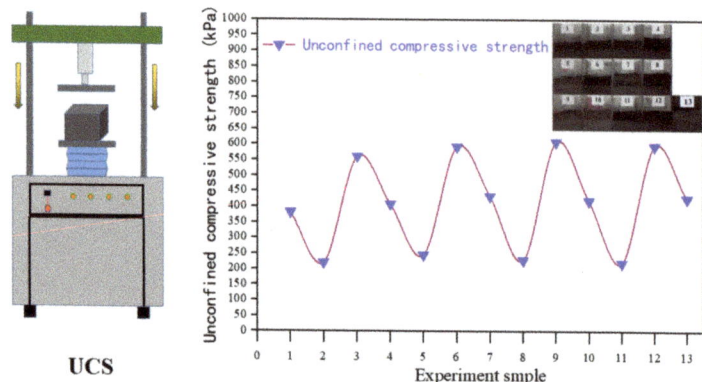

**Figure 12.** Unconfined compressive strength of the solidified specimen.

In the process of curing and stabilization, the compressive strength of the cured body is positively correlated with the curing time. The results show that the selection of reasonable curing time and cement mix proportion is the key to the sludge curing process.

## 4. Conclusions

(1) The number of oxygen-containing functional groups on the surface of BC modified by β-CD increased. β-CD-BC still has well-developed void structure and specific surface area.

(2) The compressive strength test shows that the strength is positively correlated with the amount of PC and the curing time, and the compressive strength is stable after 14 days. After β-CD-BC/PC curing, the speciation of HMs changed significantly, the residual Cu content increased from 56% to 75%; the residual Cr content increased from 54% to 82%; the residual CD content increased from 37% to 77%; residual Zn content increased from 61% to 87%; residual Pb increased from 59% to 79%.

(3) β-CD-BC achieves the adsorption of PHe through hydrophobic effect and π-π interaction, and enhances the adsorption of Pb, Cu, Cr, Cd, Zn through complexation and electrostatic attraction. This avoids the competition between the two pollutants for adsorption sites and improves the adsorption performance and adsorption capacity of BC.

**Author Contributions:** Conceptualization, G.L. and H.L.; methodology, G.L. and Y.Z.; software, X.L. and W.Z.; validation, H.L., X.C. and Y.L.; formal analysis, G.L.; investigation, L.W.; resources, H.L.; data curation, G.L.; writing—original draft preparation, G.L. and H.L.; writing—review and editing, G.L. and H.L.; supervision, H.L.; project administration, H.L. and Y.L.; funding acquisition, H.L. and Y.L. All authors have read and agreed to the published version of the manuscript.

**Funding:** This research was funded by the Key Technologies Research and Development Program (grant number 2019YFC1803804), the Fundamental Research Funds for the Central Universities (grant numbers N2001016 and N2001012), and National College Students Innovation and Entrepreneurship Training Program (grant number 210055).

**Institutional Review Board Statement:** Not applicable.

**Informed Consent Statement:** Not applicable.

**Data Availability Statement:** Data sharing not applicable. No new data were created or analyzed in this study. Data sharing is not applicable to this article.

**Conflicts of Interest:** The authors declare no conflict of interest. The funders had no role in the design of the study; in the collection, analyses, or interpretation of data; in the writing of the manuscript, or in the decision to publish the results.

## References

1. Li, L.; Zhu, C.; Liu, X.; Li, F.; Li, H.; Ye, J. Biochar amendment immobilizes arsenic in farmland and reduces its bioavailability. *Environ. Sci. Pollut. Res.* **2018**, *25*, 34091–34102. [CrossRef] [PubMed]
2. Ma, Y.; Liu, Z.; Xu, Y.; Zhou, S.; Wu, Y.; Wang, J.; Huang, Z.; Shi, Y. Remediating Potentially Toxic Metal and Organic Co-Contamination of Soil by Combining InSitu Solidification/Stabilization and Chemical Oxidation: Efficacy, Mechanism, and Evaluation. *Int. J. Environ. Res. Public Health* **2018**, *15*, 2595. [CrossRef]
3. Pu, H.; Mastoi, A.K.; Chen, X.; Song, D.; Qiu, J.; Yang, P. An integrated method for the rapid dewatering and solidification/stabilization of dredgedcontaminated sediment with a high water content. *Front. Environ. Sci. Eng.* **2021**, *15*, 67. [CrossRef]
4. Eisa, H.M.; Vaezi, I.; Ardakani, A.M. Evaluation of solidification/stabilization in arsenic-contaminated soils using lime dust and cement kiln dust. *Bull. Eng. Geol. Environ.* **2020**, *79*, 1683–1692. [CrossRef]
5. Gao, R.L.; Tang, M.; Fu, Q.L.; Guo, G.G.; Li, X.; Hu, H.Q. Fractions Transformation of Heavy Metals in Compound Contaminated Soil Treated with Biochar, Montmorillonite and Mixed Addition. *Environ. Sci.* **2017**, *38*, 361–367.
6. Ouhadi, V.R.; Yong, R.N.; Deiranlou, M. Enhancement of cement-based solidification/stabilization of a lead-contaminated smectite clay. *J. Hazard. Mater.* **2021**, *403*, 123969. [CrossRef]
7. Huang, G.; Jiang, X.; Wanyan, H.; Chen, C.; Jin, X.; Zhang, Y. Effects of Different Materials on the Stabilization of Heavy Metals in Sediments. *Res. Environ. Sci.* **2012**, *25*, 563–567.
8. Raja, R.; Pal, S. Remediation of heavy metal contaminated soils by solidification/stabilization with fly ash, quick lime and blast furnace slag. *J. Indian Chem. Soc.* **2019**, *96*, 481–486.
9. Ahmad, M.; Usman, A.R.; Al-Faraj, A.S.; Ahmad, M.; Sallam, A.; Al-Wabel, M.I. Phosphorus-loaded biochar changes soil heavy metals availability and uptakepotential of maize (*Zea mays* L.) plants. *Chemosphere* **2017**, *194*, 327–339. [CrossRef] [PubMed]
10. Ghasemi, S.; Gitipour, S.; Ghazban, F.; Hedayati, H. Treatment of Petroleum Drill Cuttings Using Stabilization/Solidification Method by Cement and Modified Clay Mixes. *Iran. J. Health Saf. Environ.* **2017**, *4*, 781–797.
11. Sora, I.N.; Pelosato, R.; Botta, D.; Dotelli, G. Chemistry and microstructure of cement pastes admixed with organic liquids. *J. Eur. Ceram. Soc.* **2002**, *22*, 1463–1473. [CrossRef]
12. Wang, F.; Wang, H.; Jin, F.; Al-Tabbaa, A. The performance of blended conventional and novel binders in the in-situ tabilization/solidification of a contaminated site soil. *J. Hazard. Mater.* **2015**, *285*, 46–52. [CrossRef] [PubMed]
13. Kogbara, R.B.; Al-Tabbaa, A.; Yi, Y.; Stegemann, J.A. Cement–fly ash stabilisation/solidification of contaminated soil: Performance properties and initiation of operating envelopes. *Appl. Geochem.* **2013**, *33*, 64–75. [CrossRef]
14. Al-Wabel, M.I.; Usman, A.R.A.; El-Naggar, A.H.; Aly, A.A.; Ibrahim, H.M.; Elmaghraby, S.; Al-Omran, A. Conocarpus biochar as a soil amendment for reducing heavymetal availability and uptake by maize plants. *Saudi J. Biol. Sci.* **2015**, *22*, 503–511. [CrossRef] [PubMed]
15. Alsbaiee, A.; Smith, B.J.; Xiao, L.; Ling, Y.; Helbling, D.E.; Dichtel, W.R. Rapid removal of organic micropollutants from water by a porous β-cyclodextrin polymer. *Nature* **2016**, *529*, 190–194. [CrossRef]
16. Liu, H.H.; Cai, X.Y.; Wang, Y.; Chen, J.W. Adsorption mechanism-based screening of cyclodextrin polymers for adsorption and separation of pesticides from water. *Water Res.* **2011**, *45*, 3499–3511. [CrossRef] [PubMed]
17. Zhao, E.P.; Repo, E.; Yin, D.L.; Meng, Y.; Jafari, S.; Sillanpaa, M. EDTA-cross-linked beta-cyclodextrin: An environmentally friendly bifunctional adsorbent for simultaneous adsorption of metals and cationic dyes. *Environ. Sci. Technol.* **2015**, *49*, 10570–10580. [CrossRef] [PubMed]
18. Li, B.Y.; Li, K.Q. Effect of nitric acid pre-oxidation concentration on pore structure and nitrogen/oxygen active decoration sites of ethylenediamine -modified biochar for mercury (II) adsorption and the possible mechanism. *Chemosphere* **2019**, *220*, 28–39. [CrossRef]
19. Liang, J.; Yang, Z.; Tang, L.; Zeng, G.; Yu, M.; Li, X.; Wu, H.; Qian, Y.; Li, X.; Luo, Y. Changes in heavy metal mobility and availability from contaminatedwetland soil remediated with combined biochar-compost. *Chemosphere* **2017**, *181*, 281–288. [CrossRef]
20. Xiao, X.; Chen, B.L.; Chen, Z.M.; Zhu, L.Z.; Schnoor, J.L. Insight into multiple and multilevel structures of biochars and their potential environmental applications: A critical review. *Environ. Sci. Technol.* **2018**, *52*, 5027–5047. [CrossRef]
21. Jiang, B.N.; Lin, Y.Q.; Mbog, J.C. Biochar derived from swine manure digestate and applied on the removals of heavy metals and antibiotics. *Bioresour. Technol.* **2018**, *270*, 603–611. [CrossRef] [PubMed]
22. Borah, H.J.; Gogoi, M.; Das, D.B.; Hazarika, S. Cyclodextrins-glutaraldehyde cross-linked nanofiltration membrane for recovery of resveratrol from plant extract. *J. Environ. Chem. Eng.* **2020**, *8*, 103620. [CrossRef]
23. He, J.; Li, Y.; Wang, C.; Zhang, K.; Lin, D.; Kong, L.; Liu, J. Rapid adsorption of Pb, Cu and Cd from aqueous solutions by β-cyclodextrin polymers. *Appl. Surf. Sci.* **2017**, *426*, 29–39. [CrossRef]
24. Pei, Z.; Li, L.; Sun, L.; Zhang, S.; Shao, X.; Yang, S.; Wen, B. Adsorption characteristics of 1,2,4-trichlorobenzene, 2,4,6-trichlorophenol, 2-naphthol and naphthalene on graphene and graphene oxide. *Carbon* **2013**, *51*, 156–163. [CrossRef]
25. Chen, Q.; Zheng, J.; Zheng, L.; Dang, Z.; Zhang, L. Classical theory and electron-scale view of exceptional Cd(II) adsorption onto mesoporous cellulose biochar via experimental analysis coupled with DFT calculations. *Chem. Eng. J.* **2018**, *350*, 1000–1009. [CrossRef]
26. Wang, J.; Chen, Z.; Chen, B. Adsorption of polycyclic aromatic hydrocarbons by graphene and graphene oxide nanosheets. *Environ. Sci. Technol.* **2014**, *48*, 4817–4825. [CrossRef]

27. Choong, C.E.; Ibrahim, S.; Yoon, Y.; Jang, M. Removal of lead and bisphenol A using magnesium silicate impregnated palm-shell waste powdered activated carbon: Comparative studies on single and binary pollutant adsorption. *Ecotoxicol. Environ. Saf.* **2018**, *148*, 142–151. [CrossRef] [PubMed]
28. Hu, X.; Hu, Y.; Xu, G.; Li, M.; Zhu, Y.; Jiang, L.; Tu, Y.; Zhu, X.; Xie, X.; Li, A. Green synthesis of a magnetic β-cyclodextrin polymer for rapid removal of organic micro-pollutants and heavy metals from dyeing wastewater. *Environ. Res.* **2020**, *180*, 108796. [CrossRef]
29. Dong, P.H.; Hayano, K.; Kikuchi, Y.; Takahashi, H.; Morikawa, Y. Deformation and crushing of particles of cement treat granulate soil. *Soils Found.* **2011**, *51*, 611–624. [CrossRef]
30. Wang, D.; Zentar, R.; Abriak, N.E.; Di, S. Long-term mechanical performance of marine sediments solidified with cement, lime, and fly ash. *Mar. Geores. Geotechnol.* **2018**, *36*, 123–130. [CrossRef]

Article

# Study on the Enhanced Remediation of Petroleum-Contaminated Soil by Biochar/g-C₃N₄ Composites

Hongyang Lin [1], Yang Yang [1], Zhenxiao Shang [2], Qiuhong Li [3], Xiaoyin Niu [2], Yanfei Ma [2,*] and Aiju Liu [2,*]

[1] School of Agricultural Engineering and Food Science, Shandong University of Technology, Zibo 255049, China; linhongyang2021@163.com (H.L.); 21403010279@stumail.sdut.edu.cn (Y.Y.)
[2] School of Resources and Environmental Engineering, Shandong University of Technology, Zibo 255049, China; alphazx@sdut.edu.cn (Z.S.); zbnxy@sdut.edu.cn (X.N.)
[3] School of Materials Science and Engineering, Shandong University of Technology, Zibo 255049, China; qhli@sdut.edu.cn
* Correspondence: mayanfei@sdut.edu.cn (Y.M.); aijvliu@sdut.edu.cn (A.L.)

**Abstract:** This work developed an environmentally-friendly soil remediation method based on BC and g-C₃N₄, and demonstrated the technical feasibility of remediating petroleum-contaminated soil with biochar/graphite carbon nitride (BC/g-C₃N₄). The synthesis of BC/g-C₃N₄ composites was used for the removal of TPH in soil via adsorption and photocatalysis. BC, g-C₃N₄, and BC/g-C₃N₄ have been characterized by scanning electron microscopy (SEM), Brunauer–Emmett–Teller surface area analyzer (BET), FT-IR, and X-ray diffraction (XRD). BC/g-C₃N₄ facilitates the degradation due to reducing recombination and better electron-hole pair separation. BC, g-C₃N₄, and BC/g-C₃N₄ were tested for their adsorption and photocatalytic degradation capacities. Excellent and promising results are brought out by an apparent synergism between adsorption and photocatalysis. The optimum doping ratio of 1:3 between BC and g-C₃N₄ was determined by single-factor experiments. The removal rate of total petroleum hydrocarbons (TPH) by BC/g-C₃N₄ reached 54.5% by adding BC/g-C₃N₄ at a dosing rate of 0.08 g/g in a neutral soil with 10% moisture content, which was 2.12 and 1.95 times of BC and g-C₃N₄, respectively. The removal process of TPH by BC/g-C₃N₄ conformed to the pseudo-second-order kinetic model. In addition, the removal rates of different petroleum components in soil were analyzed in terms of gas chromatography–mass spectrometry (GC-MS), and the removal rates of $nC_{13}$-$nC_{35}$ were above 90% with the contaminated soil treated by BC/g-C₃N₄. The radical scavenger experiments indicated that superoxide radical played the major role in the photocatalytic degradation of TPH. This work definitely demonstrates that the BC/g-C₃N₄ composites have great potential for application in the remediation of organic pollutant contaminated soil.

**Keywords:** biochar; adsorption; photocatalysis; synergy; TPH; soil

## 1. Introduction

In recent years, with the development of society, China's demand for petroleum and other products is increasing [1], but the problems of leakage in the process of petroleum exploitation, transportation, refining, processing, and use are becoming more and more serious [2,3]. A large amount of petroleum accumulates in soil, which inhibits the activity of indigenous microorganisms in soil, and destroys the ecological balance of the soil [4,5]. The petroleum hydrocarbons in soil will seep into the groundwater by migration and cause the pollution of groundwater, which poses a threat to human health [6,7].

At present, the remediation methods of petroleum-contaminated soil mainly include adsorption [8], incineration [9], chemical oxidation [10], leaching [11], phytoremediation [12], and the microbial remediation method [13–15]. Among them, the adsorption method is widely preferred because of its simple operation, low price, and relatively good adsorption effect [16,17]. BC is a pyrogenic carbonaceous material produced through the pyrolysis (350–700 °C) of agriculture and forestry biomass, food processing waste, animal manure, and other industrial waste biomass under an oxygen-free environment [18–20], which has a wide range of sources and a cheaper cost [21]. It has good adsorption and special photoelectric properties [22]. However, BC has some disadvantages in practical application, such as poor selective adsorption, easy saturation, easy desorption, insufficient long-acting performance, and so on [23]. Thus, some researchers focused on enhancing the removal capacity of BC by modifying its surface properties. For instance, Gurav et al. [20] modified pinewood BC with coconut oil to improve the hydrophobicity of BC, thereby increasing the adsorption capacity of BC on crude oil. The maximum adsorption capacity of the modified BC to crude oil reached 5315 mg/g in 60 min. Nguyen et al. [24] modified BC with $FeCl_3$, $AlCl_3$, and $CaCl_2$ to increase the number of surface functional groups, and the maximum chromium removal rate was 96.8%. Liu et al. [25] pretreated camellia oleifera shell with aminosulfonic acid to obtain the modified BC with a larger specific surface area and more functional groups, and the maximum adsorption capacity of tetracycline reached 412.95 mg/g.

In recent years, the photocatalytic degradation technology based on visible light irradiation has been diffusely recognized as a highly-efficient, low-cost, and environmentally-friendly method. However, the majority of traditional photocatalysts are metal-based materials, which can cause secondary pollution of the soil when they enter it [26]. $g-C_3N_4$ is a new type of non-metallic semiconductor photocatalytic material [27]. It has a large specific surface area, suitable energy band, and high photosensitivity, and can be regarded as a promising photocatalyst in the remediation of soil [28–30]. For example, Luo et al. [31] prepared $g-C_3N_4$ by three different precursors, including urea, dicyandiamide, and melamine, via the conventional thermal polymerization method. Under visible light or natural sunlight, $g-C_3N_4$ synthesized by three different precursors can remarkably reduce the toxicity of phenanthrene-contaminated soil. However, $g-C_3N_4$ also has the defect that the photogenerated charge carriers are recombined rapidly, which, consequently, seriously affects the application of $g-C_3N_4$ in the field of photocatalysis [32]. The $BC/g-C_3N_4$ composites were prepared by loading $g-C_3N_4$ onto the surface of BC, which could greatly improve the photocatalytic ability of $g-C_3N_4$. $BC/g-C_3N_4$ has the potential application in the remediation of soil because of its enhanced photocatalytic activities and lower toxicity. The enhanced photocatalytic activities of $BC/g-C_3N_4$ should be attributed to two aspects. On the one hand, BC could be used as effective electron transfer channels and acceptors to improve the separation of photogenerated electron–hole pairs. On the other hand, the large surface area of biochar can enrich pollutants while providing sufficient catalytic sites for photocatalytic degradation [33]. Li and Lin [34] successfully prepared the biochar-supported K-doped $g-C_3N_4$ composites, which displayed an enhanced optical absorption in the visible region and a wider photocatalytic application scope compared to pure $g-C_3N_4$. Therefore, it is of great significance to prepare $BC/g-C_3N_4$ composites by using BC with $g-C_3N_4$ from a wide range of sources.

In this experiment, $g-C_3N_4$ was prepared by high temperature pyrolysis with urea as the precursor, and a novel decontamination material $BC/g-C_3N_4$ was synthesized by the impregnation method, which is expected to have both adsorptive and photocatalytic capabilities, contributing to the cleanup of pollutants in soil media. The surface morphology and chemical structure of $BC/g-C_3N_4$ were characterized by means of SEM, BET, FT-IR, and XRD. Meanwhile, the influence of various external factors on the removal efficiency of petroleum hydrocarbons and the possible reaction mechanism were discussed. Aside from the intrinsic adsorptive property of BC materials, the $BC/g-C_3N_4$ composites were thought to have a light-responsive capability due to the existence of graphitic $g-C_3N_4$. We expect that the $BC/g-C_3N_4$ can be a value-added biomass-derived material for the removal of petroleum hydrocarbons from soil by synergistic adsorption and photocatalysis. The aim of this study is to demonstrate the feasibility of preparing the $BC/g-C_3N_4$ composites using the proposed method, and to evaluate the unknown performances of the resultant materials for the remediation of petroleum-contaminated soil.

## 2. Materials and Methods

### 2.1. Materials

Anhydrous sodium sulfate, urea, petroleum ether, calcium chloride, isopropanol, ascorbic acid, and ethylene diamine tetraacetic acid were all analytical reagents. Coconut shell carbon was purchased from Henan Gongyi Wanjiajing Environmental Protection Material Co., Ltd. (Henan, China). It was obtained by pyrolyzing coconut shells in a quartz tube furnace at 600 °C for 6 h at a heating rate of 10 °C/min in a nitrogen ($N_2$) atmosphere, and passing through a 120-mesh sieve. The soil samples were taken from the petrochemical production area of Shandong Haihua Group Co., Ltd. (Shandong, China). After collection, the soil samples were dried naturally for a week and then sieved through a 20-mesh sieve to remove large particles. The total petroleum hydrocarbon (TPH) content of the soil samples was 16,000 mg/kg, and the pH was about 7.

### 2.2. Preparation of Materials

Preparation of $g-C_3N_4$: 10g of urea was placed in a ceramic crucible with a cover, which was heated in a muffle furnace from 20 °C to 550 °C at a rate of 5 °C/min and held constant for 4 h. The samples were ground after the crucible cooling to room temperature, and $g-C_3N_4$ was obtained after passing through a 150-mesh standard sieve.

Preparation of $BC/g-C_3N_4$: the BC and $g-C_3N_4$ were mixed in proportion, and the deionized water was added into the mixed samples. Afterward, the mixed samples were ultrasonically cleaned for 30 min and stirred for 2 h at 40 °C to form the suspension. Then, the suspension was dried at 105 °C to a constant weight, and the $BC/g-C_3N_4$ composites were obtained. The doping ratios of BC to $g-C_3N_4$ were 2:1, 1:1, 1:2, 1:3, 1:4, and 1:5, and the different proportions of $BC/g-C_3N_4$ samples were obtained using the above methods.

### 2.3. Characterization Analysis of BC and $BC/g-C_3N_4$

The surface morphologies of $BC/g-C_3N_4$ and BC were obtained by a Zeiss Ultra 55 Scanning Electron Microscope (SEM, Zeiss, Sigma, Oberkochen, Germany). The specific surface areas of the samples were taken by $N_2$ adsorption-desorption using the Brunauer–Emmett–Teller method (BET, Micromeritics, asap246, Norcross, GA, USA). Fourier transform infrared spectroscopy (FT-IR) spectra was obtained by a Nicolet 5700 Series infrared spectrometer (Thermofisher, Nicolet 5700, Waltham, MA, USA). Samples were characterized by X-ray diffraction spectroscopy (XRD, BrukerAXS, D8-02, Karlsruhe, Germany) for phase identification.

### 2.4. Adsorption and Photodegradation Experiments

The adsorption experiments were conducted in the dark. Petroleum-contaminated soil samples (16 g) and 1.28 g BC or $g-C_3N_4$ were mixed in a petri dish with a diameter of 9 cm. The thickness of the mixed samples was about 1.4 mm, and the moisture content of the mixed samples were maintained at 10% by regularly adding the deionized water. The adsorption experiments were maintained at 25 °C, and the soil samples without the absorbent were used as a blank control. The soil samples were taken periodically, and then passed through a 60-mesh sieve to remove BC or $g-C_3N_4$, and the TPH content of the soil was detected by an ultraviolet spectrophotometer (UV-5100B, Shanghai Leewen Scientific Instrument Co., Ltd., Shanghai, China) with petroleum ether as the extraction agent.

The photodegradation experiments were carried out with the $g-C_3N_4$ as the photocatalyst under the irradiation of a LED light. The light was placed at the top of the incubator, and the intensity of the light was 8000 Lux. One blank sample was treated with light avoidance. The procedure of the photodegradation experiments was the same as the adsorption experiments, except for the LED light irradiation.

The adsorption and photodegradation experiments of $BC/g-C_3N_4$ were carried out under light irradiation, and the experiment procedure was the same as the photodegradation experiments. The soil samples without $BC/g-C_3N_4$ were used as a blank control group. To investigate the optimal conditions of $BC/g-C_3N_4$ treating the contaminated soil, the doping ratio, dosage, soil moisture content, and soil acidity–alkalinity were chosen as the influencing factors.

Setting of influencing factors: (1) Doping ratio: 2:1, 1:1, 1:2, 1:3, 1:4, 1:5; (2) The dosage of BC/g-$C_3N_4$: 0.01, 0.02, 0.04, 0.08, 0.12, 0.16 g/g; (3) Soil acidity and alkalinity: acidic, neutral, alkaline; (4) Soil moisture content: 2%, 5%, 10%, 20%, 30%; (5) Reaction time: 3, 5, 7, 9, 11, 13, 15, 17, 19, 21, 23, 25, 27, 29 d.

### 2.5. Free Radical Trapping Experiments

To investigate the mechanism of the photodegradation, inhibition tests on pure BC/g-$C_3N_4$ samples were carried out using different radical scavengers. Isopropanol (IPA), ascorbic acid (AA), and ethylene diamine tetraacetic acid (EDTA) were chosen as the scavengers for superoxide radical ($\cdot O_2^-$), hydroxyl radical ($\cdot OH$), and hole ($h^+$), respectively. The concentration of the scavengers was 1 mM.

### 2.6. Analysis of n-Alkanes and PAHs in Soil

Changes of petroleum components in soil samples were analyzed after allowing adsorption and photodegradation for 28 d. Samples were compared with those treated by BC or g-$C_3N_4$ alone.

In order to analyze the changes of $nC_{10}$-$nC_{40}$ in soil, the soil sample was mixed with the diatomite to dewater, and transferred into the extraction tank for extraction with the extraction repeated three times. All of the extraction liquid was collected into a concentration cup and concentrated to 1.0 mL. Then, the concentrated solution was passed through a magnesium silicate column, and the column was eluted with n-hexane of 12 mL. The concentrate and eluent were collected together and concentrated to 1 mL. The sample was then analyzed with a gas chromatography method (GC, Agilent Technologies 7890B, Palo Alto, CA, USA).

In order to analyze the changes of TPH in soil, the soil sample was mixed with anhydrous sodium sulfate and extracted by carbon tetrachloride for 0.5 h in the ultrasound instrument and 1 h in the oscillator. The extraction liquid was condensed to 1 mL in a rotary evaporator and separated by the silica gel column chromatography using dichloromethane as an eluent. The eluent was evaporated to dryness under a nitrogen atmosphere. The sample was then analyzed with a gas chromatography–mass spectrometry method (GC–MS, Agilent 7890B-5977B, Palo Alto, CA, USA).

### 2.7. Statistical Analysis

Each treatment was carried out in triplicate. The data are represented by the averages of three values, and the statistical analysis was performed using SPSS 22.0. One-way ANOVA ($p < 0.05$) was used for statistical analysis of the data obtained from the experiment, and the results of the experiments were presented with the standard deviation (SD) shown by an error bar. The letters are used to indicate whether there is a significant difference between the two data sets.

## 3. Results and Discussion

### 3.1. Characterization Analysis

#### 3.1.1. SEM Analysis

The surface morphology and microstructure of the prepared samples were investigated by SEM. As shown in Figure 1a, the microscopic morphology of BC is a tightly-arranged, banded, porous structure. The SEM image of BC/g-$C_3N_4$ is shown in Figure 1b, and the g-$C_3N_4$ was evenly distributed on the BC surface. In addition, the roughness and porosity of BC/g-$C_3N_4$ were increased, which was due to the g-$C_3N_4$ being well stacked on the BC matrix. This was beneficial for increasing the adsorption and photocatalytic efficiency of the material.

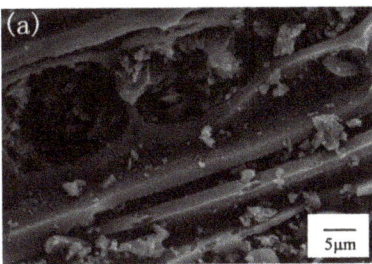 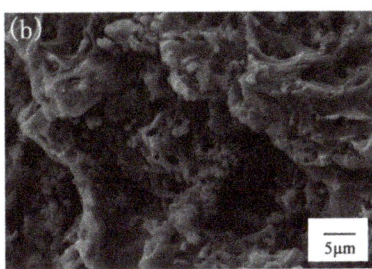

**Figure 1.** SEM images of (**a**) BC and (**b**) BC/g-$C_3N_4$ (1:3).

### 3.1.2. BET Analysis

The $N_2$ adsorption–desorption isotherms and pore size distribution for the BC and BC/g-$C_3N_4$ are shown in Figure 2. The $N_2$ BET surface area of the BC/g-$C_3N_4$ was as high as 612.69 $m^2/g$, which was higher than that of BC (541.61 $m^2/g$). After loading g-$C_3N_4$ on the surface of BC, the specific surface area of BC approximately doubled, indicating that the BC/g-$C_3N_4$ had more pore structures, which was also consistent with the result of SEM. The larger specific surface area was beneficial for the enrichment of pollutants, and provided sufficient catalytic sites for photocatalytic degradation [35]. However, the average pore size did not vary much. The average pore sizes of BC/g-$C_3N_4$ and BC were 2.03 nm and 3.30 nm, respectively. The adsorption isotherms of the BC/g-$C_3N_4$ showed type I isotherms (IUPCA classification), which was mainly monolayer adsorption. The adsorption isotherms of the BC showed type II isotherms, which was mainly multilayer absorption [36]. The loading of g-$C_3N_4$ on the surface of BC increased the specific surface area and pore structure, which facilitated the adsorption and photocatalytic degradation of TPH.

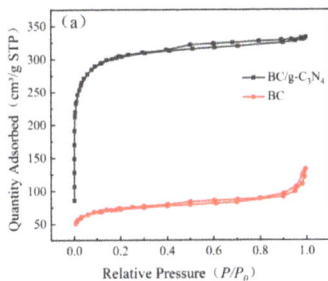 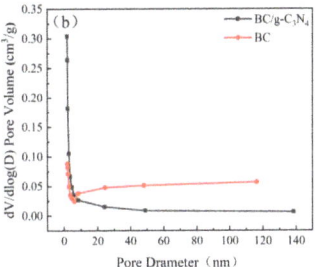

**Figure 2.** (**a**) $N_2$ adsorption–desorption isotherms; (**b**) pore size distribution curves of BC and BC/g-$C_3N_4$ (1:3).

### 3.1.3. FTIR and XRD Analysis

Figure 3a represents the FTIR spectra of BC, g-$C_3N_4$, and BC/g-$C_3N_4$. For the BC sample, the band at 1520 $cm^{-1}$ was due to the bending vibration of amino. The band at 1140 $cm^{-1}$ was attributed to the stretching vibration of C-O in cellulose, hemicellulosic and lignin or C-O-C in cellulose and hemicellulosic, and the one at 665 $cm^{-1}$ was attributed to the bending vibration peak of C-H of the aromatic ring. For the g-$C_3N_4$ and BC/g-$C_3N_4$ samples, an adsorption peak at 3200 $cm^{-1}$ was observed, which originated from the stretching vibrations of N-H and O-H. In addition, the g-$C_3N_4$ and BC/g-$C_3N_4$ samples presented a series of bands of the typical stretching vibration modes of C-N heterocycles (1600, 1480, and 1260 $cm^{-1}$) [37] and the intense bending vibration mode of the tri-s-triazine unit (800 $cm^{-1}$) [38]. The peak shape of BC/g-$C_3N_4$ was basically the same as those of g-$C_3N_4$, and the characteristic peaks of BC/g-$C_3N_4$ were weakened. This phenomenon indicated that the BC-loaded g-$C_3N_4$ was not a simple physical mixing but formed a compact structure with lower energy [28].

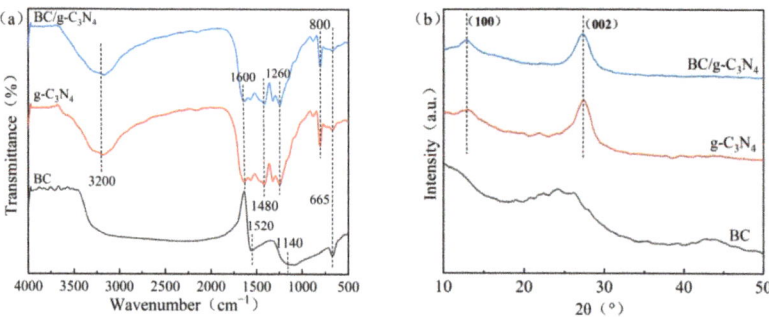

**Figure 3.** (**a**) FTIR and (**b**) XRD of BC, g-$C_3N_4$, and BC/g-$C_3N_4$ (1:3).

Figure 3b shows the XRD diffraction pattern of BC, g-$C_3N_4$, and BC/g-$C_3N_4$. The diffractograms of g-$C_3N_4$ and BC/g-$C_3N_4$ showed two peaks at $2\theta = 13.5°$ and $27°$, corresponding to two lattice planes ((100) and (002)) of g-$C_3N_4$, respectively, indicating that g-$C_3N_4$ was successfully loaded on the surface of BC.

### 3.2. Analysis of Adsorption and Photocatalytic Capacity

The removal efficiency of TPH in the contaminated soil samples by BC, g-$C_3N_4$, and BC/g-$C_3N_4$ is shown in Figure 4. The results showed that the BC removed the TPH from the soil mainly by adsorption, and 25.7% of TPH was removed in the 28 d experiment. The TPH removal rate of g-$C_3N_4$ was only 5% under dark conditions, whereas the TPH removal rate of g-$C_3N_4$ reached 27.9% under light conditions, indicating that g-$C_3N_4$ degraded TPH in soil mainly by photocatalysis. It is clearly evident that BC/g-$C_3N_4$ showed a higher removal potential than the BC and g-$C_3N_4$ ($p < 0.05$), as 54.5% of TPH was removed in the 28-d experiment, and the removal efficiency of TPH by BC/g-$C_3N_4$ was 2.12 times and 1.95 times of the BC and g-$C_3N_4$, respectively. This is because the BC/g-$C_3N_4$ had a larger specific surface area and a wider energy band, which facilitated the adsorption of TPH on its surface and the separation of electrons and holes, thus greatly improving the TPH removal efficiency of BC/g-$C_3N_4$ [39,40].

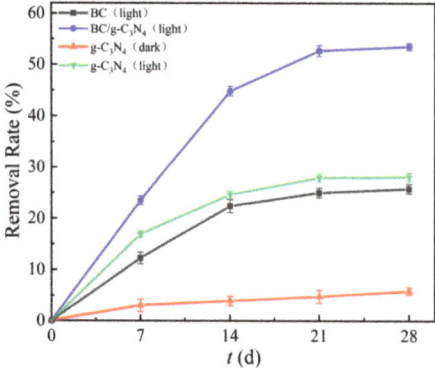

**Figure 4.** TPH removal efficiency by BC, BC/g-$C_3N_4$ (1:3), g-$C_3N_4$ (dark), g-$C_3N_4$ (light).

## 3.3. Removal of TPH in Soil by BC/g-C$_3$N$_4$

### 3.3.1. Effect of the BC to g-C$_3$N$_4$ Doping Ratio on TPH Removal Efficiency

Figure 5a shows the changes in TPH removal efficiency with different doping ratios of BC to g-C$_3$N$_4$. The results show that the TPH removal rate increased from 35% to 54% when the doping ratio changed from 2:1 to 1:3. However, as the g-C$_3$N$_4$ dosage continued to increase, the TPH removal efficiency gradually decreased. This may be due to the initially large amount of g-C$_3$N$_4$ adsorbed on the surface of BC, which increased the surface area of BC and provided more adsorption sites for TPH. At the same time, g-C$_3$N$_4$ played a role in the photocatalytic degradation of TPH in soil, and the removal rate of TPH by BC/g-C$_3$N$_4$ was greatly improved. When the g-C$_3$N$_4$ continued to increase, excessive g-C$_3$N$_4$ accumulated on the BC/g-C$_3$N$_4$ surface and occupied adsorption sites on the BC/g-C$_3$N$_4$ surface, showing competition with the adsorption of TPH. In addition, excessive g-C$_3$N$_4$ agglomerated on the surface of BC, reducing the contact area between the pollutant and the catalyst [41]. This was not conducive to the removal of TPH in the soil by BC/g-C$_3$N$_4$.

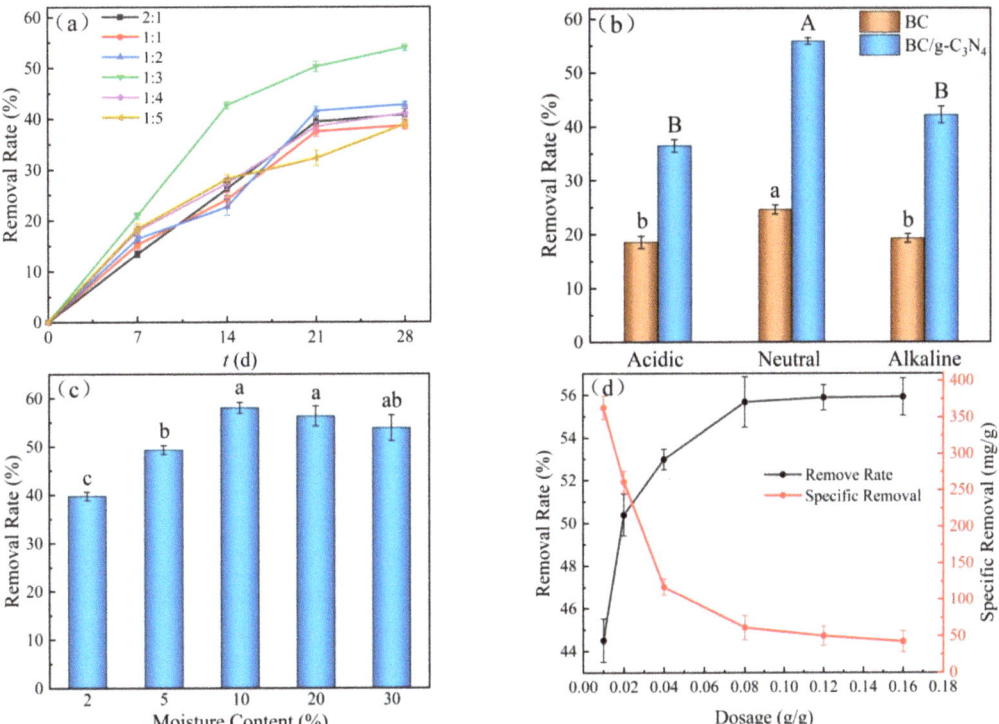

**Figure 5.** The TPH removal efficiency by BC/g-C$_3$N$_4$ under various conditions: (**a**) doping ratio, (**b**) pH, (**c**) moisture content, (**d**) dosage.

### 3.3.2. Effect of Soil Acidity and Alkalinity on TPH Removal Efficiency

Figure 5b shows the effect of soil acidity and alkalinity on the TPH removal efficiency by BC/g-C$_3$N$_4$. Both acidic soil and alkaline soil were not conducive to the TPH removal efficiency by BC/g-C$_3$N$_4$. A neutral environment was the most conducive to the removal of TPH in soil by BC/g-C$_3$N$_4$. The higher the concentration for H$^+$ in soil under acidic conditions, the more H$^+$ ions competed in the active sites of BC/g-C$_3$N$_4$ with TPH, which resulted in the reduction of TPH removal efficiency [42]. Under alkaline conditions, the organic acids and acidic substances of the petroleum reacted with the OH$^-$ ions in soil to produce surface-active substances, making TPH easily desorbed from BC/g-C$_3$N$_4$, which was not favorable to the adsorption and enrichment of TPH by BC/g-$C_3N_4$ [43]. Therefore, under neutral soil conditions, BC/g-$C_3N_4$ was more beneficial to the removal of TPH in soil.

### 3.3.3. Effect of Moisture Content on TPH Removal Efficiency

Figure 5c shows the effect of moisture content of soil on the TPH removal efficiency. It can be seen from the figure that the TPH removal efficiency gradually increased with the increasing of moisture content. Low moisture content was not conducive to the functioning of BC/g-$C_3N_4$, and the TPH removal rate of air-dried soil was only 40%. On the one hand, the moisture content of soil was so low that the electrons excited by visible light could only be transferred by air as the medium, and the transfer efficiency was not as good as in water. Thus, the TPH removal efficiency was low. The mass transfer rate of the solid–solid phase was much lower than that of the solid–liquid phase, so the low moisture content also inhibited the removal of TPH by BC/g-$C_3N_4$. When the moisture content of soil reached 10%, the TPH removal rate reached a maximum value of about 58% and then remained constant.

### 3.3.4. Effect of BC/g-$C_3N_4$ Dosage on TPH Removal Efficiency

Figure 5d shows the effect of the dosage of BC/g-$C_3N_4$ on the TPH removal efficiency. It can be seen from the figure that the dosage of BC/g-$C_3N_4$ increased from 0.01 g/g to 0.02 g/g, and the TPH removal rate increased rapidly from 44.5% to 50.8%. More adsorption sites were provided for TPH in the soil with the increase in the dosage of BC/g-$C_3N_4$, and more oxidation active intermediates were produced for the photocatalysis of TPH in the reaction system. However, with the excessive increase of BC/g-$C_3N_4$, the adsorption sites and oxidation active substances were surplus, which made the specific removal of TPH decrease. As shown in Figure 5d, when the dosage of BC/g-$C_3N_4$ increased from 0.01 g/g to 0.16 g/g, the specific removal amount of TPH decreased rapidly from 354.1 mg/g to 32.1 mg/g. Therefore, the optimal dosage of 0.08 g/g can not only achieve a better removal effect, but also control the reasonable production cost.

### 3.4. Kinetic Analysis

The experimental data were fitted by pseudo-first-order kinetic and pseudo-second-order kinetic [44]. The fitting results are shown in Figure 6 and Table 1.

$$\text{The pseudo-first-order kinetic model: } q_t = q_e\left(1 - e^{-K_1 t}\right) \tag{1}$$

$$\text{The pseudo-second-order kinetic model: } q_t = \frac{q_e^2 K_2 t}{1 + q_e K_2 t} \tag{2}$$

where $q_e$ is the equilibrium removal capacity of TPH, mg/g; $t$ is the reaction time, d; $K_1$ is the rate constant of the pseudo-first-order kinetic model, $d^{-1}$; $K_2$ is the rate constant of the pseudo-second-order kinetic model, mg·$g^{-1}·d^{-1}$.

**Table 1.** Kinetic parameters of TPH removal by BC/g-$C_3N_4$.

| Pseudo-First-Order Kinetics Model | | | Pseudo-Second-Order Kinetics Model | | |
|---|---|---|---|---|---|
| $q_e$/(mg·$g^{-1}$) | $K_1$/$d^{-1}$ | $R^2$ | $q_e$/(mg·$g^{-1}$) | $K_2$/(mg·$g^{-1}$·$d^{-1}$) | $R^2$ |
| 102.51 | 0.0620 | 0.959 | 101.72 | 0.0161 | 0.981 |

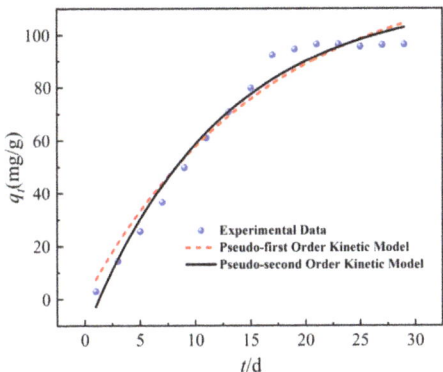

**Figure 6.** Kinetics model of TPH removal by BC/g-C$_3$N$_4$.

Figure 6 and Table 1 show the pseudo-first-order kinetic model, pseudo-second-order kinetic model, and fitting parameters of the TPH removal by BC/g-C$_3$N$_4$, respectively. As shown in Figure 6, the removal rate of TPH from soil by BC/g-C$_3$N$_4$ increased gradually with time at first. After a period of time, the removal rate of TPH gradually reached the equilibrium state. It can be seen from Table 1 that the $R^2$ of the pseudo-second-order kinetic model was larger than that of the pseudo-first-order kinetic model, and the theoretical removal capacity calculated by the pseudo-second-order kinetic model was closer to the actual value, so the pseudo-second-order kinetic model was more suitable for the process of TPH removal. Therefore, the removal of TPH in soil by BC/g-C$_3$N$_4$ was not only a simple physical adsorption, but also included a chemical process [45].

### 3.5. Gas Chromatography Analysis

The contents of n-alkanes and PAHs in soil were determined by GC and GC-MS methods. According to the test results, the total concentration of PAHs in the original soil sample was 0.63 mg/kg, and only benzo(g,h,i)perylene, benzo(k)fluoranthene, dibenz(a,h)anthracene, and indeno(1,2,3-cd)pyrene were detected. There were no PAHs that could be detected after the soil samples were treated by BC, g-C$_3$N$_4$ or BC/g-C$_3$N$_4$. Figure 7a shows the GC chromatogram of n-alkanes in soil samples treated by different materials. As observed from the chromatogram of the original soil sample, the n-alkanes were composed of a wide range of hydrocarbons from nC$_{13}$ to nC$_{35}$. The relative concentrations of nC$_{17}$-nC$_{28}$ were higher than the other n-alkanes, and that of C$_{28}$ was the highest. The peaks of nC$_{13}$ and nC$_{14}$ almost disappeared with the contaminated soil treated by BC, g-C$_3$N$_4$, or BC/g-C$_3$N$_4$. In addition, the peaks of nC$_{29}$-nC$_{35}$ also almost disappeared with the contaminated soil treated by BC/g-C$_3$N$_4$. As shown in Figure 7b, the removal rates of nC$_{13}$-nC$_{15}$ were above 95% with the contaminated soil treated by BC or g-C$_3$N$_4$. The removal rates of nC$_{16}$-nC$_{35}$ were around 50% for BC treatment, which reached around 80% for g-C$_3$N$_4$ treatment. When BC/g-C$_3$N$_4$ was used to remove n-alkanes from soils, the nC$_{13}$-nC$_{15}$ were completely removed, and the removal rates of the nC$_{16}$-nC$_{35}$ increased to more than 90%. The BC/g-C$_3$N$_4$ synergistically worked to improve the removal of petroleum hydrocarbons from the soil significantly, which was consistent with the previous results shown in Figure 4.

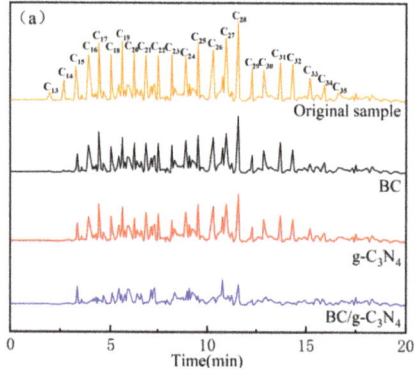
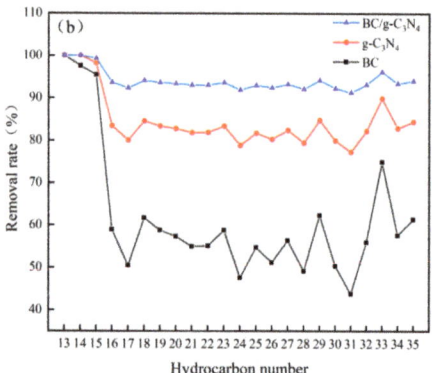

**Figure 7.** (a) GC chromatogram of n-alkanes in soil samples treated with different materials at 28 d; (b) The removal rate of each n-alkane in soil samples treated with different materials.

In addition to PAHs and n-alkanes, the TPH in contaminated soil also included branched alkanes and cycloalkanes. The removal of n-alkanes by g-$C_3N_4$ photocatalysis was higher than that by BC adsorption, but the difference in the removal of TPH was not significant, indicating that the removal of branched alkanes and cycloalkanes by BC adsorption was more effective. This is because the structure of cycloalkanes was stable and it was difficult to open their rings, so the g-$C_3N_4$ photocatalytic degradation was difficult to perform. However, the adsorption of BC was not selective and was little-influenced by the nature of the pollutants. Thus, the removal rates of different components in TPH by BC were more average. The removal rates of n-alkanes treated by g-$C_3N_4$ and BC/g-$C_3N_4$ were much higher than the TPH removal rates in soil samples. This indicated that g-$C_3N_4$ and BC/g-$C_3N_4$ were relatively ineffective in removing branched and cyclic alkanes from the TPH, and that the remaining constituents in the soil were mostly obstinate-branched and cyclic alkanes [46].

### 3.6. Photocatalysis Mechanism

Figure 8 shows the adsorption and photocatalytic mechanism of BC/g-$C_3N_4$. The g-$C_3N_4$ could generate electron–hole pairs under visible light. Under irradiation, electrons were excited from the valence band (VB) of g-$C_3N_4$ to its conduction band (CB), leaving a positively-charged hole ($h^+$) in the VB of g-$C_3N_4$. The disadvantage of the pure g-$C_3N_4$ photocatalytic material was that the photo-generated electrons and holes rapidly combined to fail, which affected its photocatalytic efficiency [47]. As a conductive channel, BC enhanced the electron transfer rate of g-$C_3N_4$ and provided more active sites for photocatalysis [48]. In addition, BC played the role of electron traps, and the photo-generated electrons of g-$C_3N_4$ could be transferred to BC for temporary storage, which was conducive to the separation of electron–hole pairs, so that more electrons were involved in the photocatalytic process and the photocatalytic efficiency was improved [49]. The rich pore structure and large specific surface area of BC could enrich TPH in soil, and had a synergistic effect with the photocatalytic process of g-$C_3N_4$.

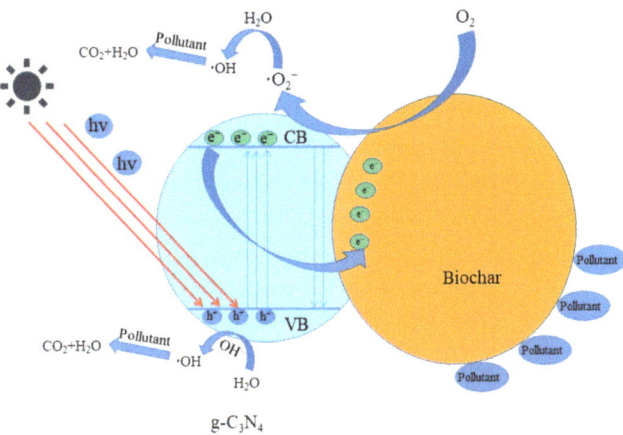

**Figure 8.** Removal mechanism of TPH by BC/g-$C_3N_4$ (1:3).

The electrons in g-$C_3N_4$ and the electrons accumulated on BC combined with $O_2$ to form superoxide radical ($\cdot O_2^-$), whereas the remaining holes in the VB combined with $H_2O$ to form hydroxyl radical ($\cdot OH$). The removal of petroleum hydrocarbons from the soil is carried out by the strong oxidizing properties of these groups. In order to understand the degradation mechanism in depth, the active radicals were identified by free radical capture experiments. Isopropanol (IPA), ascorbic acid (AA), and ethylene diamine tetraacetic acid (EDTA) were employed as the hydroxyl radical ($\cdot OH$), superoxide radical ($\cdot O_2^-$), and hole ($h^+$) scavengers. As Figure 9 shows, the removal rates of TPH decreased when the scavengers were added, indicating that $\cdot OH$, $\cdot O_2^-$, and $h^+$ were involved in the photocatalytic degradation of TPH. In particular, the removal rate of TPH greatly decreased from 55.94 to 24.87% after the addition of AA, indicating that $\cdot O_2^-$ was the main active radical in the photocatalytic process.

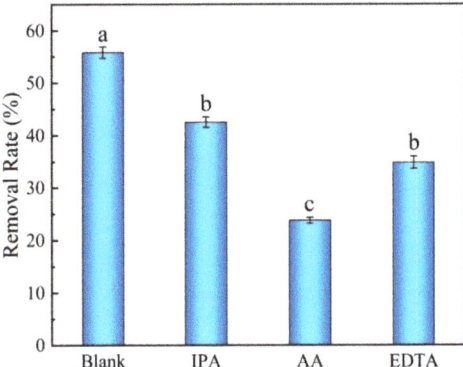

**Figure 9.** BC/g-$C_3N_4$ (1:3) toward TPH degradation in the presence of different trapping agents.

### 4. Conclusions

An environmentally-friendly soil remediation method based on the BC and g-$C_3N_4$ has been developed to remediate petroleum-contaminated soil. The novel composites BC/g-$C_3N_4$ were prepared by the impregnation method. It can be confirmed that the BC/g-$C_3N_4$ was successfully formed from the surface composition and morphological structure. The results of the adsorption and photodegradation experiments showed that the composite had superior adsorption and photocatalytic properties. Under the visible light irradiation, the BC/g-$C_3N_4$ showed excellent performance for removing the TPH from soil. The removal rate of TPH by BC/g-$C_3N_4$ was 2.12 and 1.95 times that of BC and g-$C_3N_4$. As a conductive channel, BC enhanced the electron transfer rate of g-$C_3N_4$ and provided more active sites for photocatalysis. In addition, the porous structure of BC can provide more adsorption sites for TPH. These effects synergistically improved the adsorption and photocatalytic activity of BC/g-$C_3N_4$ significantly. The optimum doping ratio of 1:3 between BC and g-$C_3N_4$ was determined by single-factor experiments, and the best removal effect could be achieved by adding BC/g-$C_3N_4$ at a dosing rate of 0.08 g/g in a neutral soil with 10% moisture content. The active species trapping experiments reveal that the $\cdot O_2^-$ plays a major role in the process of TPH degradation. The BC/g-$C_3N_4$ composites could be considered a promising material for the degradation of organic pollutants. This study provides a potential idea for the efficient degradation of TPH in soil. However, this method also faces the problem of difficult catalyst recovery in practical applications. In the future, compounding with magnetic nanomaterials should be considered, to use magnetism to recover composite materials from soil, so as to achieve the reuse of materials and to reduce costs. In addition, it is important to explore the co-doping or tri-doping of composites with certain elements to further improve their light absorption range and electron transfer capability.

**Author Contributions:** Conceptualization, Y.M. and A.L.; Data curation, H.L.; Formal analysis, H.L.; Investigation, H.L., Y.Y., Z.S., Q.L. and X.N.; Methodology, Y.M. and H.L.; Project administration, Y.M.; Resources, Y.M.; Software, Y.Y. and Z.S.; Supervision, Q.L. and A.L.; Validation, H.L., Y.Y. and X.N.; Visualization, Z.S., Q.L. and X.N.; Writing—original draft, H.L.; Writing—review and editing, H.L., Y.M. and A.L. All authors have read and agreed to the published version of the manuscript.

**Funding:** This work was funded by the Shandong Provincial Natural Science Foundation (No. ZR2020MD108, No. ZR2020ZD19), the Yellow River Delta Research Institute Innovation Research Foundation Program (No. 118171), the National Natural Science Foundation of China (No. 51804188), and the Ph.D. Program of Shandong Provincial Natural Science Foundation (No. ZR2018BEE015).

**Institutional Review Board Statement:** Not applicable.

**Informed Consent Statement:** Not applicable.

**Conflicts of Interest:** The authors declare no conflict of interest.

### Abbreviations

Biochar (BC), graphite carbon nitride (g-$C_3N_4$), biochar/graphite carbon nitride (BC/g-$C_3N_4$), total petroleum hydrocarbon (TPH), scanning electron microscopy (SEM), Brunauer–Emmett–Teller surface area analyzer (BET), Fourier transform infrared spectroscopy (FTIR), X-ray diffraction (XRD), gas chromatography–mass spectrometry (GC-MS), valence band (VB), conduction band (CB), isopropanol (IPA), ascorbic acid (AA), ethylene diamine tetraacetic acid (EDTA).

### References

1. Varjani, S.J.; Upasani, V.N. Core Flood study for enhanced oil recovery through ex-situ bioaugmentation with thermo- and halo-tolerant rhamnolipid produced by *Pseudomonas aeruginosa* NCIM 5514. *Bioresour. Technol.* **2016**, *220*, 175–182. [CrossRef] [PubMed]
2. Lee, Y.Y.; Seo, Y.; Ha, M.; Lee, J.; Yang, H.; Cho, K.S. Evaluation of rhizoremediation and methane emission in diesel-contaminated soil cultivated with tall fescue (*Festuca arundinacea*). *Environ. Res.* **2021**, *194*, 110606. [CrossRef] [PubMed]
3. Zhang, K.; Zhang, Z.-H.; Wang, H.; Wang, X.-M.; Zhang, X.-H.; Xie, Y.F. Synergistic effects of combining ozonation, ceramic membrane filtration and biologically active carbon filtration for wastewater reclamation. *J. Hazard. Mater.* **2020**, *382*, 121091. [CrossRef] [PubMed]
4. Bulai, I.S.; Adamu, H.; Umar, Y.A.; Sabo, A. Biocatalytic remediation of used motor oil-contaminated soil by fruit garbage enzymes. *J. Environ. Chem. Eng.* **2021**, *9*, 105465. [CrossRef]
5. Guo, H.; Yao, J.; Cai, M.; Qian, Y.; Guo, Y.; Richnow, H.H.; Blake, R.E.; Doni, S.; Ceccanti, B. Effects of petroleum contamination on soil microbial numbers, metabolic activity and urease activity. *Chemosphere* **2012**, *87*, 1273–1280. [CrossRef]

6. Kwon, M.J.; O'Loughlin, E.J.; Ham, B.; Hwang, Y.; Shim, M.; Lee, S. Application of an in-situ soil sampler for assessing subsurface biogeochemical dynamics in a diesel-contaminated coastal site during soil flushing operations. *J. Environ. Manag.* **2018**, *206*, 938–948. [CrossRef]
7. Pinedo, J.; Ibañez, R.; Lijzen, J.P.; Irabien, A. Assessment of soil pollution based on total petroleum hydrocarbons and individual oil substances. *J. Environ. Manag.* **2013**, *130*, 72–79. [CrossRef]
8. Zahed, M.A.; Salehi, S.; Madadi, R.; Hejabi, F. Biochar as a sustainable product for remediation of petroleum contaminated soil. *Curr. Res. Green Sustain. Chem.* **2021**, *4*, 100055. [CrossRef]
9. Samolada, M.C.; Zabaniotou, A.A. Comparative assessment of municipal sewage sludge incineration, gasification and pyrolysis for a sustainable sludge-to-energy management in Greece. *Waste Manag.* **2014**, *34*, 411–420. [CrossRef]
10. Desalegn, B.; Megharaj, M.; Chen, Z.L.; Naidu, R. Green mango peel-nanozerovalent iron activated persulfate oxidation of petroleum hydrocarbons in oil sludge contaminated soil. *Environ. Technol. Innov.* **2018**, *11*, 142–152. [CrossRef]
11. Chen, S.L.; Yi, Z.Y.; Wang, J.; Pan, C.Y.; Chang, S.; Guo, Q.W.; Zhou, J.G.; Sun, L. Case study on remediation of diesel contaminated soil and groundwater by eluent-extraction technology. *Environ. Eng.* **2020**, *38*, 178–182. [CrossRef]
12. Steliga, T.; Kluk, D. Application of *Festuca arundinacea* in phytoremediation of soils contaminated with Pb, Ni, Cd and petroleum hydrocarbons. *Ecotoxicol. Environ. Saf.* **2020**, *194*, 110409. [CrossRef]
13. Chen, C.-H.; Liu, P.G.; Whang, L.-M. Effects of natural organic matters on bioavailability of petroleum hydrocarbons in soil-water environments. *Chemosphere* **2019**, *233*, 843–851. [CrossRef]
14. Dickson, U.J.; Coffey, M.; Mortimer, R.J.G.; Smith, B.; Ray, N.; Di Bonito, M. Investigating the potential of sunflower species, fermented palm wine and *Pleurotus ostreatus* for treatment of petroleum-contaminated soil. *Chemosphere* **2020**, *240*, 124881. [CrossRef]
15. Jasmine, J.; Mukherji, S. Impact of bioremediation strategies on slurry phase treatment of aged oily sludge from a refinery. *J. Environ. Manag.* **2019**, *246*, 625–635. [CrossRef]
16. Yuan, M.J.; Tong, S.T.; Zhao, S.Q.; Jia, C.Q. Adsorption of polycyclic aromatic hydrocarbons from water using petroleum coke-derived porous carbon. *J. Hazard. Mater.* **2010**, *181*, 1115–1120. [CrossRef]
17. Esmaeili, A.; Saremnia, B. Comparison study of adsorption and nanofiltration methods for removal of total petroleum hydrocarbons from oil-field wastewater. *J. Pet. Sci. Eng.* **2018**, *171*, 403–413. [CrossRef]
18. Sajjad, A.; Jabeen, F.; Farid, M.; Fatima, Q.; Akbar, A.; Ali, Q.; Hussain, I.; Iftikhar, U.; Farid, S.; Ishaq, H.K. Biochar: A Sustainable Product for Remediation of Contaminated Soils. In *Plant Ecophysiology and Adaptation under Climate Change: Mechanisms and Perspectives II*; Hasanuzzaman, M., Ed.; Springer: Singapore, 2020; pp. 787–799. [CrossRef]
19. Aziz, S.; Ali, M.I.; Farooq, U.; Jamal, A.; Liu, F.-J.; He, H.; Guo, H.; Urynowicz, M.; Huang, Z. Enhanced bioremediation of diesel range hydrocarbons in soil using biochar made from organic wastes. *Environ. Monit. Assess.* **2020**, *192*, 569. [CrossRef]
20. Gurav, R.; Bhatia, S.K.; Choi, T.-R.; Choi, Y.-K.; Kim, H.J.; Song, H.-S.; Park, S.L.; Lee, H.S.; Lee, S.M.; Choi, K.-Y.; et al. Adsorptive removal of crude petroleum oil from water using floating pinewood biochar decorated with coconut oil-derived fatty acids. *Sci. Total Environ.* **2021**, *781*, 146636. [CrossRef]
21. Sun, Y.W.; Zeng, B.Y.; Dai, Y.T.; Liang, X.J.; Zhang, L.J.; Ahmad, R.; Su, X.T. Modification of sludge-based biochar using air roasting-oxidation and its performance in adsorption of uranium(VI) from aqueous solutions. *J. Colloid Interface Sci.* **2022**, *614*, 547–555. [CrossRef]
22. Leichtweis, J.; Silvestri, S.; Carissimi, E. New composite of pecan nutshells biochar-ZnO for sequential removal of acid red 97 by adsorption and photocatalysis. *Biomass-Bioenergy* **2020**, *140*, 105648. [CrossRef]
23. Wang, H.X.; Teng, H.W.; Wang, X.Y.; Xu, J.L.; Sheng, L.X. Physicochemical modification of corn straw biochar to improve performance and its application of constructed wetland substrate to treat city tail water. *J. Environ. Manag.* **2022**, *310*, 114758. [CrossRef]
24. Nguyen, D.L.T.; Binh, Q.A.; Nguyen, X.C.; Nguyen, T.T.H.; Vo, Q.N.; Nguyen, T.D.; Tran, T.C.P.; Kim, S.Y.; Nguyen, T.P.; Bae, J.; et al. Metal salt-modified biochars derived from agro-waste for effective congo red dye removal. *Environ. Res.* **2021**, *200*, 111492. [CrossRef]
25. Liu, Y.N.; Li, F.M.; Deng, J.Q.; Wu, Z.Q.; Lei, T.Z.; Tan, M.J.; Wu, Z.J.; Qin, X.L.; Li, H. Mechanism of sulfamic acid modified biochar for highly efficient removal of tetracycline. *J. Anal. Appl. Pyrolysis* **2021**, *158*, 105247. [CrossRef]
26. Wang, J.; Luo, Z.J.; Song, Y.Y.; Zheng, X.R.; Qu, L.L.; Qian, J.C.; Wu, Y.W.; Wu, X.Y.; Wu, Z.R. Remediation of phenanthrene contaminated soil by g-$C_3N_4$/$Fe_3O_4$ composites and its phytotoxicity evaluation. *Chemosphere* **2019**, *221*, 554–562. [CrossRef]
27. Hao, Q.; Chen, T.; Wang, R.T.; Feng, J.R.; Chen, D.M.; Yao, W.Q. A separation-free polyacrylamide/bentonite/graphitic carbon nitride hydrogel with excellent performance in water treatment. *J. Clean. Prod.* **2018**, *197*, 1222–1230. [CrossRef]
28. Rajalakshmi, N.; Barathi, D.; Meyvel, S.; Sathya, P. S-scheme $Ag_2CrO_4$/g-$C_3N_4$ photocatalyst for effective degradation of organic pollutants under visible light. *Inorg. Chem. Commun.* **2021**, *132*, 108849. [CrossRef]
29. Sharma, A.; Kanth, S.K.; Xu, S.S.; Han, N.; Zhu, L.; Fan, L.L.; Liu, C.; Zhang, Q.F. Visible light driven g-$C_3N_4$/$Bi_4NbO8X$ (X=Cl, Br) heterojunction photocatalyst for the degradation of organic pollutants. *J. Alloys Compd.* **2022**, *895*, 162576. [CrossRef]
30. Fan, G.; Ma, Z.Y.; Li, X.B.; Deng, L.J. Coupling of $Bi_2O_3$ nanoparticles with g-$C_3N_4$ for enhanced photocatalytic degradation of methylene blue. *Ceram. Int.* **2021**, *47*, 5758–5766. [CrossRef]

31. Luo, Z.J.; Song, Y.Y.; Wang, M.J.; Zheng, X.R.; Qu, L.L.; Wang, J.; Wu, X.Y.; Wu, Z.R. Comparison of g-$C_3N_4$ synthesized by different precursors in remediation of phenanthrene contaminated soil and ecotoxicity. *J. Photochem. Photobiol. A Chem.* **2020**, *389*, 112241. [CrossRef]
32. Luo, S.Y.; Li, S.P.; Zhang, S.; Cheng, Z.Y.; Nguyen, T.T.; Guo, M.H. Visible-light-driven Z-scheme protonated g-$C_3N_4$/wood flour biochar/$BiVO_4$ photocatalyst with biochar as charge-transfer channel for enhanced RhB degradation and Cr(VI) reduction. *Sci. Total Environ.* **2022**, *806*, 150662. [CrossRef] [PubMed]
33. Tang, R.D.; Gong, D.X.; Deng, Y.C.; Xiong, S.; Zheng, J.F.; Li, L.; Zhou, Z.P.; Su, L.; Zhao, J. π-π stacking derived from graphene-like biochar/g-$C_3N_4$ with tunable band structure for photocatalytic antibiotics degradation via peroxymonosulfate activation. *J. Hazard. Mater.* **2021**, *423*, 126944. [CrossRef] [PubMed]
34. Li, F.Y.; Lin, M.X. Synthesis of Biochar-Supported K-doped g-$C_3N_4$ Photocatalyst for Enhancing the Polycyclic Aromatic Hydrocarbon Degradation Activity. *Int. J. Environ. Res. Public Health* **2020**, *17*, 2065. [CrossRef] [PubMed]
35. Shi, L.; Liang, L.; Ma, J.; Wang, F.; Sun, J. Remarkably enhanced photocatalytic activity of ordered mesoporous carbon/g-$C_3N_4$ composite photocatalysts under visible light. *Dalton Trans.* **2014**, *43*, 7236–7244. [CrossRef]
36. Muttakin, M.; Mitra, S.; Thu, K.; Ito, K.; Saha, B.B. Theoretical framework to evaluate minimum desorption temperature for IUPAC classified adsorption isotherms. *Int. J. Heat Mass Transf.* **2018**, *122*, 795–805. [CrossRef]
37. Ramli, A.Z.; Sabudin, S.; Batcha, M.F.M.; Waehahyee, M. Analysis of Tar Properties Produced During Co-Gasification of Empty Fruit Bunch Pellet and Oil Palm Shell. *Int. J. Eng. Sci.* **2020**, *12*, 52–59. Available online: https://publisher.uthm.edu.my/ojs/index.php/ijie/article/view/2930 (accessed on 15 May 2022).
38. Shekardasht, M.B.; Givianrad, M.H.; Gharbani, P.; Mirjafary, Z.; Mehrizad, A. Preparation of a novel Z-scheme g-$C_3N_4$/RGO/$Bi_2Fe_4O_9$ nanophotocatalyst for degradation of Congo Red dye under visible light. *Diam. Relat. Mater.* **2020**, *109*, 108008. [CrossRef]
39. Li, H.; Mahyoub, S.A.A.; Liao, W.J.; Xia, S.Q.; Zhao, H.C.; Guo, M.Y.; Ma, P.S. Effect of pyrolysis temperature on characteristics and aromatic contaminants adsorption behavior of magnetic biochar derived from pyrolysis oil distillation residue. *Bioresour. Technol.* **2017**, *223*, 20–26. [CrossRef]
40. Kumar, A.; Kumar, R.; Sharma, G.; Naushad, M.; Stadler, F.J.; Ghfar, A.A.; Dhiman, P.; Saini, R.V. Sustainable nano-hybrids of magnetic biochar supported g-$C_3N_4$/$FeVO_4$ for solar powered degradation of noxious pollutants—Synergism of adsorption, photocatalysis & photo-ozonation. *J. Clean. Prod.* **2017**, *165*, 431–451. [CrossRef]
41. Lin, M.X. Photocatalysis and Microwave Degradation of Petroleum Hydrocarbons in Soil by Biochar-Supported Modified g-$C_3N_4$. Master's Thesis, Liaoning Petrochemical University, Fushun, China, 2019. [CrossRef]
42. Fan, S.S.; Wang, Y.; Wang, Z.; Tang, J.; Tang, J.; Li, X.D. Removal of methylene blue from aqueous solution by sewage sludge-derived biochar: Adsorption kinetics, equilibrium, thermodynamics and mechanism. *J. Environ. Chem. Eng.* **2017**, *5*, 601–611. [CrossRef]
43. Pradubmook, T.; O'Haver, J.H.; Malakul, P.; Harwell, J.H. Effect of pH on adsolubilization of toluene and acetophenone into adsorbed surfactant on precipitated silica. *Colloids Surf. A Physicochem. Eng. Asp.* **2003**, *224*, 93–98. [CrossRef]
44. Bulut, E.; Özacar, M.; Şengil, I.A. Adsorption of malachite green onto bentonite: Equilibrium and kinetic studies and process design. *Microporous Mesoporous Mater.* **2008**, *115*, 234–246. [CrossRef]
45. Ji, X.Q.; Lv, L.; Chen, F.; Yang, C.P. Sorption properties and mechanisms of organic dyes by straw biochar. *Acta Sci. Circumstantiae* **2016**, *36*, 1648–1654. [CrossRef]
46. Li, Y.-T.; Zhang, J.-J.; Li, Y.-H.; Chen, J.-L.; Du, W.-Y. Treatment of soil contaminated with petroleum hydrocarbons using activated persulfate oxidation, ultrasound, and heat: A kinetic and thermodynamic study. *Chem. Eng. J.* **2021**, *428*, 131336. [CrossRef]
47. Li, Y.; Xing, X.Y.; Pei, J.Z.; Li, R.; Wen, Y.; Cui, S.C.; Liu, T. Automobile exhaust gas purification material based on physical adsorption of tourmaline powder and visible light catalytic decomposition of g-$C_3N_4$/$BiVO_4$. *Ceram. Int.* **2020**, *46*, 12637–12647. [CrossRef]
48. Ma, Z.-P.; Zhang, L.N.; Ma, X.; Zhang, Y.-H.; Shi, F.-N. Design of Z-scheme g-$C_3N_4$/BC/$Bi_{25}FeO_{40}$ photocatalyst with unique electron transfer channels for efficient degradation of tetracycline hydrochloride waste. *Chemosphere* **2022**, *289*, 133262. [CrossRef]
49. Yin, H.F.; Cao, Y.; Fan, T.L.; Li, P.F.; Liu, X.H. Construction of AgBr/β-$Ag_2WO_4$/g-$C_3N_4$ ternary composites with dual Z-scheme band alignment for efficient organic pollutants removal. *Sep. Purif. Technol.* **2021**, *272*, 118251. [CrossRef]

Article

# Antimony Immobilization in Primary-Explosives-Contaminated Soils by Fe–Al-Based Amendments

Ningning Wang [1], Yucong Jiang [2], Tianxiang Xia [1,*], Feng Xu [3], Chengjun Zhang [4], Dan Zhang [1] and Zhiyuan Wu [1]

[1] Beijing Key Laboratory for Risk Modeling and Remediation of Contaminated Sites, National Engineering Research Center of Urban Environmental Pollution Control, Beijing Municipal Research Institute of Eco-Environmental Protection, Beijing 100037, China; wangningning@cee.cn (N.W.); zhangdan@cee.cn (D.Z.); wuzhiyuan@cee.cn (Z.W.)

[2] Beijing Institute of Mineral Resources and Geology, Beijing 101500, China; jiangyu_cong@163.com

[3] Technical Centre for Soil, Agriculture and Rural Ecology and Environment, Ministry of Ecology and Environment, Beijing 100012, China; xufeng@tcare-mee.cn

[4] State Key Laboratory of Water Environment Simulation, School of Environment, Beijing Normal University, Beijing 100875, China; zhangcj@mail.bnu.edu.cn

* Correspondence: xiatianxiang@cee.cn; Tel.: +86-010-88385869

**Abstract:** Soils at primary explosives sites have been contaminated by high concentrations of antimony (Sb) and co-occurring heavy metals (Cu and Zn), and are largely overlooked and neglected. In this study, we investigated Sb concentrations and species and studied the effect of combined Fe- and Fe–Al-based sorbent application on the mobility of Sb and co-occurring metals. The content of Sb in soil samples varied from 26.7 to 4255.0 mg/kg. In batch experiments, $FeSO_4$ showed ideal Sb sorption (up to 97% sorption with 10% $FeSO_4 \cdot 7H_2O$), whereas the sorptions of 10% $Fe^0$ and 10% goethite were 72% and 41%, respectively. However, Fe-based sorbents enhanced the mobility of co-occurring Cu and Zn to varying levels, especially $FeSO_4 \cdot 7H_2O$. $Al(OH)_3$ was required to prevent Cu and Zn mobilization. In this study, 5% $FeSO_4 \cdot 7H_2O$ and 4% $Al(OH)_3$ mixed with soil was the optimal combination to solve this problem, with Sb, Zn, and Cu stabilizations of 94.6%, 74.2%, and 82.2%, respectively. Column tests spiked with 5% $FeSO_4 \cdot 7H_2O$, and 4% $Al(OH)_3$ showed significant Sb (85.85%), Zn (83.9%), and Cu (94.8%) retention. The pH-regulated results indicated that acid conditioning improved Sb retention under alkaline conditions. However, no significant difference was found between the acidification sets and those without pH regulation. The experimental results showed that 5% $FeSO_4 \cdot 7H_2O$ + 4% $Al(OH)_3$ without pH regulation was effective for the stabilization of Sb and co-occurring metals in primary explosive soils.

**Keywords:** primary explosives site; heavy metal contamination; antimony; co-occurring metal; Fe–Al-based amendment; immobilization

## 1. Introduction

Antimony (Sb) is an element of growing environmental concern because of the widespread use and uncontrolled release of Sb compounds into the environment [1–4]. In addition, antimony has a carcinogenic potential [5] and Sb(III) has been shown to potentially cause lung cancer in female rats [6]. Background concentrations of Sb in soil tend to be lower than 1 mg/kg [3]. However, the major smelting and general use of Sb has led to severe site contamination, which poses a significant risk to the local environment [7–9] and humans [10]. An increasing number of studies have shown that antimony pollution is a global issue [11,12] because of its toxicity to humans and its role in causing liver, skin, and respiratory and cardiovascular diseases [10]. Soil is an important medium for Sb concentration and migration. In recent years, increased attention has been given to Sb-contaminated soil in mining areas and shooting ranges [10,13–16]. However, few studies have investigated Sb-contaminated soils by primary explosives because of confidentiality and sensitivity. A significant input of Sb into the environment occurs through the production and use of primary explosives because Sb was historically used as a combustible agent for classical primary explosives, which contain 33.4% $Sb_2S_3$ [17]. The weathering and corrosion of combustion residue lead to the mobilization of metalloid Sb in anionic form [18]. Primary explosives sites, which are often characterized by critical concentrations of co-occurring copper (Cu) and zinc (Zn) [19–21], can be of particular environmental concern because they represent hazardous multi-element contamination sources for sites zones. The leachate of Sb and co-occurring metals from primary explosive production and use areas poses a serious long-term threat to the environment and human health. Thus, the immobilization or a reduction in the mobility and bioavailability of Sb and co-occurring metals in primary explosives sites is critical.

Sb enters the soil environment primarily as $Sb_2S_3$ (stibnite) and $Sb_2O_3$ (senarmontite, valentinite) [22]. Chemical/physical weathering and naturally occurring oxidation and microbial processes are often responsible for converting primary Sb mineral phases, predominantly sulfides and oxides, to secondary Sb minerals, which are soluble in soil pore water and more mobile in the environment [23]. Most of the antimony is confined to the top 30 cm layer of the soil and is bound tightly to soil-derived humic acid molar mass fractions that are extracted from the top 10 cm layer [12]. The formation of stable secondary Sb minerals by precipitation and adsorption on metal oxyhydroxides (for example Fe-based sorbents [24,25]) are the most prominent naturally occurring processes that can control the mobility and transformation of Sb species in soil systems [26]. In the natural environment, the mobility, bioavailability, and toxicity of Sb are primarily dependent on its chemical speciation [3,27]. Antimony exists in a variety of oxidation states (−III, 0, III, V), with oxidation states III and V being predominant in aqueous environments across a wide pH range (4–10) [28]. In the natural environment, Sb(III) occurs primarily as $Sb(OH)_3$ under anaerobic conditions between pH 2 and 10 [3,27], whereas Sb(V) is a predominant species that exists as $Sb(OH)_6^-$ in aerated environments [29] and has a high affinity to amorphous and crystalline Fe-(hydr)oxides, with which it can form stable bidentate inner-sphere complexes [30]. These interactions are favored by goethite in the pH range of 7.5–9.0, by hematite at pH 8.5, by ferrihydrite in the pH range of 7.0–7.9, and by akaganeite in the pH range of 9.5–10 [31]. However, Sb with different valence states has different properties; for instance, Sb(III) can adsorb strongly on goethite over a wide pH range from 3 to 12, whereas the maximum adsorption of Sb(V) occurs only below pH 7 [32]. These Fe-based metals adsorb Sb strongly and act as oxidants to transform Sb(III) to Sb(V) [33,34]. Laboratory-scale testing indicated that $Fe_2(SO_4)_3$ may be applicable to Sb immobilization in soils [15]. The sorption effect is based on the reaction of Sb(V) with the surface hydroxyl group of Fe-based materials. However, the mobility of co-occurring Cu and Zn was enhanced after Fe-based sorbent addition [15]. In the pH range of 5–9, co-occurring metals behave differently, and are commonly present in soil solutions as cations at acidic and circumneutral pH or as soluble soil organic matter (SOM)-metal(II) complexes at higher pH [31]. At neutral and alkaline pH, substantial amounts of heavy metals are immobilized as Me-hydroxides, Me-caralbonates or Me-hydroxycarbonates (Me indicates heavy metals). Soluble heavy metals show a limited affinity for hydroxyl groups because of their cationic nature, but they interact more strongly with negatively charged components [31]. Thus, the different speciation, mobility, and bioavailability between Sb and co-occurring metals make the identification of suitable amendments challenging. Aluminum (hydr)oxides show important sorption properties for Pb, Cd, and Zn [35]. Aluminum (hydr)oxides can be protonated, which makes their surface positively charged and generates electrostatic interactive forces with negatively charged Sb(V) [36]. Only a few amendments, which are based mostly on Fe- and Al-containing materials, have been tested with variable success as Sb-immobilizing agents. Limited studies have examined the mobility of Sb and co-occurring metals in soil and the selection of ideal sorbents for their immobilization.

The focus of this work was on soil that was contaminated with Sb in primary explosives production sites with the main goals being: first, to investigate the mobility and speciation of Sb in primary explosive sites; second, to evaluate the effect of the combined application of Fe–Al mixed amendments for primary explosive sites using batch and column tests; and third, investigate the pH effect on Sb immobilization.

## 2. Materials and Methods

*2.1. Sb-Contaminated Soils and Adsorbents*

Sb-contaminated soils were collected from primary explosives production workshops from a primary explosives site in Heilongjiang, China. The specific coordinates have not been provided because of the confidentiality and sensitivity of this military enterprise. This site produced primary explosives for more than 60 years and remains operational. Based on previous environmental site investigation by soil boreholes, it was found that Sb was enriched mostly on the surface soil layer, and thus, nine soil samples (S1–S9) were collected at the surface (0–20 cm depth). Samples were air-dried at room temperature, crushed with a wooden roller and sieved through a 150-µm mesh, which is thought to better estimate human exposure than bulk soil [37]. The samples were mixed and prepared for soil analysis.

The soil sample with the highest concentration of Sb was used for soil immobilization studies. Four adsorbents were used: ferrous sulfate ($FeSO_4·7H_2O$, powder); goethite ($HFeO_2$, powder); $Fe^0$ (powder); and aluminum hydroxide ($Al(OH)_3$, powder).

*2.2. Soil Analysis*

Soil physicochemical properties were measured on triplicate samples with three blanks. Soil pH was measured in a 1:5 ($w/v$) soil/deionized soil suspension by using a pH meter (PHSJ-4A, INESA Scientific Instrument Co., Ltd., Shanghai, China) after 1 h of equilibration according to ISO10390:2005. After microwave-assisted digestion with $HCl + HNO_3 + HClO_4$ (3:1:1) at 190 °C for 15 min, the mixture was cooled, filtered (<0.45 µm), and diluted with ultrapure deionized water. Total Sb ($Sb_{tot}$) concentrations were determined by hydride generation atomic fluorescence spectrometry (HG-AFS) (AFS 9700, Titan Instrument Co. Ltd., Beijing, China) [2]. The total contents of Fe, Mn, Zn, and Cu were determined by using inductively coupled plasma atomic emission spectrometry (ICP-AES) (NexION300x, PerkinElmer, Waltham, MA, USA) [38]. The modified European Community Bureau of Reference (BCR) sequential extraction method [39] was used as a sequential extraction process for Sb. Sb speciation in soil is divided into a soluble/exchangeable fraction (F1), reducible fraction (F2), oxidizable fraction (F3), and residual fraction (F4) based on the BCR technique [40]. The processes were as follows—step 1: 0.5 g soil + 20 mL of 0.11 mol/L HAc was agitated continuously at ambient temperature for 16 h, and supernatant was used to determine the acid-soluble fraction; step 2: the soil in step 1 + 20 mL of 0.5 mol/L $NH_2OH·HCl$ was agitated at ambient temperature for 16 h, and the supernatant was used to determine the reducible fraction; step 3: the soil in step 2 + 5 mL of 8 mol/L $H_2O_2$ was agitated continuously at ambient temperature for 1 h, placed in a water bath at 85 °C for 1 h and treated with 25 mL of 1 mol/L $NH_4Oac$ for 16 h at ambient temperature, with supernatant being used to determine the oxidizable fraction; step 4: the residual soil was digested with aqua regia to obtain a residue fraction.

*2.3. Immobilization Experiment*

2.3.1. Batch Experiments

Batch experiments and column leachate experiments were performed on the soils that were most polluted with Sb (S9). Experiments were performed in a 50 mL centrifuge tube with four grams of untreated soil or soil amendment mix (0%, 2%, 5%, and 10% iron-based adsorbent; 0%, 2%, and 4% aluminum-based adsorbent) and the corresponding volume of deionized water was treated with a Milli-Q water purification device (Millipore Corp, Billerica, MA, USA) with a liquid-to-solid ratio of 10 (L/S). Centrifuge tubes were shaken at 100 rpm/min for 10 days at room temperature (25 ± 1 °C) in an incubator shaker. Based on the earlier sorption studies of Sb to Fe-based materials, 10 days reaction time was sufficient to reach equilibrium [15,41]. All experiments were performed in triplicate. After reaction, the samples were centrifuged, and the supernatant was filtered with a 0.45 μm filter membrane (PES, ReLAB). The filtrate was used for Sb, Cu, and Zn analysis.

2.3.2. Column Leachate Experiments

Column leachate experiments were carried out in triplicate with untreated soil and in quintuplicate with 5% $FeSO_4$ and 4% $Al(OH)_3$ stabilized soil without or with pH regulators (i) 4% sodium bisulfate and (ii) 4% sodium carbonate. Column leachate experiments, as shown in Figure 1, were carried out in polyethylene columns (length 300 mm × internal diameter 22 mm) packed with 160 g thoroughly mixed soil material of untreated soil (Group A), stabilized soil + 4% $NaHSO_4$ (Group B), stabilized soil (Group C), and stabilized soil + 4% $Na_2CO_3$ (Group D).

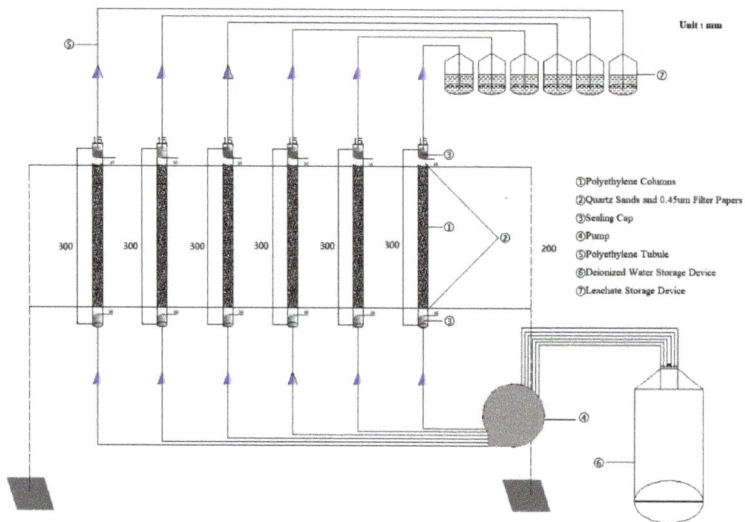

**Figure 1.** Structure of column experiments.

Double filter papers (0.45 um) and 50 mm long quartz sands were installed at the top and bottom of the column for particle retention. Deionized water was added to the columns in an upward flow direction at approximately 0.38 mL/min by a peristaltic pump over an 18-day period. The calculated pore volume (PV) in the soil column was 0.043 L, based on an assumed soil porosity of 0.4 (total 550–600 PV) [15]. During the column experiment, samples were collected regularly and filtered for immediate analysis.

## 3. Results and Discussions

### 3.1. Soil Characteristics and Risk

The main soil physicochemical properties are summarized in Table 1. All nine soil samples were calcareous, with pH values varying from 7.69 to 8.37, which is consistent with the local soil pH range [42]. The concentrations of Fe and Mn in the soil were $2 \times 10^4$–$3.22 \times 10^4$ mg/kg and 396–719 mg/kg, respectively. The total concentration of Sb, Cu, and Zn in the studied area varied from 26.73 to 4255, 24.29 to 312.3, and 67.62 to 1330 mg/kg, respectively. Antimony was the main contaminant in this primary explosives site and was 0.34–211.8 times higher than the Chinese screening value of the first land use category (20 mg/kg), based on the document of "Soil Environmental Quality Risk Control Standard for Soil Contamination of Development Land", which was issued by the Ministry of Ecology and Environment of China. The highest concentration of Sb in soil sample S9 (4255 mg/kg) was 211.8 times and 105.4 times higher than the corresponding risk screening value (20 mg/kg) and control value (40 mg/kg), respectively.

Table 1. Main physicochemical characteristics of soils.

| Soil | pH | Concentrations (mg/kg) | | | | |
|---|---|---|---|---|---|---|
| | | Mn | Fe | Cu | Zn | $Sb_{tot}$ |
| S1 | 8.37 ± 1.70 | 499 ± 66.54 | 32,200 ± 876.43 | 25.37 ± 0.17 | 71.19 ± 14.81 | 26.73 ± 2.31 |
| S2 | 7.8 ± 0.27 | 573 ± 36.67 | 30,000 ± 156.47 | 24.29 ± 3.12 | 67.62 ± 0.52 | 61.32 ± 4.27 |
| S3 | 8.11 ± 0.75 | 545 ± 20.82 | 28,900 ± 488.27 | 33.55 ± 1.57 | 83.80 ± 3.50 | 108.01 ± 6.43 |
| S4 | 7.96 ± 0.19 | 413 ± 41.93 | 24,200 ± 893.30 | 232.85 ± 5.87 | 402.75 ± 95.65 | 216.60 ± 9.64 |
| S5 | 7.69 ± 0.40 | 628 ± 16.05 | 31,200 ± 366.70 | 186.74 ± 2.75 | 868.02 ± 78.23 | 267.26 ± 172.5 |
| S6 | 7.93 ± 1.41 | 396 ± 14.83 | 23,000 ± 259.53 | 103.76 ± 5.44 | 331.36 ± 11.83 | 512.09 ± 19.52 |
| S7 | 8.15 ± 1.58 | 513 ± 37.53 | 25,000 ± 456.95 | 312.30 ± 14.83 | 981.60 ± 10.42 | 719.36 ± 23.34 |
| S8 | 7.91 ± 0.08 | 554 ± 52.51 | 23,600 ± 99.28 | 35.26 ± 1.56 | 215.88 ± 7.25 | 953.91 ± 39.72 |
| S9 | 8.25 ± 1.87 | 719 ± 189.57 | 20,000 ± 127.71 | 80.16 ± 3.05 | 1330.05 ± 36.67 | 4255.03 ± 231.50 |

The BCR method is used extensively as a sequential extraction process for heavy metal [38]; thus, BCR sequential analysis was used to measure the main contaminant (Sb). The fraction of Sb in the primary-explosives-contaminated soil in Figure 2 followed the order (average values): F4 fraction (38.05–94.22%) > F2 fraction (0.01–31.80%) > F3 fraction (0.32–21.55%) > F1 fraction (0.76–12.92%). The F4 fraction of the heavy metals is considered to be associated with stable minerals with the lowest mobility [40]; thus, Sb in S2 soil sample was the most stable. As a result, S2 may pose least risk. The F4 fraction of Sb was slightly different in S1, S2, and S3, and decreased significantly in the other soils, especially in S9 (only 38.05%). The Sb of the F1 and F2 fractions can be combined as directly available fractions [43] with a direct toxicity [44] because the Sb in these two fractions is highly mobile when environmental conditions (such as pH and Eh) change. The F3 fraction of Sb is easily mobilized and transformed into the F1 and F2 fractions, and potential eco-toxicity should not be ignored [44]. The total concentrations of F1, F2, and F3 fractions are a bioavailable fraction because of their direct and potential eco-toxicity [40]. Figure 2 shows that Sb had the highest proportion of F1 + F2 + F3 phase (61.95%) in S9, which indicates that S9 had a high migration potential and the greatest biological impact [36]. The amount of exchangeable Sb is negatively correlated with the concentration of Fe ($r = -0.867, p < 0.01$) for all soil samples.

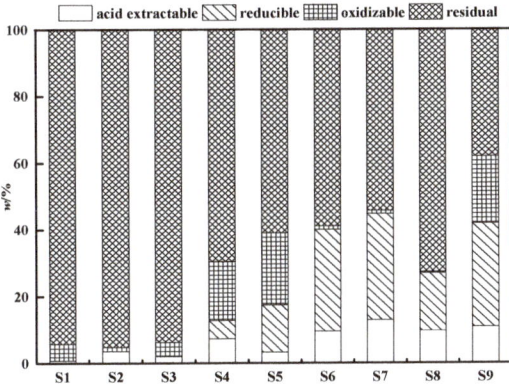

Figure 2. Sequential extraction of Sb from primary-explosives-contaminated soils.

## 3.2. Batch Experiments

### 3.2.1. Changes in pH and Sb Fractions

We evaluated the application of ferrous sulfate (FeSO$_4$·7H$_2$O), goethite (HFeO$_2$), and Fe$^0$ at three different percentages (2%, 5%, and 10%), as potential amendments for the remediation of Sb and soil contaminated with co-occurring metals. To obtain stabilizing results, Al(OH)$_3$ was applied (0%, 2%, and 4%) as combined amendments with Fe-based amendments. As shown in Figure 3, after the equilibration period of 10 d, FeSO$_4$·7H$_2$O addition significantly decreased the water extract pH (21%, 24%, and 28% decreased at 2%, 5%, and 10% amendment percentages, respectively) compared with the control, whereas HFeO$_2$ and Fe$^0$ had little effect on the water extract pH. This behavior is a result of the hydrolysis reaction of FeSO$_4$ (Fe$^{2+}$ + 2H$_2$O $\rightleftharpoons$ Fe(OH)$_2$ +2H$^+$) in the pore water of soil [45,46]. The application of Al(OH)$_3$ increased the water extract pH in a small range (less than one unit) regardless of the combined application with FeSO$_4$·7H$_2$O, HFeO$_2$, or Fe$^0$. This slight increase in pH is attributed mainly to the weak alkalinity of Al(OH)$_3$ and partial dissolution of amendments and adsorption/precipitation [47].

Sequential BCR extraction procedures were used for 5%Fe + 4%Al-based modified soils (with the best modified result) and are presented in Figure 4. The Sb proportions of F1, F2, F3, and F4 fractions in the control soil were 8.7%, 4.5%, 27.0%, and 59.7%, respectively. The F1 fraction [40] decreased significantly in modified soils compared with the control. Approximately 66.67%, 44.44%, and 66.67% reductions (vs. control) were observed for FeSO$_4$ + Al(OH)$_3$-, goethite + Al(OH)$_3$-, and Fe$^0$ + Al(OH)$_3$-modified soil, respectively. This result is important because this fraction of Sb has the most biological impact with a direct toxicity [44]. With F1, F2 fractions are considered the most available to soil biota and the most easily leached to groundwater [43]. However, only FeSO$_4$ + Al(OH)$_3$ addition reduced the F2 fraction from 5% to 1%, despite a decrease in soil pH. The decrease in soil pH was likely caused by an increase in exchange sites (Fe/Al oxides and oxyhydroxides), which has a great affinity for Sb [48]. Fe–Al-based amendment additions induced a shift of F1, F2, and F3 fractions towards to F4 fraction, which was more strongly retained by the Fe and Al (hydr)oxides [29,49] by adsorption. The concentration decrease in the F1, F2, and F3 fractions resulted in a subsequent low bioavailability and eco-toxicity [40]. However, this result was not clear for the goethite- and Al(OH)$_3$-modified soil with only a 2% bioavailable fraction shift to a stable residual fraction. FeSO$_4$ + Al(OH)$_3$ addition caused a significant increase in F4 (71%) compared with goethite + Al(OH)$_3$ and Fe$^0$ + Al(OH)$_3$ groups, which could explain its high sorption capacity of Sb [50]. Overall, the BCR extract results showed that the combined addition of 5%FeSO$_4$ and 4%Al(OH)$_3$ induced a significant redistribution of Sb with a reduction in its more labile and bioavailable fraction and an increase in the residual fraction.

### 3.2.2. Effects of Fe- and Al-Based Sorbents on Sb

The results of the batch experiments (Figure 3) showed that the leachate concentration of Sb in the control groups exceeded 14 mg/L, which is higher than the fifth category water limit (0.01 mg/L) of the "Standard for groundwater quality" issued by the General Administration of Quality Supervision, Inspection and Quarantine of the People's Republic of China (AQSIQ, 2017). This result proved a high liquidity of antimony in primary-explosives-contaminated soils and occurs mainly because of the high percentage of soluble fraction (11.03%) and reducible fraction (30.12%) of Sb in S9 (Figure 2). Thus, it is necessary to immobilize or reduce the mobility and bioavailability of Sb in explosives-contaminated soils to reduce plant and human bioavailability and Sb leachate in groundwater [51].

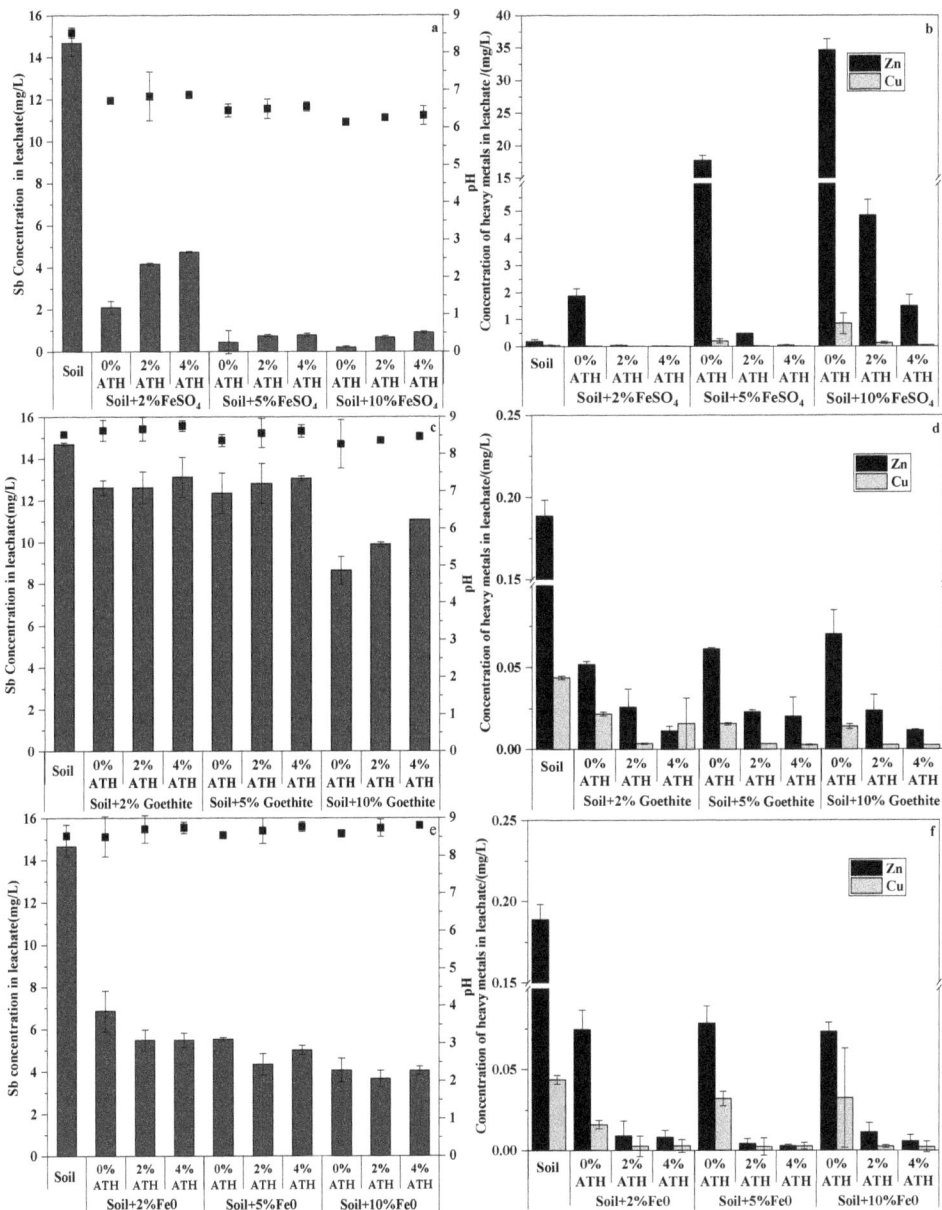

**Figure 3.** pH and Sb and co-occurring heavy metals concentration in water extracts ((**a**,**b**)—FeSO$_4$ + ATH; (**c**,**d**)—Goethite + ATH; (**e**,**f**)—Fe$^0$ + ATH).

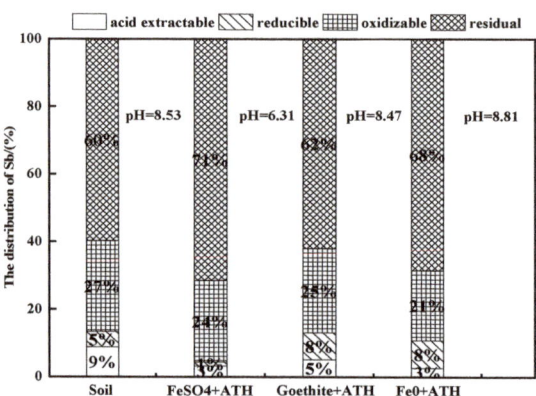

**Figure 4.** Sequential extraction of Sb in amended soil with Fe and Al mixed adsorbent.

The amendment additions did not change the total Sb concentrations. However, Fe-based amendments prevented Sb mobility and reduced Sb bioavailability by a strong preference of Sb binding to Fe hydroxides [52]. $FeSO_4 \cdot 7H_2O$ possessed the strongest sorption properties for Sb with 85.64%, 97.21%, and 98.50% sorptions with 2%, 5%, and 10% additions, respectively (Figure 3a). The sorption property was enhanced by the addition of various concentrations of $FeSO_4 \cdot 7H_2O$. The sorption increased gradually with an increase in $FeSO_4 \cdot 7H_2O$ concentration from 0% to 10% and was moderate with $FeSO_4 \cdot 7H_2O$ concentrations from 5% to 10%. The amendments of $Fe^0$ and $HFeO_2$ showed a lower Sb sorption compared with $FeSO_4 \cdot 7H_2O$. The sorption increased gradually with an increase in $Fe^0$ or $HFeO_2$ concentration from 0% to 10%; 10% $Fe^0$ was retained up to 72.34% Sb (Figure 3c), and the $HFeO_2$ only retained up to 41.05% Sb (Figure 3e). In the severely Sb-contaminated primary explosives site, $FeSO_4 \cdot 7H_2O$ was more effective for Sb immobilization than $Fe^0$ or $HFeO_2$. The results are consistent with the immobilization of Fe-based amendments for Sb in shooting range soil [15]. On the basis of the thermodynamics principle, Sb(V) should be the main form in oxic environments [15,53]. In addition, widely distributed dissolved iron in the environment impacts rapid Sb(III) oxidation [54]. Ferric ion and iron oxyhydroxides in the environment are strong Sb adsorbents [30]. Sb(III) can be oxidized simultaneously into Sb(V) once adsorbed on the iron oxyhydroxide surface [54]. The immobilization mechanism of iron oxide for Sb(V) was summarized as direct precipitation, co-precipitation, and adsorption [55,56]. The direct precipitation mechanism could lead to secondary Fe–Sb mineral tripuhyite ($FeSbO_4$) formation, which is an important and ultimate sink for Sb in an environment with a low solubility (log Kso = −13.41) [57]. Sb(V) adsorption was a predominant mechanism, and co-precipitation was important in $FeSO_4 \cdot 7H_2O$-modified soil [58]. FeOOH adsorption and the hydrolysis product of Fe(II) and Fe(III), rather than co-precipitation, was predominant in the coagulation mechanism [30]. The Fe(II) in the solution was oxidized rapidly to Fe(III), which improved the antimony removal efficiency [46]. Fe(II) oxidation and ferric-hydroxide formation may lead to an increase in the adsorption between ferric flocs and Sb [59,60]. $FeSO_4 \cdot 7H_2O$ addition can reduce the soil solution pH, and at pH < 7, iron oxide showed a strong affinity to Sb. When pH < 3, this level [$H^+$] could inhibit the extent of Fe(II) and Fe(III) hydrolysis, which limited the Sb(V) immobilization efficiency. For pH 3–6, the isoelectric point of $FeSO_4$-produced iron flocs was 7.5. Iron flocs with a positive charge had a better capture of negatively charged $Sb(OH)_6^-$ at a weak acid condition and for pH 5 to 6 [58]. As a result, $FeSO_4 \cdot 7H_2O$ amendment could promote Sb(V) immobilization by reducing the soil solution pH. The maximum Sb(V) adsorption on $HFeO_2$ existed below pH 7 [32] and was considered pH-dependent [61]. In this study, the soil extract solution pH exceeded 7; thus, $HFeO_2$ immobilized Sb in these primary-explosives-contaminated soils to a certain extent. When in contact with oxygenated water, $Fe^0$ converts to activated $Fe^0$ that contains ferrihydrite and goethite, which are distributed on the surface [4]. The large surface coverage makes activated $Fe^0$ to be recognized as a suitable adsorbent for Sb adsorption and immobilization [62]. However, the immobilization efficiency of $Fe^0$ for Sb was lower than that for $FeSO_4 \cdot 7H_2O$. According to batch experiments of Fe-based amendments, $FeSO_4 \cdot 7H_2O$ was most appropriate for high concentrations of Sb-contaminated soil.

$Al(OH)_3$ addition slightly increased Sb mobility regardless of $FeSO_4$-, goethite-, or $Fe^0$-modified soils by increasing the soil pH [58]. The sorption of Sb in $FeSO_4$- and $Al(OH)_3$-modified soil decreased slightly from 85.82% to 67.77% in 2% $FeSO_4$, from 97.09% to 94.69% in 5% $FeSO_4$, and from 98.60% to 93.88% in 10% $FeSO_4$ with $Al(OH)_3$ addition from 0% to 4%. Approximately 2.47–21.03% reductions were observed with $Al(OH)_3$ addition. Sb sorption in goethite- and $Al(OH)_3$-modified soil decreased from 14.09% to 10.68% in 2% goethite, from 15.91% to 11.02% in 5% goethite, and from 41.04% to 24.55% in 10% goethite with an $Al(OH)_3$ addition from 0% to 4%. Approximately 24.19–40.20% reductions were observed. However, a change in sorption was not clear for $Fe^0$- and $Al(OH)_3$-modified soil, even with substantial amounts of $Al(OH)_3$ [48]. In the Fe–Al mixed-addition cases, a reduction in sorption may have resulted from the change in pH [63]. $Al(OH)_3$ shows a lower adsorption efficiency than iron oxides, especially in neutral and alkaline environments [64]. However, $FeSO_4$- and $Al(OH)_3$-modified soil still showed a more efficient sorption, which is attributed to the ion-exchange ability of Fe–Al double hydroxides and $SbO_3^-$ adsorption on the FeO(OH) surface [65].

3.2.3. Effect of Fe- and Al-Based Sorbents on Zn and Cu

$FeSO_4 \cdot 7H_2O$ amendment addition to the soil resulted in a high concentration decrease in Sb; however, higher concentrations of Zn and Cu were detected in the soil extract solution, especially for Zn (Figure 3). Zn release increased gradually with an increase in $FeSO_4 \cdot 7H_2O$ concentration from 0% to 10%. After $FeSO_4 \cdot 7H_2O$ addition, the highest leachate concentration of Zn was 34.57 mg/L, which was 181.94 times the original soil leachate concentration (0.19 mg/L). Similar promotion leachate effects of $FeSO_4 \cdot 7H_2O$ were found for Cu. The leachate concentration of Cu was 0.84 mg/L with 10% $FeSO_4 \cdot 7H_2O$ addition, which was 21 times the leachate concentration compared with the original soil (0.04 mg/L). $FeSO_4 \cdot 7H_2O$ addition decreased the water extract pH by the hydrolysis of $FeSO_4$ [45,46]. The acid environment promoted Zn and Cu release, which resulted in a high leachate concentration of Zn and Cu [66,67]. In contrast to $FeSO_4$, goethite and $Fe^0$ application significantly reduced the concentration of Zn and Cu. No further reduction was caused by goethite and $Fe^0$ addition. As shown in Figure 3, the maximum adsorption efficiencies of Zn and Cu were 72.60% and 68.05%, respectively, in goethite-treated soil. The maximum adsorption efficiencies of Zn and Cu were 61.29% and 63.31%, respectively, in $Fe^0$-modified soil. Zn and Cu fixation by goethite could be attributed mainly to metal diffusion into the structural lattice of goethite by the following reactions: $\equiv$Fe-OH + $Me^{2+}$ + H2O $\rightleftharpoons$ $\equiv$Fe-O-MeOH$_2^+$; $\equiv$Fe-O-MeOH$_2^+$ + $Me^{2+}$ + 2H2O $\rightleftharpoons$ Fe-O-MeOH$_2^+$ + Me(OH)$_2$(s) + $2H^+$ [68]. With $Al(OH)_3$ addition, the Zn and Cu concentration in the leachate decreased significantly, especially in $FeSO_4 \cdot 7H_2O$-treated soil. $Al(OH)_3$ (4%) decreased the extractable Zn by 99.72%, 83.47%, and 96.26% in $FeSO_4 \cdot 7H_2O$, goethite, and $Fe^0$-modified soil, respectively (Figure 3), whereas 4% $Al(OH)_3$ addition decreased the extractable Cu by 96.11%, 83.61%, and 93.21% in $FeSO_4 \cdot 7H_2O$, goethite, and $Fe^0$-modified soil, respectively. The reduction in extractable Zn and Cu in Fe–Al-based modified soil may indicate that $Al(OH)_3$ was important in Zn and Cu immobilization. These results are similar to those reported previously [36] and are somewhat expected in Fe–Al-based modified soil, where a substantial amount of the Sb and co-occurring metals are likely to form stable surface complexes or precipitates with iron and aluminum hydroxide [36]. In soil that is amended with 5% ferrous sulfate and 4% aluminum hydroxide, the leachate concentrations of Sb, Zn, and Cu decreased significantly, and the maximum stabilization efficiencies were 94.69%, 74.17%, and 82.15%, respectively. Goethite (10%) and aluminum hydroxide (2%, pH = 8.36) addition yielded a 32.60% stabilization efficiency of Sb and an decreases of 87.45% and 93.73% in Zn and Cu, respectively. The mixture of 10% $Fe^0$ and 2% aluminum hydroxide (pH = 8.74) stabilized Sb, Zn, and Cu in soil with efficiencies of 75.00%, 94.01%, and 94.14%, respectively. For high Sb contamination and light pollution with Zn and Cu, 5% $FeSO_4 \cdot 7H_2O$ and 4% $Al(OH)_3$ addition was shown to be the most suitable amendment course.

### 3.3. Column Experiments

#### 3.3.1. Column Experiments for Sb Immobilization

In column experiments, the initial Sb concentration in leachate of untreated soil was 16.5 mg/L, which exceeds the GB/T 14848—2017 (Standard for Groundwater Quality) type-V leachate limit (0.01 mg/L) (AQSIQ, 2017) (Figure 5a). When the pore volume reached 3.7, the concentration of Sb in leachate increased substantially to a peak of 17.73 mg/L. Similar leachate results were found in shooting range soil [15]. Okkenhaug et al. [15] assumed that this initial increase in Sb occurred because of the depletion of $H^+$ in the soil during leachate, which induced a decrease in the positive surface charge of minerals such as iron oxyhydroxides. Consequently, the sorption capacity of negatively charged anions decreased. In this study, no significant pH increase was recorded. Sb mobilization may have increased in untreated soil through the depletion of organic and inorganic components and through the partial dissolution of reducible and oxidizable fractions [1,69]. After an initial increase, the Sb concentration decreased exponentially up to a PV of 74.4. At this stage, the pH decreased rapidly to 7.03. Subsequently, the pH stabilized at ~7. As a result, the Sb concentration stabilized at 0.99 mg/L. The trend in Sb leachate was consistent with the pH trend, which meant that Sb dissolution was affected by soil pH [70]. Hockmann et al. [71] found a similar effect for Sb in large-range soil under large seepage conditions. The initial Sb concentration of leachate in a 5% $FeSO_4 \cdot 7H_2O$- and 4% $Al(OH)_3$-modified column was 0.024 mg/L, which was 675 times lower than that of untreated soil, and the stabilization efficiency was 99.86%. The pH was adjusted to 3–10 by the addition of pH regulators, $NaHSO_4$ or $Na_2CO_3$. The leachate concentration of Sb in the stabilized soil is shown in Figure 5a. The leachate concentrations of Sb in descending order was as follows: Group A, Group D, Group B, Group C. Group D showed a high initial Sb concentration (16.82 mg/L) compared with Group B (0.29 mg/L) and Group C (0.02 mg/L) as a result of the rapid increase in pH that was induced by large amounts of $OH^-$, which decreased the positive surface charge of iron and aluminum oxyhydroxide. Consequently, Group B decreased the retention capacity of negatively charged anions [67,72]. The leachate concentration of Sb in Group B decreased to levels similar to Group C with an increase in PV and a decrease in pH.

The calculation results of the cumulative leachate amount of Sb in the dynamic process of each group are shown in Figure 5b. When the PV reached 223.3, the total amounts of Sb that was leached in Groups A, B, C, and D were 577.61, 79.75, 81.71, and 101.01 mg/kg, respectively. Soils that were amended with $FeSO_4 \cdot 7H_2O$ mixed with $Al(OH)_3$ showed a good Sb stabilization effect [36], with stabilization efficiencies of 86.19%, 85.85%, and 82.51% in Groups B, C, and D, respectively. Group B decreased the soil pH significantly, but did not increase the retention effect of Sb. The pH of Group C was consistent with untreated soil when the PV reached 37.2. The 5% $FeSO_4 \cdot 7H_2O$ and 4% $Al(OH)_3$ mixed addition showed the best stabilization performance of Sb with less pH disturbance.

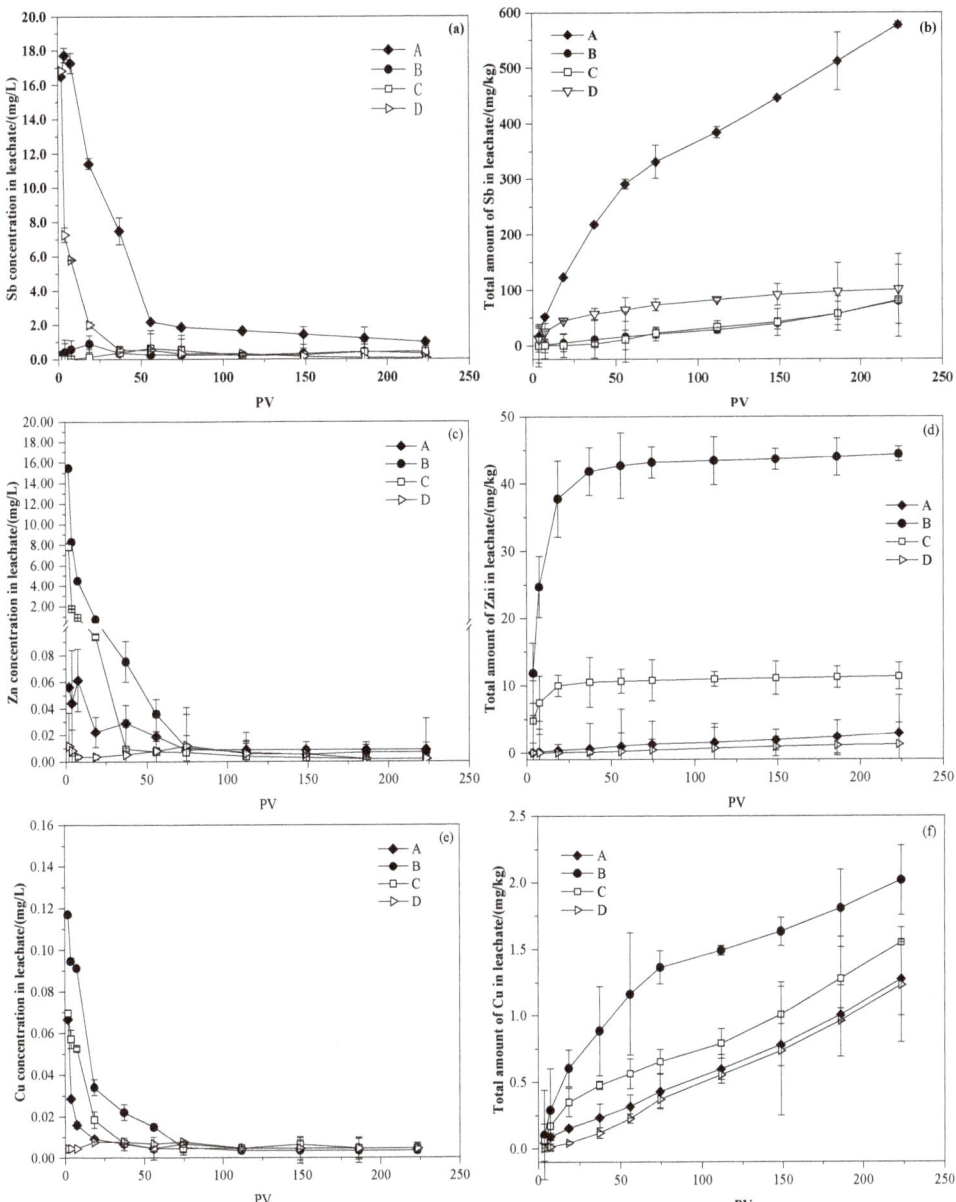

**Figure 5.** Changes in leachate concentration and accumulation of heavy metals in column experiment: (♦) untreated soil, (●) Fe–Al-stabilized soil with acidification, (□) Fe–Al-stabilized soil, (▷) Fe–Al-stabilized soil with alkalization ((**a**,**c**,**e**)—Sb, Zn, Cu concentration in leachate respectively; (**b**,**d**,**f**)—total leaching amount of Sb, Zn, Cu/ weight of soil).

3.3.2. Column Experiments for Zn and Cu Immobilization

The initial Zn (0.056 mg/L) and Cu (0.067 mg/L) concentrations in the leachate of untreated soil in Figure 5c,e, exceeded the GB/T 14848–2017 (Standard for Groundwater Quality) (AQSIQ, 2017) type-I and -II leachate limit, respectively. As shown in Figure 6c–f, the leachate concentration of Zn and Cu decreased significantly when PV < 74.4, showed a slower decrease and tended to stabilize when PV ≥ 74.4. The final leachate concentration of Zn and Cu stabilized at ~0.009 and 0.0035 mg/L, which was 6.2 and 19.1 times lower than that of the initial concentration, respectively. The initial Zn and Cu concentrations in the leachate of soil, amended with 5% $FeSO_4 \cdot 7H_2O$ and 4% $Al(OH)_3$, were 7.83 and 0.069 mg/L, which were 139 and 1.04 times higher than those of untreated soil, respectively. After Fe–Al-based adsorbent addition, the soil became transitory acidic, which promoted Zn and Cu leachate [66]. A significant increase in the leachate concentration of Zn and Cu was observed in stabilized soil with acidification, but a significant decrease in the leachate concentration of Zn and Cu was observed in stabilized soil with alkalization, which was the result of the surface charge variation in minerals that were induced by pH and the positively charged property of Zn and Cu [67]. For PV ≥ 74.4, the leachate concentrations of Zn and Cu in the three groups of modified soil were consistent with untreated soil as the pH eventually equilibrated.

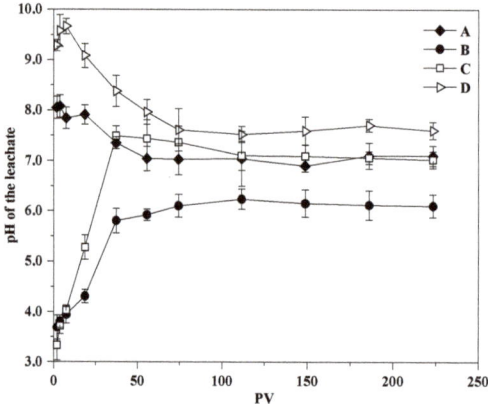

**Figure 6.** Leach solution pH with change in PV.

The calculation results of the cumulative leachate amounts of Zn and Cu in the dynamic process of each group are shown in Figure 5d, f. The total leachate amount of Zn in Groups A, B, C, and D was 2.88, 44.37, 11.34, and 1.19 mg/kg, respectively. The amount of released Zn increased slowly when PV ≥ 74.4 and depended on the soil pH [73]. The total amounts of Cu leached in Groups A, B, C, and D were 1.27, 2.02, 1.55, and 1.23 mg/kg, respectively. The highest release resulted in Fe–Al-stabilized soil with the acidification group. In the four groups, even a moderate change in pH could induce persistent Cu release.

3.3.3. pH Effects on Sb and Co-Occurring Heavy Metals Immobilization

The initial leachate pH for a 5% $FeSO_4 \cdot 7H_2O$ + 4% $Al(OH)_3$ addition with or without a pH regulator differed significantly from the pH of the control groups; however, pH differences between the groups decreased to PVs up to 55.4 (Figure 5). The initial pH values of Groups A and D were 8.05 and 9.30; these values decreased with an increase in PV, and the pH of the column leachate tended towards stability when the PV reached 74.4. In contrast, the initial pH values in the leachate of Groups B and C were 3.70 and 3.30, which increased with the increase in PV to 37.2 and 74.4. The dynamic column experiment results show that the leachate solution pH stabilized at 6–8. The addition of 5% $FeSO_4$ and 4% $Al(OH)_3$ without a pH regulator had less effect on the final column soil pH.

The change in pH with PV affected Sb, Zn, and Cu stabilization in the soil. Sb could be amended by Fe–Al-based mixed amendments in a wide pH range (pH = 3–10) by strong binding to Fe oxyhydroxides as inner-sphere surface complexes [74] and the optimization pH range for adsorption was less than 7. This result agreed with the findings of Leuz et al. [32], who found that $H^+$ could increase the positive surface charge of iron oxyhydroxides and enhance the sorption capacity of $Sb(OH)_6^-$. However, when the pH exceeded 8, Sb(V) desorption was observed in the study of Leuz et al. [32]. This desorption reaction occurred mainly because the surface electronegativity of alkaline soil exceeded that of acidic soil, and it had a strong repulsive force to the anion $Sb(OH)_6^-$, which was not conducive to adsorption [75]. Therefore, the retention and stabilization efficiency of Sb decreased with an increase in pH. In contrast to Sb, the stabilization efficiency of Zn and Cu in soil increased significantly with an increase in pH. The stabilization of Zn and Cu under alkaline conditions was better than that under acidic conditions. This reduced retention effectiveness in acidic conditions could be caused by the following: First, $H^+$ replaces $Zn^{2+}$ and $Cu^+$ in the complex that is formed by the adsorption reaction, which induces Zn and Cu ions to migrate into the soil pore water [76]. Second, the partial dissolution of the sorbent, especially $Al(OH)_3$, in the soil leads to reduced sorption sites and potentially a lower retention of Zn and Cu. However, because of limited Sb sorption capacity, the partial dissolution of $Al(OH)_3$ had little effect on Sb adsorption under acidic conditions [4]. This pH contrast between Sb stabilization and co-occurring metals highlights the immobilization advantages of $FeSO_4$-mixed $Al(OH)_3$ addition.

## 4. Conclusions

This work shows that the studied Fe–Al-based amendments are suitable for the simultaneous stabilization of Sb and co-occurring metals, such as Cu and Zn in primary explosives-sites soil that is contaminated with a high concentration of Sb. Experimental results show that $FeSO_4 \cdot 7H_2O$ had an ideal retention of Sb and $Al(OH)_3$ and could prevent Cu and Zn mobilization. Batch and column leachate results showed that a 5% $FeSO_4 \cdot 7H_2O$ + 4% $Al(OH)_3$ application mixed with soil could immobilize Sb and retain co-occurring metals in highly contaminated Sb soils. The results of pH-regulated column tests indicated that acid conditioning favored Sb retention. Amendment addition had a positive impact on Sb retention and reduced the labile fractions of Sb. However, the retention behavior and mechanism, ion-exchange mechanism involved in these processes, and possible changes in remediation effect with time must be determined from further experiments. Amendment selection for primary-explosives-contaminated soils is based primarily on its ability to reduce the concentration of labile Sb, Cu, and Zn. In contrast, additional effects, such as the impact of amendments on the soil microbial abundance, community, diversity, and soil functionality are often neglected. Hydrolytic reactions of iron salt can cause soil acidification, and consequently limit site utilization. Sulfate addition to soil from $FeSO_4 \cdot 7H_2O$ likely had a substantial influence on the microbial community. These changes may affect key soil ecosystems and result in an ecological risk in the amended soil. The soil ecological risk must be investigated further after the addition of these efficient chemical amendments.

**Author Contributions:** Conceptualization, T.X.; writing—original draft preparation, N.W. and Y.J.; writing—review and editing, N.W.; validation, F.X., resources, C.Z. and Z.W.; supervision, D.Z.; funding acquisition, N.W. All authors have read and agreed to the published version of the manuscript.

**Funding:** This research was funded by [Beijing Municipal Natural Science Foundation] grant number [8204063].

**Acknowledgments:** This work was supported financially by Beijing Municipal Natural Science Foundation (8204063). The authors acknowledge the support of the Beijing Municipal Research Institute of Eco-Environment Protection, Beijing Institute of Mineral Resources and Geology, Beijing Normal University, and Technical Centre for Soil, Agriculture and Rural Ecology and Environment, Ministry of Ecology and Environment. These four institutions support resource sharing and an experimental platform. We thank Laura Kuhar, from Liwen Bianji (Edanz) (www.liwenbianji.cn/; 28 January 2022), for editing the English text of a draft of this manuscript.

**Conflicts of Interest:** The authors declare no conflict of interest.

## References

1. Hu, X.; Kong, L.; He, M. Kinetics and mechanism of photopromoted oxidative dissolution of antimony trioxide. *Environ. Sci. Technol.* **2014**, *48*, 14266–14272. [CrossRef]
2. Kong, L.; Hu, X.; He, M. Mechanisms of Sb(III) oxidation by pyrite-induced hydroxyl radicals and hydrogen peroxide. *Environ. Sci. Technol.* **2015**, *49*, 3499–3505. [CrossRef]
3. Filella, M.; Belzile, N.; Chen, Y.W. Antimony in the environment: A review focused on natural waters I. occurrence. *Earth-Sci. Rev.* **2002**, *57*, 125–176. [CrossRef]
4. He, M.; Wang, N.; Long, X.; Zhang, C.; Ma, C.; Zhong, Q.; Wang, A.; Wang, Y.; Pervaiz, A.; Shan, J. Antimony speciation in the environment: Recent advances in understanding the biogeochemical processes and ecological effects. *J. Environ. Sci.* **2019**, *75*, 14–39. [CrossRef] [PubMed]
5. Gebel, T. Arsenic and antimony: Comparative approach on mechanistic toxicology. *Chem.-Biol. Interact.* **1997**, *107*, 131–144. [CrossRef]
6. Gerhardsson, L.; Brune, D.; Nordberg, G.F.; Wester, P.O. Antimony in lung, liver and kidney from deceased smelter workers. Scandinavian. *J. Work. Environ. Health* **1982**, *8*, 201–208. [CrossRef]
7. Wang, N.; Wang, A.; Kong, L.; He, M. Calculation and application of Sb toxicity coefficient for potential ecological risk assessment. *Sci. Total Environ.* **2018**, *610-611*, 167–174. [CrossRef]
8. Wang, N.; Wang, A.; Xie, J.; He, M. Responses of soil fungal and archaeal communities to environmental factors in an ongoing antimony mine area. *Sci. Total Environ.* **2019**, *652*, 1030–1039. [CrossRef]
9. Wang, X.; He, M.; Xi, J.; Lu, X.; Xie, J. Antimony, arsenic and other toxic elements in the topsoil of an antimony mine area. In *Molecular Environmental Soil Science at the Interfaces in the Earth's Critical Zone*; Springer: Berlin/Heidelberg, Germany, 2010; pp. 58–61.
10. Wu, F.; Fu, Z.; Liu, B.; Mo, C.; Chen, B.; Corns, W.; Liao, H. Health risk associated with dietary co-exposure to high levels of antimony and arsenic in the world's largest antimony mine area. *Sci. Total Environ.* **2011**, *409*, 3344–3351. [CrossRef]
11. Filella, M.; Philippo, S.; Belzile, N.; Chen, Y.W.; Quentel, F. Natural attenuation processes applying to antimony: A study in the abandoned antimony mine in Goesdorf, Luxembourg. *Sci. Total Environ.* **2009**, *407*, 6205–6216. [CrossRef]
12. Amarasiriwardena, D.; Wu, F. Antimony: Emerging toxic contaminant in the environment Preface. *Microchem. J.* **2011**, *97*, 1–3. [CrossRef]
13. Wang, N.; Zhang, S.; He, M. Bacterial community profile of contaminated soils in a typical antimony mining site. *Environ. Sci. Pollut. Res.* **2018**, *25*, 141–152. [CrossRef]
14. Ahmad, M.; Lee, S.S.; Lim, J.E.; Lee, S.E.; Cho, J.S.; Moon, D.H.; Hashimoto, Y.; Ok, Y.S. Speciation and phytoavailability of lead and antimony in a small arms range soil amended with mussel shell, cow bone and biochar: EXAFS spectroscopy and chemical extractions. *Chemosphere* **2014**, *95*, 433–441. [CrossRef] [PubMed]
15. Okkenhaug, G.; Amstatter, K.; Bue, H.L.; Cornelissen, G.; Breedveld, G.D.; Henriksen, T.; Mulder, J. Antimony (Sb) Contaminated Shooting Range Soil: Sb Mobility and Immobilization by Soil Amendments. *Environ. Sci. Technol.* **2013**, *47*, 6431–6439. [CrossRef]
16. Wang, X.; He, M.; Xie, J.; Xi, J.; Lu, X. Heavy metal pollution of the world largest antimony mine-affected agricultural soils in Hunan province (China). *J. Soils Sediments* **2010**, *10*, 827–837. [CrossRef]
17. Matyáš, R.; Pachman, B. *Primary Explosives*; Springer: Berlin/Heidelberg, Germany, 2013; p. 350.
18. Hu, X.; He, M.; Kong, L. Photopromoted oxidative dissolution of stibnite. *Appl. Geochem.* **2015**, *61*, 53–61. [CrossRef]
19. Huynh, M.H.V.; Coburn, M.D.; Meyer, T.J.; Wetzler, M. Green primary explosives: 5-Nitrotetrazolato-N2-ferrate hierarchies. *Proc. Natl. Acad. Sci. USA* **2006**, *103*, 10322–10327. [CrossRef]
20. Brede, U.; Hagel, R.; Redecker, K.H.; Weuter, W. Primer compositions in the course of time: From black powder and SINOXID to SINTOX compositions and SINCO booster. *Propellants Explos. Pyrotech.* **1996**, *21*, 113–117. [CrossRef]
21. Jiang, Y.; Jia, X.; Xia, T.; Li, N.; Zhang, W. Environmental Risk Assessment of Antimony in Contaminated Soil by Primary Explosives. *Res. Environ. Sci.* **2020**, *33*, 485–493.
22. Okkenhaug, G.; Zhu, Y.G.; Luo, L.; Lei, M.; Li, X.; Mulder, J. Distribution, speciation and availability of antimony (Sb) in soils and terrestrial plants from an active Sb mining area. *Environ. Pollut.* **2011**, *159*, 2427–2434. [CrossRef]
23. Herath, I.; Vithanage, M.; Bundschuh, J. Antimony as a global dilemma: Geochemistry, mobility, fate and transport. *Environ. Pollut.* **2017**, *223*, 545–559. [CrossRef] [PubMed]
24. Yuan, Z.; Zhang, G.; Lin, J.; Zeng, X.; Ma, X.; Wang, X.; Wang, S.; Jia, Y. The stability of Fe(III)-As(V) co-precipitate in the presence of ascorbic acid: Effect of pH and Fe/As molar ratio. *Chemosphere* **2019**, *218*, 670–679. [CrossRef] [PubMed]
25. Zhang, D.; Wang, S.; Gomez, M.A.; Wang, Y.; Jia, Y. Long-term stability of the Fe(III)-As(V) coprecipitates: Effects of neutralization mode and the addition of Fe(II) on arsenic retention. *Chemosphere* **2019**, *237*, 124503. [CrossRef] [PubMed]
26. Tandy, S.; Meier, N.; Schulin, R. Use of soil amendments to immobilize antimony and lead in moderately contaminated shooting range soils. *J. Hazard. Mater.* **2017**, *324*, 617–625. [CrossRef] [PubMed]
27. Filella, M.; Belzile, N.; Chen, Y.W. Antimony in the environment: A review focused on natural waters II. Relevant solution chemistry. *Earth-Sci. Rev.* **2002**, *59*, 265–285. [CrossRef]
28. Ilgen, A.G.; Majs, F.; Barker, A.J.; Douglas, T.A.; Trainor, T.P. Oxidation and mobilization of metallic antimony in aqueous systems with simulated groundwater. *Geochim. Cosmochim. Acta* **2014**, *132*, 16–30. [CrossRef]

29. Filella, M.; Williams, P.A.; Belzile, N. Antimony in the environment: Knowns and unknowns. *Environ. Chem.* **2009**, *6*, 95–105. [CrossRef]
30. Guo, X.; Wu, Z.; He, M.; Meng, X.; Jin, X.; Qiu, N.; Zhang, J. Adsorption of antimony onto iron oxyhydroxides: Adsorption behavior and surface structure. *J Hazard. Mater.* **2014**, *276*, 339–345. [CrossRef] [PubMed]
31. Garau, G.; Silvetti, M.; Vasileiadis, S.; Donner, E.; Diquattro, S.; Deiana, S.; Lombi, E.; Castaldi, P. Use of municipal solid wastes for chemical and microbiological recovery of soils contaminated with metal(loid)s. *Soil Biol. Biochem.* **2017**, *111*, 25–35. [CrossRef]
32. Leuz, A.-K.; Moench, H.; Johnson, C.A. Sorption of Sb(III) and Sb(V) to goethite: Influence on Sb(III) oxidation and mobilization. *Environ. Sci. Technol.* **2006**, *40*, 7277–7282. [CrossRef]
33. Kong, L.; He, M. Mechanisms of Sb(III) photooxidation by the excitation of organic Fe(III) complexes. *Environ. Sci. Technol.* **2016**, *50*, 6974–6982. [CrossRef] [PubMed]
34. Almas, A.R.; Pironin, E.; Okkenhaug, G. The partitioning of Sb in contaminated soils after being immobilization by Fe-based amendments is more dynamic compared to Pb. *Appl. Geochem.* **2019**, *108*, 104378. [CrossRef]
35. Wang, Y.; Michel, F.M.; Levard, C.; Choi, Y.; Eng, P.J.; Brown, G.E., Jr. Competitive Sorption of Pb(II) and Zn(II) on Polyacrylic Acid-Coated Hydrated Aluminum-Oxide Surfaces. *Environ. Sci. Technol.* **2013**, *47*, 12131–12139. [CrossRef] [PubMed]
36. Garau, G.; Silvetti, M.; Castaldi, P.; Mele, E.; Deiana, P.; Deiana, S. Stabilising metal(loid)s in soil with iron and aluminium-based products: Microbial, biochemical and plant growth impact. *J. Environ. Manag.* **2014**, *139*, 146–153. [CrossRef] [PubMed]
37. James, K.; Peters, R.E.; Laird, B.D.; Ma, W.K.; Wickstrom, M.; Stephenson, G.L.; Siciliano, S.D. Human Exposure Assessment: A Case Study of 8 PAH Contaminated Soils Using in Vitro Digestors and the Juvenile Swine Model. *Environ. Sci. Technol.* **2011**, *45*, 4586–4593. [CrossRef] [PubMed]
38. Zhang, S.; Wang, Y.; Pervaiz, A.; Kong, L.; He, M. Comparison of diffusive gradients in thin-films (DGT) and chemical extraction methods for predicting bioavailability of antimony and arsenic to maize. *Geoderma* **2018**, *332*, 1–9. [CrossRef]
39. Rauret, G.; Lopez-Sanchez, J.F.; Sahuquillo, A.; Rubio, R.; Davidson, C.; Ure, A.; Quevauviller, P. Improvement of the BCR three step sequential extraction procedure prior to the certification of new sediment and soil reference materials. *J. Environ. Monit.* **1999**, *1*, 57–61. [CrossRef]
40. Weng, H.-X.; Ma, X.-W.; Fu, F.-X.; Zhang, J.-J.; Liu, Z.; Tian, L.-X.; Liu, C. Transformation of heavy metal speciation during sludge drying: Mechanistic insights. *J. Hazard. Mater.* **2014**, *265*, 96–103. [CrossRef]
41. Liu, Y.; Yan, J.; Liu, F.; Shen, C.; Li, F.; Yang, B.; Huang, M.; Sang, W. Nanoscale iron (oxyhydr)oxide-modified carbon nanotube filter for rapid and effective Sb(III) removal. *Rsc Adv.* **2019**, *9*, 9.
42. Gao, F.; Ju, T.; Wu, X.; Wang, Y.; Li, X.; Fan, P.; Luan, T.; Zhou, J. Spatial Variability and Autocorrelation Analysis of pH in a Mollisol Tillage Area of Northeast China. *Soils* **2018**, *50*, 566–573.
43. Shen, F.; Liao, R.; Ali, A.; Mahar, A.; Guo, D.; Li, R.; Xining, S.; Awasthi, M.K.; Wang, Q.; Zhang, Z. Spatial distribution and risk assessment of heavy metals in soil near a Pb/Zn smelter in Feng County, China. *Ecotoxicol. Environ. Saf.* **2017**, *139*, 254–262. [CrossRef]
44. Chen, M.; Li, X.-M.; Yang, Q.; Zeng, G.-M.; Zhang, Y.; Liao, D.-X.; Liu, J.-J.; Hu, J.-M.; Guo, L. Total concentrations and speciation of heavy metals in municipal sludge from Changsha, Zhuzhou and Xiangtan in middle-south region of China. *J. Hazard. Mater.* **2008**, *160*, 324–329. [CrossRef] [PubMed]
45. Tresintsi, S.; Simeonidis, K.; Vourlias, G.; Stavropoulos, G.; Mitrakas, M. Kilogram-scale synthesis of iron oxy-hydroxides with improved arsenic removal capacity: Study of Fe(II) oxidation-precipitation parameters. *Water Res.* **2012**, *46*, 5255–5267. [CrossRef]
46. Zhang, T.; Zhao, Y.; Bai, H.; Wang, W.; Zhang, Q. Enhanced arsenic removal from water and easy handling of the precipitate sludge by using $FeSO_4$ with $CaCO_3$ to $Ca(OH)_2$. *Chemosphere* **2019**, *231*, 134–139. [CrossRef] [PubMed]
47. Jang, J.-H.; Dempsey, B.A.; Burgos, W.D. Solubility of hematite revisited: Effects of hydration. *Environ. Sci. Technol.* **2007**, *41*, 7303–7308. [CrossRef] [PubMed]
48. Garau, G.; Silvetti, M.; Deiana, S.; Deiana, P.; Castaldi, P. Long-term influence of red mud on As mobility and soil physico-chemical and microbial parameters in a polluted sub-acidic soil. *J. Hazard. Mater.* **2011**, *185*, 1241–1248. [CrossRef]
49. Wilson, S.C.; Lockwood, P.V.; Ashley, P.M.; Tighe, M. The chemistry and behaviour of antimony in the soil environment with comparisons to arsenic: A critical review. *Environ. Pollut.* **2010**, *158*, 1169–1181. [CrossRef] [PubMed]
50. Manzano, R.; Silvetti, M.; Garau, G.; Deiana, S.; Castaldi, P. Influence of iron-rich water treatment residues and compost on the mobility of metal(loid)s in mine soils. *Geoderma* **2016**, *283*, 1–9. [CrossRef]
51. Palansooriya, K.N.; Shaheen, S.M.; Chen, S.S.; Tsang, D.C.W.; Hashimoto, Y.; Hou, D.; Bolan, N.S.; Rinklebe, J.; Ok, Y.S. Soil amendments for immobilization of potentially toxic elements in contaminated soils: A critical review. *Environ. Int.* **2020**, *134*, 105046. [CrossRef] [PubMed]
52. Scheinost, A.C.; Rossberg, A.; Vantelon, D.; Xifra, I.; Kretzschmar, R.; Leuz, A.-K.; Funke, H.; Johnson, C.A. Quantitative antimony speciation in shooting-range soils by EXAFS spectroscopy. *Geochim. Cosmochim. Acta* **2006**, *70*, 3299–3312. [CrossRef]
53. Serafimovska, J.M.; Arpadjan, S.; Stafilov, T.; Tsekova, K. Study of the antimony species distribution in industrially contaminated soils. *J. Soils Sediments* **2013**, *13*, 294–303. [CrossRef]
54. Kong, L.; He, M.; Hu, X. Rapid photooxidation of Sb(III) in the presence of different Fe(III) species. *Geochim. Cosmochim. Acta* **2016**, *180*, 214–226. [CrossRef]
55. Guo, X.; Wu, Z.; He, M. Removal of antimony(V) and antimony(III) from drinking water by coagulation-flocculation-sedimentation (CFS). *Water Res.* **2009**, *43*, 4327–4335. [CrossRef] [PubMed]

56. Wu, Z.; He, M.; Guo, X.; Zhou, R. Removal of antimony (III) and antimony (V) from drinking water by ferric chloride coagulation: Competing ion effect and the mechanism analysis. *Sep. Purif. Technol.* **2010**, *76*, 184–190. [CrossRef]
57. Leverett, P.; Reynolds, J.K.; Roper, A.J.; Williams, P.A. Tripuhyite and schafarzikite: Two of the ultimate sinks for antimony in the natural environment. *Mineral. Mag.* **2012**, *76*, 891–902. [CrossRef]
58. Liu, Y.; Lou, Z.; Yang, K.; Wang, Z.; Zhou, C.; Li, Y.; Cao, Z.; Xu, X. Coagulation removal of Sb(V) from textile wastewater matrix with enhanced strategy: Comparison study and mechanism analysis. *Chemosphere* **2019**, *237*, 124494. [CrossRef] [PubMed]
59. Jarvis, P.; Sharp, E.; Pidou, M.; Molinder, R.; Parsons, S.A.; Jefferson, B. Comparison of coagulation performance and floc properties using a novel zirconium coagulant against traditional ferric and alum coagulants. *Water Res.* **2012**, *46*, 4179–4187. [CrossRef]
60. Dehghani, M.H.; Karimi, B.; Rajaei, M.S. The effect of aeration on advanced coagulation, flotation and advanced oxidation processes for color removal from wastewater. *J. Mol. Liq.* **2016**, *223*, 75–80. [CrossRef]
61. Essington, M.E.; Stewart, M.A. Influence of Temperature and pH on Antimonate Adsorption by Gibbsite, Goethite, and Kaolinite. *Soil Sci.* **2015**, *180*, 54–66. [CrossRef]
62. Mishra, S.; Dwivedi, J.; Kumar, A.; Sankararamakrishnan, N. Removal of antimonite (Sb(III)) and antimonate (Sb(V)) using zerovalent iron decorated functionalized carbon nanotubes. *RSC Adv.* **2016**, *6*, 95865–95878. [CrossRef]
63. Dai, C.; Zhou, Z.; Zhou, X.; Zhang, Y. Removal of Sb (III) and Sb(V) from Aqueous Solutions Using nZVI. *Water Air Soil Pollut.* **2014**, *225*, 1799. [CrossRef]
64. Daoyong, Z.; Xiangliang, P.A.N.; Guijin, M.U.; Huiming, G.U. Removal of antimony from water by adsorption to red mud. *Technol. Water Treat.* **2008**, *34*, 34–37.
65. Kameda, T.; Yagihashi, N.; Park, K.-S.; Grause, G.; Yoshioka, T. Preparetation of Fe-Al layered double hydroxide and its application in Sb removal. *Fresenius Environ. Bull.* **2009**, *18*, 1006–1010.
66. Hai-ying, Z.; You-cai, Z.; Guo-xin, Z.; Jing-yu, Q.I.; Cui-xiang, G.U.O.; Li-jie, S. Leaching Behavior of Heavy Metals from MSWI Fly Ash. *Environ. Sci. Technol.* **2010**, *33*, 130–132.
67. Li, J.; Chen, J.; Cao, H.; He, P. Influence of pH on leaching of pollutants from sewage sludge. *Chin. J. Environ. Eng.* **2013**, *7*, 4983–4989.
68. Cui, H.; Yang, X.; Xu, L.; Fan, Y.; Yi, Q.; Lia, R.; Zhou, J. Effects of goethite on the fractions of Cu, Cd, Pb, P and soil enzyme activity with hydroxyapatite in heavy metal-contaminated soil. *Rsc. Adv.* **2017**, *7*, 45869–45877. [CrossRef]
69. Hu, X.; He, M. Organic ligand-induced dissolution kinetics of antimony trioxide. *J. Environ. Sci.* **2017**, *56*, 87–94. [CrossRef] [PubMed]
70. Hu, X.; Guo, X.; He, M.; Li, S. pH-dependent release characteristics of antimony and arsenic from typical antimony-bearing ores. *J. Environ. Sci.* **2016**, *44*, 171–179. [CrossRef] [PubMed]
71. Hockmann, K.; Tandy, S.; Lenz, M.; Reiser, R.; Conesa, H.M.; Keller, M.; Studer, B.; Schulin, R. Antimony retention and release from drained and waterlogged shooting range soil under field conditions. *Chemosphere* **2015**, *134*, 536–543. [CrossRef]
72. Klitzke, S.; Lang, F. Mobilization of Soluble and Dispersible Lead, Arsenic, and Antimony in a Polluted, Organic-rich Soil—Effects of pH Increase and Counterion Valency. *J. Environ. Qual.* **2009**, *38*, 933–939. [CrossRef]
73. Ke, X.; Zhang, F.J.; Zhou, Y.; Zhang, H.J.; Guo, G.L.; Tian, Y. Removal of Cd, Pb, Zn, Cu in smelter soil by citric acid leaching. *Chemosphere* **2020**, *255*, 126690. [CrossRef] [PubMed]
74. Mitsunobu, S.; Takahashi, Y.; Terada, Y.; Sakata, M. Antimony(V) Incorporation into Synthetic Ferrihydrite, Goethite, and Natural Iron Oxyhydroxides. *Environ. Sci. Technol.* **2010**, *44*, 3712–3718. [CrossRef]
75. Ma, X.; Qin, J.; Zhang, Y. A Comparison of Desorption Behaviors of Sb in Different Soils. *J. Agro-Environ. Sci.* **2015**, *34*, 1528–1534.
76. Du, X.; Cui, S.; Wang, Q.; Han, Q.; Liu, G. Non-competitive and competitive adsorption of Zn (II), Cu (II), and Cd (II) by a granular Fe-Mn binary oxide in aqueous solution. *Environ. Prog. Sustain. Energy* **2021**, *40*, e13611. [CrossRef]

Article

# A Comprehensive Evaluation Method for Soil Remediation Technology Selection: Case Study of Ex Situ Thermal Desorption

Shuang Li [1,2], Liao He [1], Bo Zhang [1], Yan Yan [1,2], Wentao Jiao [1] and Ning Ding [1,*]

1 State Key Laboratory of Urban and Regional Ecology, Research Center for Eco-Environmental Sciences, Chinese Academy of Sciences, Beijing 100085, China; lishuang_1216@126.com (S.L.); 2018520089@bipt.edu.cn (L.H.); zhangbo@rcees.ac.cn (B.Z.); yyan@rcees.ac.cn (Y.Y.); wtjiao@rcees.ac.cn (W.J.)
2 College of Resources and Environment, University of Chinese Academy of Sciences, Beijing 101408, China
* Correspondence: ningding@rcees.ac.cn

**Abstract:** Quantitative evaluation of different contaminated soil remediation technologies in multiple dimensions is beneficial for the optimization and comparative selection of technology. Ex situ thermal desorption is widely used in remediation of organic contaminated soil due to its excellent removal effect and short engineering period. In this study, a comprehensive evaluation method of soil remediation technology, covering 20 indicators in five dimensions, was developed. It includes the steps of constructing an indicator system, accounting for the indicator, normalization, determining weights by analytic hierarchy process, and comprehensive evaluation. Three ex situ thermal desorption technology—direct thermal desorption, indirect thermal desorption, and indirect thermal heap—in China were selected for the model validation. The results showed that the direct thermal desorption had the highest economic and social indicator scores of 0.068 and 0.028, respectively. The indirect thermal desorption had the highest technical and environmental indicator scores of 0.118 and 0.427, respectively. The indirect thermal heap had the highest resource indicator score of 0.175. With balanced performance in five dimensions, the indirect thermal desorption had the highest comprehensive score of 0.707, which is 1.6 and 1.4 times higher than the direct thermal desorption and indirect thermal heap, respectively. The comprehensive evaluation method analyzed and compared the characteristics of the ex situ thermal desorption technology from different perspectives, such as specific indicators, multiple dimensions, and single comprehensive values. It provided a novel evaluation approach for the sustainable development and application of soil remediation technology.

**Keywords:** comprehensive evaluation method; contaminated soil; ex situ thermal desorption; environmental impact; resource utilization

## 1. Introduction

Establishing a comprehensive and practical evaluation system is of critical importance to the sustainable development of technologies. Comprehensive evaluation refers to the use of a systematic and standardized method that includes simultaneous multiple indicators for evaluation. Comprehensive evaluation can analyze the whole process of technology implementation, and provide information for process optimization in terms of technological, economical, and social aspects [1–3]. Therefore, comprehensive evaluation is very important for process optimization of technology, and the comparison and selection among different technologies.

Contaminated soil remediation is an important issue in the environmental field [4]. In past decades, a variety of soil remediation technologies have been developed [5]. To evaluate different soil remediation technologies, one first needs to focus on the characteristic indicators such as efficiency, stability, and applicability. The technology consumes raw and auxiliary materials, and energy during implementation, resulting in the consumption of natural resources. At the same time, emissions from energy consumption or process physicochemical reactions can result in environmental impacts. The economic cost, benefit, and technical value of the technology are also important factors of concern to investors and decision makers. In addition, the implementation of such pollutant removal technologies can have certain social impacts, such as job opportunities for local residents, but also negative social effects, such as concerns from adjacent residents and potential impact on workers' health. A comprehensive evaluation can avoid the transfer of technological loads between different dimensions.

Ex situ thermal desorption has become one of the most effective remediation technologies for organic contaminated soil [6–9]. Ex situ soil remediation usually is the second choice after in situ technology, which are more sustainable and less costly; thus, the effort to analyze the impact of the ex situ remediation processes is necessary. Since the 1980s, scholars from the United States, France, Canada, Argentina, South Korea, and other countries have carried out thermal desorption remediation research on a variety of organic contaminated soils [10]. In Europe, thermal desorption has also been widely used in engineering practice [6,11–14]. In America, among the 571 ex situ soil remediation projects carried out during 1982 to 2014, 77 used ex situ thermal desorption remediation technology, accounting for 13.5% of the total number of projects [14]. The independent research, and the development and application of the equipment for ex situ thermal desorption technology in China started late. The first patent on thermal desorption remediation technology was granted in 2009, and the first related article was published in 2011 [15,16]. As of 2017, a total of 23 ex situ thermal desorption remediation projects for contaminated sites have been carried out [15].

At present, the evaluation of carbon emission and environmental impact of ex situ thermal desorption technology has been carried out [17–20], but there are few literature reports on its quantitative evaluation at the levels of different dimensions, such as technical characteristics, resources, environment, economy, and society. A comprehensive study can provide a theoretical basis for the directional selection of ex situ thermal desorption technology in terms of specific indicators, and further provide scientific support for the overall development of ex situ thermal desorption technology. In this paper, multilevel comprehensive evaluation is carried out for direct and indirect ex situ thermal desorption technology, and its key influencing factors and advantageous indicators are determined through comparative analysis, which further reflects the importance of technology evaluation methods in selecting appropriate technology. The establishment of a comprehensive evaluation model is conducive to the optimization, improvement, and comparative selection of technology, and can provide a new analytical method for the quantitative comparison between different ex situ thermal desorption technologies.

## 2. Method and Data

### 2.1. Methodological Framework

A comprehensive evaluation method for ex situ thermal desorption technology was constructed in this study, and its framework is shown in Figure 1. The main steps of technology evaluation include: (1) determining the evaluation object and the technology involved in the evaluation; (2) describing the remediation technology; (3) determining the evaluation indicator set and collecting the evaluation indicator parameters; (4) determining the weight and quantification method of the evaluation indicator; (5) comprehensively analyzing and weighting each indicator, and calculating the score of each evaluated dimension; and (6) obtaining the comprehensive evaluation result.

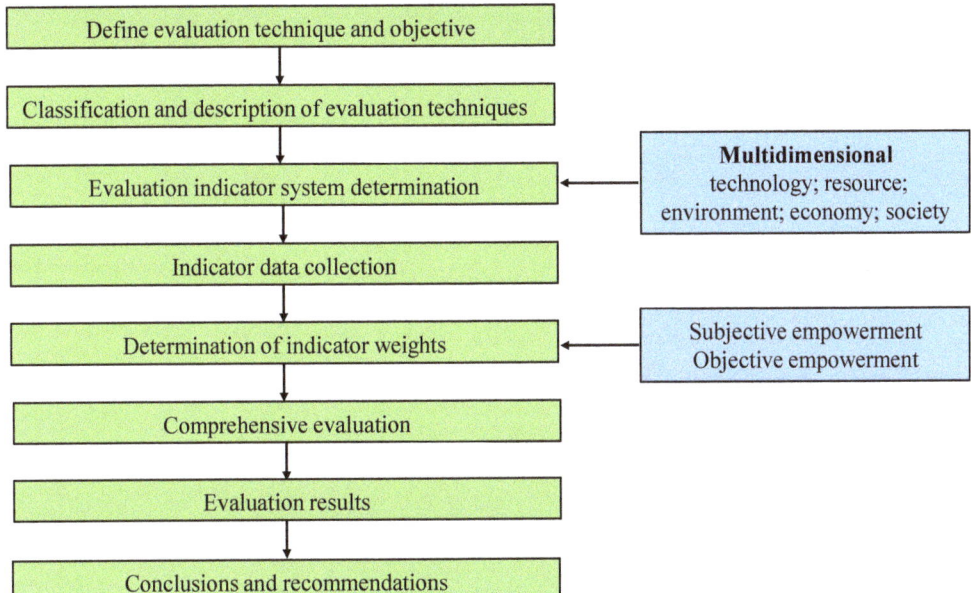

**Figure 1.** Comprehensive evaluation method framework.

*2.2. Comprehensive Evaluation Indicator System*

2.2.1. Evaluation Indicators

To comprehensively evaluate the performance of ex situ thermal desorption technology, an evaluation indicator system was constructed by referring to a sustainable development indicator, a green development indicator, and an environmental pollution prevention and control technology evaluation method. It contains five dimensions, which are technology, resource, environment, economy, and society, and has a total of 20 indicators. The dimensions and the indicators are shown in Table 1.

(1) Technical indicators

In terms of technical indicators, characteristics of efficiency and operation are constructed. The efficiency indicators reflect the characteristics of the technology in pollutant removal and thermal efficiency; the operation indicators reflect whether the technology still has instability. The technical efficiency indicators select heat transfer efficiency and pollutant removal rate; the operational indicators select secondary pollutants, fault condition, and comprehensive energy consumption as secondary indicators.

(2) Resource indicators

The resource indicators reflect the demand for various material inputs in the process of technology implementation, which select raw material consumption, energy consumption, and water consumption as secondary indicators.

(3) Environmental indicators

Environmental indicators include two parts, the first part is the environmental impact during the implementation of technology, focusing on noise and peculiar smell, and the second part is the whole process environmental impact, in which we applied the life cycle assessment (LCA) method to calculate the typical environmental impact. This study selects global warming potential (GWP), eutrophication potential, acidification potential, and ozone layer depletion potential.

Table 1. Comprehensive evaluation indicators of ex situ thermal desorption remediation technology.

| Dimensions | Indicators | Units | Indicator Definition |
|---|---|---|---|
| Technical indicators | Heat transfer efficiency | % | Heat transfer rate per unit time |
| | Pollutant removal rate | % | Removal rate of target pollutants (removal rate to standard) |
| | Secondary pollutants | / | Whether to produce other secondary pollutants (produce exceed the standard, produce but not exceed, not produce) |
| | Fault condition | / | Whether it can operate stably and produce failure situations (no fault, minor fault, and serious fault affect the operation) |
| | Comprehensive energy consumption | MJ/t soil remediation | Energy consumption during operation |
| Resource indicators | Raw materials consumption | kg/t soil remediation | Whether to consume dehydrating agents, conditioning agents, odor inhibitors, etc. |
| | Energy consumption | kWh, $m^3$, L ect./t soil remediation | Consumption of electricity, natural gas, gasoline, etc., from life cycle perspective |
| | Water consumption | $m^3$/t soil remediation | Fresh water consumption |
| Environmental indicators | Global warming potential | kg $CO_2$-Equiv./t soil remediation | Life cycle assessment methodology indicator |
| | Eutrophication potential | kg Phosphate-Equiv./t remediation | Life cycle assessment methodology indicator |
| | Acidification potential | kg $SO_2$-Equiv./t soil remediation | Life cycle assessment methodology indicator |
| | Ozone layer depletion potential | kg R11-Equiv./t soil remediation | Life cycle assessment methodology indicator |
| | Peculiar smell | / | Peculiar smell during the implementation of technology |
| | Noise | decibel | Noise impact during implementation of technology |
| Economic indicators | Investment return period | Year | The number of years from the time the project starts production to the time when the full construction investment is recovered |
| | Direct benefit | Yuan (RMB)/t soil remediation | Net profit of remediation of unit contaminated soil |
| | Indirect benefit | Yuan (RMB)/t soil remediation | Disposal costs reduced by remediation of unit contaminated soil |
| Social indicators | Job opportunity | person/t soil remediation | Jobs created during the operation |
| | Social income | % | The income level of practitioners, the income per person per month/local average income |
| | "Not in my back yard" (NIMBY) | / | Residents or local units worry that remediation technology will bring many negative effects on health, environmental quality, and asset value |

(4) Economic indicators

Economic indicators are designed to reveal the costs and benefits of technology, which select investment return period, direct benefits, and indirect benefits as secondary indicators.

(5) Social indicators

Social indicators reflect the basic social benefits and negative effects. The social benefits include the job opportunities and social income, while the negative effects primarily consider the concerns from the neighboring residents, i.e., "not in my back yard" (NIMBY).

2.2.2. Quantification of Evaluation Indicators

Once the indicator system is defined, the indicators need to be quantified and normalized. The indicators are divided into qualitative and quantitative indicators; qualitative indicators are graded according to the severity, and quantitative indicators are calculated based on the definitions of the indicators.

The indicators, with different units and magnitudes, need to be normalized. The common methods of evaluation indicators are mainly the normalization method (quantitative indicators) and rank assignment method (qualitative indicators) [21]. These two methods are used in this study.

To ensure the accuracy of assessment, all quantitative indicators were normalized before calculating the weights, with the largest values selected as criterion 1 for positive indicators and the smallest values selected as criterion 1 for negative indicators, with all indicator values between 0 and 1.

### 2.3. Weight Determination Method

#### 2.3.1. Weighting Calculation

The determination of the weight coefficient can be performed by using the expert scoring method, target distance method, analytic hierarchy process (AHP), or entropy weight method [22,23]. The expert scoring method mainly collects experts' opinions on the importance of relevant indicators according to the relevant knowledge mastered by experts in the opinion table, and summarizes the different opinions to reach common opinions. The target distance method is widely used in the environmental field and it mainly represents the severity of the environmental impact effect based on studying the gap between the current level and the target level [24]. In this paper, we choose the AHP, which is mainly used to solve the problem of decision making, with a combination of qualitative and quantitative method. The direct participation of decision makers ensures the consistency of the model thinking process, which can provide support for various fields with complex problems [25,26].

#### 2.3.2. Judgment Matrix Construction

The weight coefficients in this study were determined by the AHP. The importance scales of different indicators in this method and their meanings are shown in Table 2 [27].

**Table 2.** Indicator importance scale.

| Importance Scale $a_{ij}$ | Description | Importance Scale $a_{ij}$ | Description |
|---|---|---|---|
| 1 | Two factors have the same importance | 9 | $i$ is more important than $j$ |
| 3 | $i$ is slightly more important than $j$ | 2,4,6,8 | scale median |
| 5 | $i$ is more important than $j$ | reciprocal | $j$ compared to $i$ |
| 7 | $i$ is extremely more important than $j$ | | |

Table 2 quantifies the relative importance of indicators in different dimensions. On this basis, the weight of each indicator can be calculated according to the root method or the sum product method. According to a number of expert opinions and literature reports [28], combined with judgment matrix construction, the importance scale of dimensions or indicators is determined, as shown in Tables 3–8.

**Table 3.** Importance scale of different dimensional layers.

| | Technology | Resources | Environment | Economy | Society |
|---|---|---|---|---|---|
| Technology | 1 | 1/3 | 1/5 | 3 | 3 |
| Resources | 3 | 1 | 1/3 | 5 | 7 |
| Environment | 5 | 3 | 1 | 7 | 9 |
| Economy | 1/3 | 1/5 | 1/7 | 1 | 5 |
| Society | 1/3 | 1/7 | 1/9 | 1/5 | 1 |
| Weight | 0.118 | 0.265 | 0.513 | 0.071 | 0.033 |

**Table 4.** Importance scale of the technical indicator.

| | Heat Transfer Efficiency | Pollutant Removal Rate | Secondary Pollutants | Failure Situation | Comprehensive Energy Consumption |
|---|---|---|---|---|---|
| Heat transfer efficiency | 1 | 1/7 | 1/3 | 1/5 | 1/3 |
| Pollutant removal rate | 7 | 1 | 5 | 3 | 5 |
| Secondary pollutants | 3 | 1/5 | 1 | 1/3 | 3 |
| Fault condition | 5 | 1/3 | 3 | 1 | 3 |
| Comprehensive energy consumption | 3 | 1/5 | 1/3 | 1/3 | 1 |

Table 5. Importance scale of the resource indicator.

|  | Raw Materials Consumption | Energy Consumption | Water Consumption |
|---|---|---|---|
| Raw materials consumption | 1 | 1 | 1/3 |
| Energy consumption | 1 | 1 | 1/3 |
| Water consumption | 3 | 3 | 1 |

Table 6. Importance scale of the environmental indicator.

|  | Greenhouse Effect | Eutrophication | Acidification Effect | Ozone Layer Destruction | Peculiar Smell | Noise |
|---|---|---|---|---|---|---|
| Global warming potential | 1 | 5 | 3 | 7 | 3 | 3 |
| Eutrophication potential | 1/5 | 1 | 1/3 | 3 | 1/3 | 1/3 |
| Acidification potential | 1/3 | 3 | 1 | 5 | 3 | 3 |
| Ozone layer depletion potential | 1/7 | 1/3 | 1/5 | 1 | 1/3 | 1/3 |
| Peculiar smell | 1/3 | 3 | 1/3 | 3 | 1 | 1 |
| Noise | 1/3 | 3 | 1/3 | 3 | 1 | 1 |

Table 7. Importance scale of the economic indicator.

|  | Investment Return Period | Direct Benefit | Indirect Income |
|---|---|---|---|
| Investment return period | 1 | 1/5 | 1/3 |
| Direct benefit | 5 | 1 | 3 |
| Indirect benefit | 3 | 1/3 | 1 |

Table 8. Importance scale of the social indicator.

|  | Job Opportunity | Social Income | Adjacent Effect |
|---|---|---|---|
| Job opportunity | 1 | 3 | 7 |
| Social income | 1/3 | 1 | 5 |
| NIMBY | 1/7 | 1/5 | 1 |

2.3.3. Weighting Coefficient Determination

Based on the analysis above, this study uses the sum product method to calculate the weight coefficient, which can be divided into two steps:

First, the judgment matrix is normalized by column, and the rows are added and summed as follows:

$$\overline{W}_i = \sum_{j=1}^{n} \frac{a_{ij}}{\sum_{i=1}^{n} a_{ij}}$$

Second, normalization is carried out, and the result is the weight coefficient of each environmental indicator, which can be obtained from the following formula:

$$W_i = \frac{\overline{W}_i}{\sum_{i=1}^{n} \overline{W}_i}$$

The weight coefficient results obtained are shown in Table 9.

Table 9. Weight coefficient.

| Primary Indicators | Secondary Indicators | Secondary Weight | Primary Weight |
|---|---|---|---|
| Technical indicator | Heat transfer efficiency | 0.045 | 0.118 |
| | Pollutant removal rate | 0.498 | |
| | Secondary pollutants | 0.129 | |
| | Fault condition | 0.245 | |
| | Comprehensive energy consumption | 0.083 | |
| Resource indicator | Raw materials consumption | 0.200 | 0.265 |
| | Energy consumption | 0.200 | |
| | Water consumption | 0.600 | |
| Environmental indicator | Global warming potential | 0.398 | 0.513 |
| | Eutrophication potential | 0.067 | |
| | Acidification potential | 0.240 | |
| | Ozone layer depletion potential | 0.041 | |
| | Peculiar smell | 0.127 | |
| | Noise | 0.127 | |
| Economic indicator | Investment return period | 0.105 | 0.071 |
| | Direct benefit | 0.637 | |
| | Indirect benefit | 0.258 | |
| Social indicator | Job opportunity | 0.649 | 0.033 |
| | Social income | 0.279 | |
| | NIMBY | 0.072 | |

2.3.4. Consistency Test of Judgment Matrix

Inconsistent judgments may derive from the comparison matrix obtained by the two-by-two comparison method used in AHP. Therefore, a consistency test is required. Additionally, the consistency test mainly refers to the fact that when variable a is relative important to variable b, and variable b is relative important to variable c, then variable a must be more important than variable c.

The consistency indicator $CI$ is

$$CI = \frac{(\lambda_{max} - n)}{(n - 1)}$$

The formula for determining consistency is

$$CR = \frac{CI}{RI} < 0.1$$

where $CR$ is the consistency ratio. $RI$ is the average random consistency indicator, and its value is shown in Table 10.

Table 10. Average random consensus indicator.

| Numerical Value $n$ | 1 | 2 | 3 | 4 | 5 | 6 | 7 | 8 | 9 |
|---|---|---|---|---|---|---|---|---|---|
| RI | 0.00 | 0.00 | 0.58 | 0.90 | 1.12 | 1.24 | 1.32 | 1.41 | 1.45 |

Based on the calculation and analysis of the formula above, the consistency test results obtained for the five dimensions are shown in Table 11.

Table 11. Consistency ratio of five dimensions.

| | Technical Indicator | Resource Indicator | Environmental Indicator | Economic Indicator | Social Indicator |
|---|---|---|---|---|---|
| CR | 0.066 | 0 | 0.051 | 0.037 | 0.064 |

As can be seen from the these tables, the consistency ratio of importance ranking of all indicators is less than 0.1, indicating that the ranking results have a satisfactory consistency and can be accepted [29].

2.4. Comprehensive Evaluation Methodology

The total score $S$ can be obtained from the weighted average of the values of each dimension:

$$S = \sum_{i=1}^{p} D_i$$

where $S$ is a single indicator of comprehensive evaluation and $D_i$ is the score of each indicator/dimension at different levels.

2.5. Data Source

The data obtained from the inquiry of remediation site staff, inspection of project reports, and test reports are used in this study, as shown in Table 12. The data of this study are divided into five categories. The first is the data of resource and energy consumption, and environment emissions, which mainly come from the statistical, recorded, and monitor data on the remediation site. The technical data mainly come from interviews of technicians at the remediation site, and most of the economic and social data come from the technology report. The data of NIMBY came from the survey and interview of nearby residents. In addition, the basic data of LCA designed in the environmental dimension mainly come from CAS RCEES 2020 developed by the Research Center for Eco-Environmental Sciences, Chinese Academy of Sciences. This database supports the publication of many related studies [21,30,31].

Table 12. Description of data types and sources.

| Data Types | Data Sources |
|---|---|
| Energy consumption and material consumption | On-site research |
| Technical specifications, failure situation, and efficiency | Provided by on-site technicians |
| Economic cost input and benefits | On-site research and project reports |
| Social employment and salary | On-site research and project reports |
| NIMBY | Survey and interview |
| Full process environmental impact base data | China localized life cycle assessment database CAS RCEES |

## 3. Case Study

3.1. Remediation Site and Technology Selection

Thermal desorption can be divided into two parts: the thermal desorption stage and the off-gas treatment stage, as shown in Figure 2. Ex situ thermal desorption technologies involve excavating and transporting contaminated soil from the original site where the pollution occurred to other sites for remediation. The principle is that through direct or indirect heating, the contaminated soil reaches a certain temperature, in which the organic pollutants are converted into a gas phase and volatilized into the desorption off-gas, and then completely removed by the gas treatment system, so as to obtain clean soil [32,33]. According to different contact modes between heat source and contaminated soil, ex situ thermal desorption technology can be divided into direct and indirect thermal desorption technology [5,34].

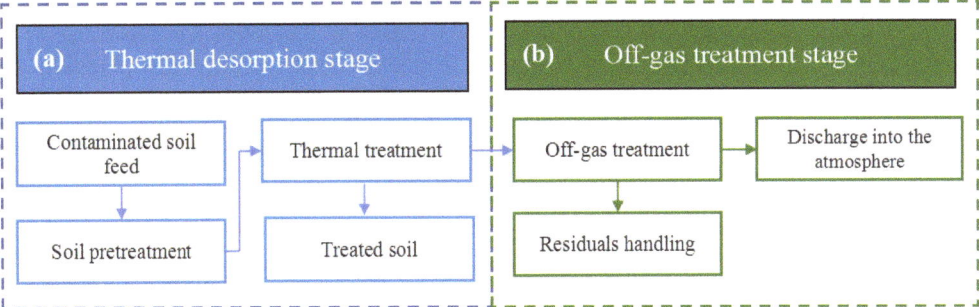

**Figure 2.** Basic process of a thermal desorption system.

In this study, ex situ thermal desorption technologies were selected at contaminated sites in three cities of China, and their process flow chart is shown in Figure 3.

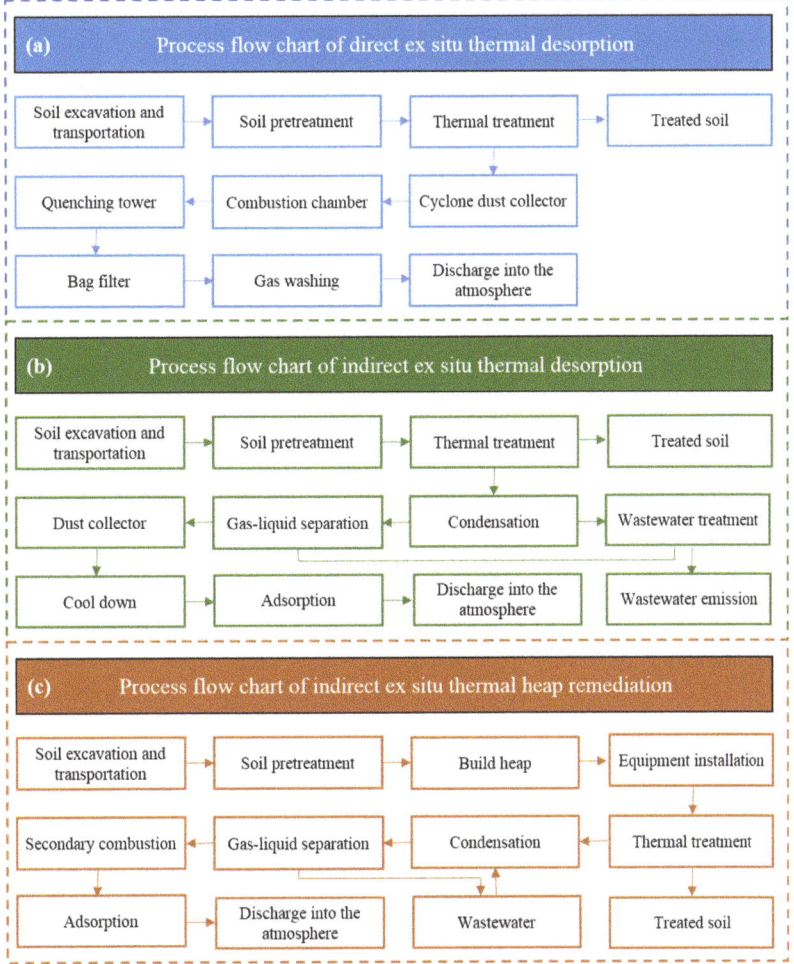

**Figure 3.** Flow chart for three ex situ thermal desorption technology.

Direct ex situ thermal desorption is adopted at the Tianjin contaminated site in China, with the process flow chart shown in Figure 3a. The soil is dehydrated, screened, and loaded in the pretreatment workshop. Rotary kilns use natural gas as raw material to heat the soil. The heated soil is humidified and cooled with water, and the off-gas generated in the treatment process is discharged into the atmosphere by cyclone dust collector, combustion chamber, quenching tower, bag filter, and gas washing. No waste water is generated during thermal desorption.

Indirect ex situ thermal desorption is adopted at the Liuzhou contaminated site in China, with the process flow chart shown in Figure 3b. The soil is dehydrated, screened, and loaded in the pretreatment workshop. Rotary kilns use natural gas as raw material to heat the soil. The heated soil is humidified and cooled with water, and the off-gas generated in the treatment process is discharged into the atmosphere by condensation, gas–liquid separation, dust collector, cool down, and adsorption. The wastewater generated is collected in the collection tank and pumped to the wastewater treatment equipment on-site before being discharged.

Indirect thermal heap remediation is adopted at the Linyi contaminated site in China, with the process flow chart shown in Figure 3c. The construction of thermal heap includes the following steps: pretreatment of soil, construction of the heap body, fuel system, heating and extraction system installation, and the auxiliary system installation. The off-gas generated in the treatment process is discharged into the atmosphere by condensation, gas–liquid separation, secondary combustion and adsorption, and the wastewater generated is collected and stored in a temporary storage system.

### 3.2. Results Analysis

The results obtained by this method can show the characteristics of technology and comparison from different angles: (1) Analyze the specific parameters and improvement hotspots of the technology from the specific indicator performance, such as the GWP, and emission sources that caused GWP. (2) Trace the improvement direction from different dimensions and show the balance characteristics of the technology between dimensions. For example, the poor performance of the resource dimension indicates that it has high demand for resources, energy, and raw and auxiliary materials. At the same time, we should comprehensively consider technology from different dimensions, not only pursue one dimension and ignore the other dimensions. For example, the performance of technical efficiency and operation process is well, but the environmental impact is high. (3) Promptly judge the comprehensive performance of different technologies under a single indicator of comprehensive evaluation.

Primary and secondary indicators of the three kinds of ex situ thermal desorption is demonstrated in Table 13. The comparison of secondary indicators of the three ex situ thermal desorption is shown in Figure 4, the comparison of primary indicators of the three ex situ thermal desorption is shown in Figure 5 and the comprehensive comparison radar chart is shown in Figure 6.

#### 3.2.1. Indicator Performance

In terms of technical indicators, the conclusion drawn from Table 13 shows that three kinds of ex situ thermal desorption have different advantages on five indicators. The scores of three technical indicators of the indirect thermal heap are lower than those of the direct thermal desorption and indirect thermal desorption, whereas the scores for heat transfer efficiency and comprehensive energy consumption of the indirect thermal desorption are higher than those of the direct thermal desorption and indirect thermal heap. Overall, the indirect thermal desorption has the best technical indicator.

The score of raw material and energy consumption of the indirect thermal desorption is higher than those of direct thermal desorption and indirect thermal heap, but the score of water consumption of the indirect thermal heap is much higher than that of direct thermal desorption and indirect thermal desorption. Overall, owing to less raw material consumption, the indirect thermal heap has the best resource indicators.

**Table 13.** Comparison of primary and secondary indicators for the three ex situ thermal desorption processes.

| Dimensions | Indicators | Secondary Indicators | | | Primary Indicators | | |
|---|---|---|---|---|---|---|---|
| | | Direct Thermal Desorption | Indirect Thermal Desorption | Indirect Thermal Heap | Direct Thermal Desorption | Indirect Thermal Desorption | Indirect Thermal Heap |
| Technical indicators | Heat transfer efficiency | 0.016 | 0.045 | 0.014 | 0.106 | 0.118 | 0.104 |
| | Pollutant removal rate | 0.498 | 0.498 | 0.493 | | | |
| | Secondary pollutants | 0.129 | 0.129 | 0.129 | | | |
| | Fault condition | 0.245 | 0.245 | 0.245 | | | |
| | Comprehensive energy consumption | 0.011 | 0.083 | 0.009 | | | |
| Resource indicators | Raw material consumption | 0.099 | 0.200 | 0.001 | 0.052 | 0.108 | 0.175 |
| | Energy consumption | 0.064 | 0.200 | 0.059 | | | |
| | Water consumption | 0.035 | 0.009 | 0.600 | | | |
| Environmental indicators | Global warming potential | 0.184 | 0.398 | 0.223 | 0.197 | 0.427 | 0.169 |
| | Eutrophication potential | 0.007 | 0.067 | 0.014 | | | |
| | Acidification potential | 0.024 | 0.240 | 0.052 | | | |
| | Ozone layer depletion potential | 0.0005 | 0.0009 | 0.041 | | | |
| | Peculiar smell | 0.042 | 0.127 | 0 | | | |
| | Noise | 0.127 | 0 | 0 | | | |
| Economic indicators | Investment return period | 0.062 | 0.105 | 0.075 | 0.068 | 0.031 | 0.037 |
| | Direct benefit | 0.637 | 0.335 | 0.335 | | | |
| | Indirect benefit | 0.258 | 0.068 | 0.118 | | | |
| Social indicators | Job opportunity | 0.649 | 0.325 | 0.464 | 0.028 | 0.022 | 0.026 |
| | Social income | 0.149 | 0.279 | 0.248 | | | |
| | NIMBY | 0.072 | 0.072 | 0.072 | | | |
| | Total score | | | | 0.452 | 0.707 | 0.511 |

The scores for GWP, eutrophication potential, acidification potential, and peculiar smell indicators of the indirect thermal desorption are higher than those of the direct thermal desorption and indirect thermal heap, and the noise of the direct thermal desorption is higher than that of the indirect thermal desorption and indirect thermal heap. The best performance of the indirect thermal desorption in the GWP is due to the low energy consumption. Overall, the indirect thermal desorption has the best environmental indicators.

The score of direct and indirect benefit of the direct thermal desorption is higher than that of the indirect thermal heap, while the indirect thermal desorption has least indirect benefit. In general, the direct thermal desorption has the best economic indicator.

The direct thermal desorption has more job opportunities than that of the indirect thermal desorption and indirect thermal heap, and the indirect thermal desorption's social income is higher than that of the direct thermal desorption and indirect thermal heap. In general, the direct thermal desorption has the best social indicators.

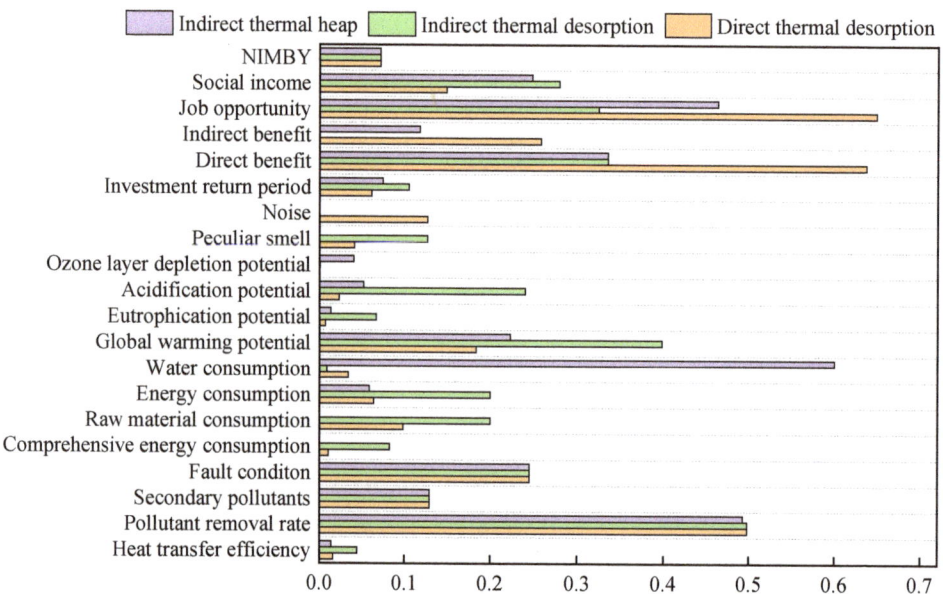

**Figure 4.** Secondary indicators of the three ex situ thermal desorption processes.

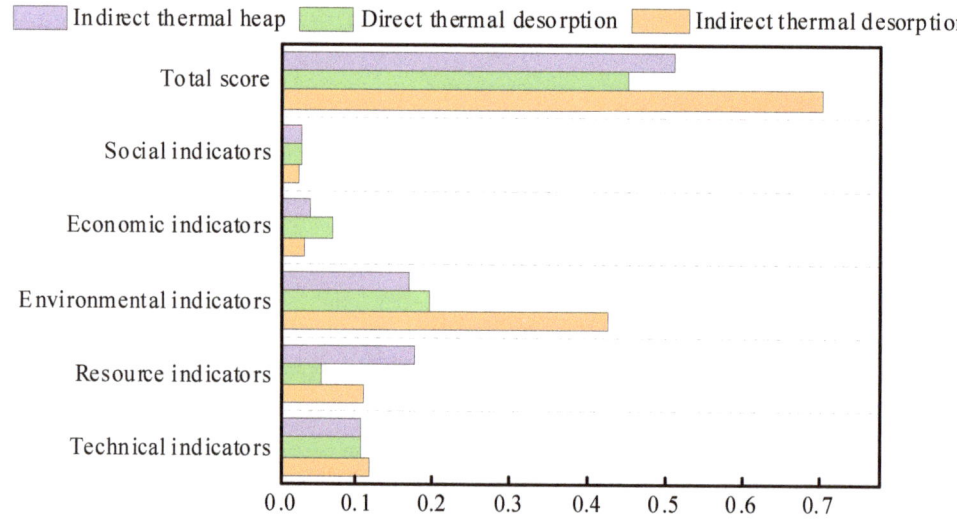

**Figure 5.** Comparison of primary indicators for the three ex situ thermal desorption processes.

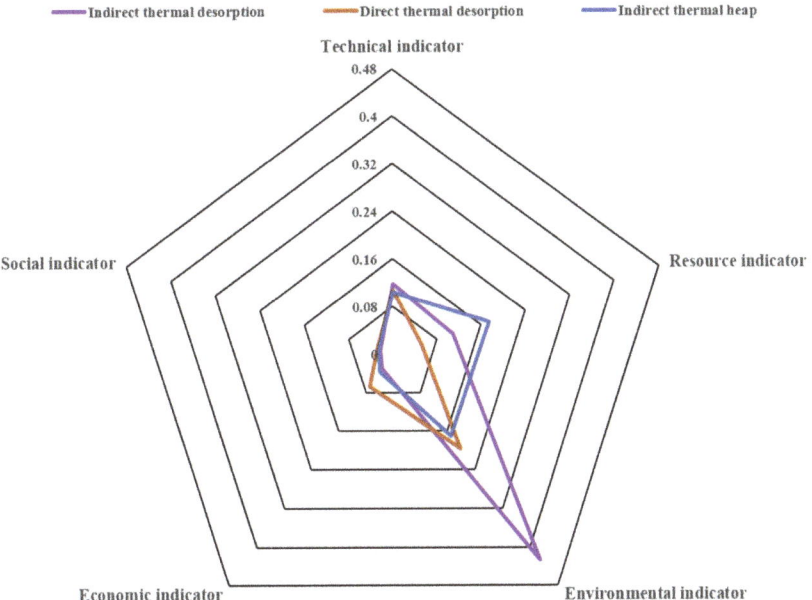

**Figure 6.** Comprehensive comparison radar map for the three ex situ thermal desorption technology.

3.2.2. Dimensional Analysis

It can be seen from Figures 4 and 5 that the indirect thermal desorption has the highest environmental indicator score among the three ex situ thermal desorption technology. Specifically, the environmental indicator of indirect thermal desorption is 2.1 times and 2.5 times higher than that of direct thermal desorption and indirect thermal heap, respectively. The primary reason is that indirect thermal desorption has much higher scores in peculiar smell, acidification potential, and GWP. The direct thermal desorption has the highest economic indicator score, 2.2 times and 1.8 times higher than the indirect thermal desorption and indirect thermal heap, respectively, mainly in the indirect and direct benefits. The indirect thermal heap has the highest resource indicator score, which is 3.3 times and 1.6 times higher than the direct thermal desorption and indirect thermal desorption, respectively, mainly in water consumption.

Combining the weights of the five first-level indicators, the indirect thermal desorption has the highest total score, 1.5 times and 1.4 times higher than the direct thermal desorption and indirect thermal heap, respectively. In general, the overall scoring order of the three ex situ thermal desorption technology is as follows: indirect thermal desorption > indirect thermal heap > direct thermal desorption. In the comprehensive evaluation and analysis of this case, when selecting a remediation method for a contaminated site, in addition to considering the site's own situation, it can also provide a technical basis for its key indicators, which is comparatively reference-valuable.

The information shown in Figure 6 suggests that the three ex situ thermal desorption processes have their own advantages in five dimensions: the indirect thermal desorption ranks highest in terms of environmental indicators and technical indicators, especially the environmental indicators. The environmental benefits of the indirect thermal desorption are far superior to the other two sites. In terms of both social indicators and economic indicators, the direct thermal desorption has a score that is higher than the other two technologies, and the economic indicators of the direct thermal desorption are more advantageous. The indirect thermal heap achieves the highest score in terms of resource indicators. During comparison of the ex situ thermal desorption, it is notable that although all of the three sites adopt ex situ thermal desorption technology, the target pollutants removed are different, therefore the types and concentrations of secondary pollutants produced are different.

## 4. Discussion

### 4.1. Indicator System

In order to construct a multidimensional indicator system for the evaluation of ex situ remediation technologies, this study developed a research framework, as shown in Figure 1. Different published indicator systems developed for other research purposes (e.g., green development) were referenced in this study. Indicators related to the evaluation of ex situ remediation technologies were selected from the relevant literature, and these indicators were employed in our study to form a system of indicators applicable to ex situ remediation technologies.

Technical indicators mainly consider the factors in the technical recommendation list. Resource indicators mainly cover the consumption of resources, energy, and materials. Environmental indicators draw on the LCA method of environmental impact assessment and focus on important factors such as GWP. The economic indicators mainly consider the cost, rationality, and benefit. The social dimension takes into account the two stakeholders, which are local communities and workers.

### 4.2. Methodological Applicability

In this study, three kinds of ex situ thermal desorption remediation technologies are selected to verify the credibility of the model. There are some studies on the environmental impact assessment of the technology, but all use the LCA method to evaluate the GWP or other environmental impacts. However, only GWP, an indicator or environmental impact dimension, cannot be used to evaluate the comprehensive sustainability of the technology [22]. Different from the above research, the evaluation model developed in this study covers five dimensions—technology, resources, environment, economy and society—which creates a more comprehensive evaluation.

In this study, the AHP and comprehensive evaluation are used to empower and aggregate the five dimensions of sustainability performance. The results show that indirect thermal desorption has the highest score of technical and environmental indicators, indirect thermal heap has the highest score of resource indicators, direct thermal desorption has the highest score of economic and social indicators. With the balanced performance of five dimensions, indirect thermal desorption has the highest comprehensive evaluation, which shows the importance of the comprehensive model.

The case study shows that the current comprehensive evaluation model developed in this study can be widely applied to various remediation technologies, and can reveal the characteristics of the technologies. Firstly, during the case study, the data are easy to collect and obtain, and the calculation of the indicators is straightforward. Secondly, in the process of weight determination and comprehensive evaluation, the methodology is mature and easy to operate. The results of the study reflect the characteristics of the technologies at three levels: specific indicators, dimensions, and single indicators of comprehensive. The analysis of different types of ex situ thermal desorption cases shows that the model can be well applied to the evaluation of this case. For other types of technologies, this model can also be used to obtain reliable results.

In addition, the process optimization conclusions of this study can also be applied to other technical cases. For example, reducing energy consumption and increasing the proportion of renewable energy, which can reduce resource and energy consumption, reduce environmental impact, improve technical performance, and reduce costs to a certain extent. Social impact is also a link that must be given attention in the use of remediation technology, such as improving employee welfare and reducing adverse effects such as NIMBY, which can promote the process optimization and market application of technology.

## 5. Conclusions

Comprehensive evaluation of different soil remediation technologies is of critical importance to the optimization and selection of proper technology. Comprehensive evaluation refers to the use of a systematic and standardized method for simultaneous evaluation of multiple indicators. It includes the steps of constructing an indicator system, accounting for the indicators, normalization, determining weights by AHP, and comprehensive evaluation.

In this study, a comprehensive evaluation method of soil remediation technology, covering 20 indicators in five dimensions, was developed. Three ex situ thermal desorption processes—direct thermal desorption, indirect thermal desorption, and indirect thermal heap—were selected for the method validation. The results showed that direct thermal desorption had the highest economic and social indicator scores. Indirect thermal desorption had the highest technical and environmental indicator scores. Indirect thermal heap had the highest resource indicator score. With balanced performance in five dimensions, the overall comprehensive score order of the three ex situ thermal desorption is indirect thermal desorption > indirect thermal heap > direct thermal desorption. Our evaluation system can provide a theoretical basis for the improvement and selection of ex situ thermal desorption remediation technology. Our study can also provide a novel evaluation approach for the sustainable development and application of soil remediation technology.

**Author Contributions:** Conceptualization, N.D. and Y.Y.; methodology, N.D.; software, S.L.; validation, W.J. and Y.Y.; formal analysis, L.H.; investigation, L.H.; resources, N.D.; data curation, N.D. and L.H.; writing—original draft preparation, S.L.; writing—review and editing, B.Z. All authors have read and agreed to the published version of the manuscript.

**Funding:** This research received no external funding.

**Institutional Review Board Statement:** Not applicable.

**Informed Consent Statement:** Not applicable.

**Data Availability Statement:** Not applicable.

**Acknowledgments:** This research is supported by the National Key Research and Development Plan (grant numbers 2018YFC1802106 and 2020YFC1807501).

**Conflicts of Interest:** The authors declare no conflict of interest.

## References

1. Caroline, V.; Antônio, T. Sustainability in Life Cycle Analysis of Nanomaterials Applied in Soil Remediation. In *The International Congress on Environmental Geotechnics*; Springer: Singapore, 2018; pp. 537–543.
2. Visentin, C.; da Silva Trentin, A.W.; Braun, A.B.; Thomé, A. Lifecycle assessment of environmental and economic impacts of nano-iron synthesis process for application in contaminated site remediation. *J. Clean. Prod.* **2019**, *231*, 307–319. [CrossRef]
3. Harbottle, M.J.; Al-Tabbaa, A.; Evans, C.W. A comparison of the technical sustainability of in situ stabilisation/solidification with disposal to landfill. *J. Hazard. Mater.* **2007**, *141*, 430–440. [CrossRef] [PubMed]
4. Dan, W. *Research Status and Trends of Soil Pollution from 1999 to 2018*; IOP Publishing: Bristol, UK, 2021.
5. Zhao, C.; Dong, Y.; Feng, Y.; Li, Y.; Dong, Y. Thermal desorption for remediation of contaminated soil: A review. *Chemosphere* **2019**, *221*, 841–855. [CrossRef] [PubMed]
6. Ding, D.; Song, X.; Wei, C.; LaChance, J. A review on the sustainability of thermal treatment for contaminated soils. *Environ. Pollut.* **2019**, *253*, 449–463. [CrossRef] [PubMed]
7. Aresta, M.; Dibenedetto, A.; Fragale, C.; Giannoccaro, P.; Pastore, C.; Zammiello, D.; Ferragina, C. Thermal desorption of polychlorobiphenyls from contaminated soils and their hydrodechlorination using Pd-and Rh-supported catalysts. *Chemosphere* **2008**, *70*, 1052–1058. [CrossRef]
8. Zhang, X.; Yao, A. Pilot experiment of oily cuttings thermal desorption and heating characteristics study. *J. Pet. Explor. Prod. Technol.* **2019**, *9*, 1263–1270. [CrossRef]
9. Falciglia, P.P.; Giustra, M.G.; Vagliasindi, F. Low-temperature thermal desorption of diesel polluted soil: Influence of temperature and soil texture on contaminant removal kinetics. *J. Hazard. Mater.* **2011**, *185*, 392–400. [CrossRef]
10. Gharibzadeh, F.; Kalantary, R.R.; Esrafili, A.; Ravanipour, M.; Azari, A. Desorption kinetics and isotherms of phenanthrene from contaminated soil. *J. Environ. Health Sci. Eng.* **2019**, *17*, 171–181. [CrossRef]

11. Biache, C.; Mansuy-Huault, L.; Faure, P.; Munier-Lamy, C.; Leyval, C. Effects of thermal desorption on the composition of two coking plant soils: Impact on solvent extractable organic compounds and metal bioavailability. *Environ. Pollut.* **2008**, *156*, 671–677. [CrossRef]
12. Troxler, W.L.; Cudahy, J.J.; Zink, R.P.; Yezzi, J.J., Jr.; Rosenthal, S.I. Treatment of nonhazardous petroleum-contaminated soils by thermal desorption technologies. *Air Waste* **1993**, *43*, 1512–1525. [CrossRef]
13. Bykova, M.V.; Alekseenko, A.V.; Pashkevich, M.A.; Drebenstedt, C. Thermal desorption treatment of petroleum hydrocarbon-contaminated soils of tundra, taiga, and forest steppe landscapes. *Environ. Geochem. Health* **2021**, *43*, 2331–2346. [CrossRef] [PubMed]
14. U.S. EPA. *Superfund Remedy Report*; Office of Land and Emergency Management: Washington, DC, USA, 2017.
15. Luo, Y.; Tu, C. *Twenty Years of Research and Development on Soil Pollution and Remediation in China*; Springer Nature Press: Singapore, 2018.
16. Shen, Z.; Chen, Y.; Li, S.; Qun, G.; Lili, G.; Wentao, J.; Peng, L.; Longjie, J.; Jia, L. Application of ex-situ thermal desorption technology and equipment in contaminated site remediation projects in China. *Chin. J. Environ. Eng.* **2019**, *13*, 2060–2073.
17. Morais, S.A.; Delerue-Matos, C. A perspective on LCA application in site remediation services: Critical review of challenges. *J. Hazard. Mater.* **2010**, *175*, 12–22. [CrossRef] [PubMed]
18. Visentin, C.; da Silva Trentin, A.W.; Braun, A.B.; Thomé, A. Application of life cycle assessment as a tool for evaluating the sustainability of contaminated sites remediation: A systematic and bibliographic analysis. *Sci. Total Environ.* **2019**, *672*, 893–905. [CrossRef]
19. Owsianiak, M.; Lemming, G.; Hauschild, M.Z.; Bjerg, P.L. Assessing environmental sustainability of remediation technologies in a life cycle perspective is not so easy. *ACS Publ.* **2013**, *47*, 1182–1183. [CrossRef]
20. Amponsah, N.Y.; Wang, J.; Zhao, L. A review of life cycle greenhouse gas (GHG) emissions of commonly used ex-situ soil treatment technologies. *J. Clean. Prod.* **2018**, *186*, 514–525. [CrossRef]
21. Zhang, L. Sustainability Assessment of Solid Waste Recycling Technologies Based on Life Cycle Thinking. Master's Thesis, University of Chinese Academy of Sciences, Beijing, China, 2021.
22. Xiong, F.; Pan, J.; Lu, B.; Ding, N.; Yang, J. Integrated technology assessment based on LCA: A case of fine particulate matter control technology in China. *J. Clean. Prod.* **2020**, *268*, 122014. [CrossRef]
23. Wen, Q.; Liu, G.; Wu, W.; Liao, S. Multicriteria comprehensive evaluation framework for industrial park-level distributed energy system considering weights uncertainties. *J. Clean. Prod.* **2021**, *282*, 124530. [CrossRef]
24. Guo, Y. Overview of weight determination methods. *Rural. Econ. Sci.—Technol.* **2018**, *29*, 252–253.
25. Saaty, R.W. The analytic hierarchy process—what it is and how it is used. *Math. Model.* **1987**, *9*, 161–176. [CrossRef]
26. Ahmed, M.; Qureshi, M.N.; Mallick, J.; Hasan, M.; Hussain, M. Decision support model for design of high-performance concrete mixtures using two-phase AHP-TOPSIS approach. *Adv. Civ. Eng.* **2019**, *2019*, 1696131. [CrossRef]
27. Mamun, M.; Howladar, M.F.; Sohail, M.A. Assessment of surface water quality using fuzzy analytic hierarchy process (FAHP): A case study of Piyain River's sand and gravel quarry mining area in Jaflong, Sylhet, 9. *Sustain. Dev.* **2019**, *9*, 100208. [CrossRef]
28. Wei, Z.L.; Liu, D.J.; Liu, W.Y. Environmental impact and comprehensive benefit evaluation of energy grass based on life cycle assessment. *J. Beijing Jiaotong Univ.* **2013**, *2*, 138–143.
29. Zhang, L.; Lavagnolo, M.C.; Bai, H.; Pivato, A.; Raga, R.; Yue, D. Environmental and economic assessment of leachate concentrate treatment technologies using analytic hierarchy process. *Resour. Conserv. Recycl.* **2019**, *141*, 474–480. [CrossRef]
30. Ding, N.; Liu, J.; Yang, J.; Yang, D. Comparative life cycle assessment of regional electricity supplies in China. *Resour. Conserv. Recycl.* **2017**, *119*, 47–59. [CrossRef]
31. Ding, N.; Liu, N.; Lu, B.; Yang, J. Life cycle greenhouse gas emissions of aluminum based on regional industrial transfer in China. *J. Ind. Ecol.* **2021**, *25*, 1657–1672. [CrossRef]
32. De Percin, P.R. Application of thermal desorption technologies to hazardous waste sites. *J. Hazard. Mater.* **1995**, *40*, 203–209. [CrossRef]
33. Khan, F.I.; Husain, T.; Hejazi, R. An overview and analysis of site remediation technologies. *J. Environ. Manag.* **2004**, *71*, 95–122. [CrossRef]
34. Wang, B.; Wu, A.; Li, X.; Ji, L.; Sun, C.; Shen, Z.; Chen, T.; Chi, Z. Progress in fundamental research on thermal desorption remediation of organic compound-contaminated soil. *Waste Dispos. Sustain. Energy* **2021**, *3*, 83–95. [CrossRef]

MDPI  
St. Alban-Anlage 66  
4052 Basel  
Switzerland  
Tel. +41 61 683 77 34  
Fax +41 61 302 89 18  
www.mdpi.com

*International Journal of Environmental Research and Public Health* Editorial Office  
E-mail: ijerph@mdpi.com  
www.mdpi.com/journal/ijerph

www.ingramcontent.com/pod-product-compliance
Lightning Source LLC
LaVergne TN
LVHW070652100526
838202LV00013B/947